3rd Thermal EOR International Workshop (ThEOR2018)

Thermal Methods for Enhanced Oil Recovery: Laboratory Testing, Simulation and Oilfields Applications

IOP Conference Series: Earth and Environmental Science
Volume 282

Chengdu, China
15 – 19 October 2018

ISBN: 978-1-5108-9328-3
ISSN: 1755-1307

Printed from e-media with permission by:

Curran Associates, Inc.
57 Morehouse Lane
Red Hook, NY 12571

CURRAN ASSOCIATES INC. proceedings .com

Some format issues inherent in the e-media version may also appear in this print version.

Institute of Physics
Dirac House, Temple Back
Bristol BS1 6BE UK

Phone: 44 1 17 929 7481
Fax: 44 1 17 920 0979

techtracking@iop.org

Additional copies of this publication are available from:

Curran Associates, Inc.
57 Morehouse Lane
Red Hook, NY 12571 USA
Phone: 845-758-0400
Fax: 845-758-2633
Email: curran@proceedings.com
Web: www.proceedings.com

TABLE OF CONTENTS

IOP Conf. Series: Earth and Environmental Science **282** (2019) 011001 doi:10.1088/1755-1315/282/1/011001

Preface

ThEOR2018 aims to bring together key researchers and professionals from the industry, R&D centers and academic institutes to share experiences, perspectives, insights and the latest developments in thermal methods for EOR. The workshop's program contains one-day training courses and three days of plenary and keynote lectures, technical and poster sessions on cutting edge issues related to the petroleum industry. At the same time, we try to keep the balance between traditional and emerging fields, fundamentals and applications, modeling and experiment, academic interests and industry needs. The Workshop was be held in Southwest Petroleum University, Chengdu, P.R. China on October 15-19, 2018."

Peer review statement

All papers published in this volume of *IOP Conference Series: Earth and Environmental Science* have been peer reviewed through processes administered by the proceedings Editors. Reviews were conducted by expert referees to the professional and scientific standards expected of a proceedings journal published by IOP Publishing.

IOP Conf. Series: Earth and Environmental Science **282** (2019) 012001 doi:10.1088/1755-1315/282/1/012001

Changes in the material composition of fluid seals of bitumen deposits during thermal effects on the productive formation

L Sitdikova, N Khasanova, E Sidorova

Kazan Federal University, 18 Kremlevskaya Str., Kazan, Russia

E-mail: sitdikova8432@mail.ru

Abstract. Fluid seals are the most important element of oil fields, because they prevent the spread and dispersion of hydrocarbons and contribute to the preservation of deposits. The role of the fluid seals increases in the later stages of oil field development, especially when using thermal methods of intensification. As a result, changes in the mechanical, mineralogical, and lithological properties of the fluid seal rocks can occur. The deposits of natural bitumen of the Volga-Ural petroleum province are a complex geological and geochemical object. The degree of preservation of hydrocarbon deposits is determined by the filtration characteristics of the fluid seal. The composition of the fluid seals is represented by dense clay components of the rock. The use of thermal or steam-thermal methods of exposure to hydrocarbon reservoirs, primarily affects the minerals-indicators (clay minerals of cement collectors). The results of thermal effects on the reservoir can lead to changes in the material composition, parameters of porosity and permeability of the fluid seal rocks: the clay component is destroyed with the active participation of organic matter.

1. Introduction

Within the Volga-Ural petroleum province in the Permian sediments numerous accumulations of natural bitumen and bitumen occurrences are known (Figure 1).

Features of distribution of bitumen accumulations in the Permian section are controlled by the spread of fluid seal horizons and reservoir rocks [1]. In the distribution and preservation of bitumen deposits, a large role is played by the peculiarities of fluid seal rocks. The quality of the fluid seal rocks can be assessed by the following criteria: the thickness of the fluid seals, their solidity, integrity, and material composition. An important criterion is areal sustainability of fluid seals [2].

The main criteria for the quality of fluid seals allowed the authors involved in the study of bituminous rock complexes to distinguish the following types of shielding fluid seals: tires of regional, zonal and local types. Rocks of fluid seals have different thickness, in some places reach up to 100 meters or more. Local tires are developed only in certain areas of bitumen deposits, however, they are thin.

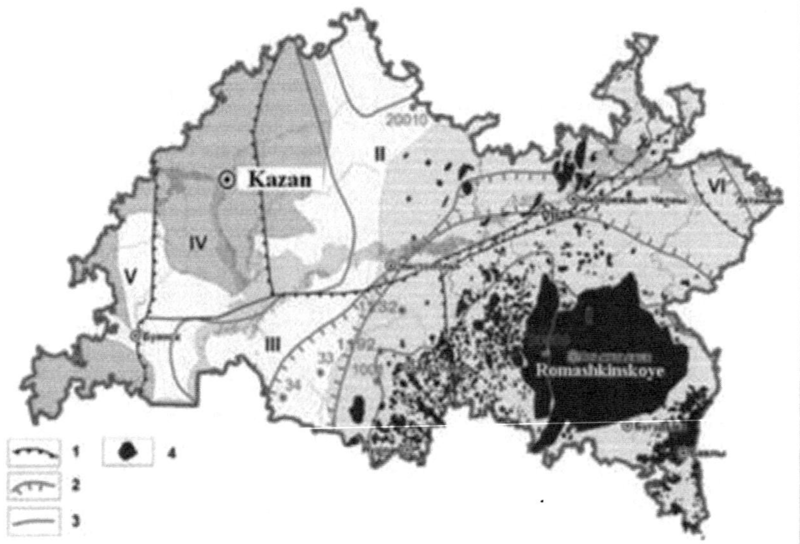

Figure 1. Schematic map of the Romashkinskoye oil field and its satellites (within the territorial boundaries of the Republic of Tatarstan) [2].

In the distribution of types of the fluid seals and reservoirs in the Permian sediments section are distinguished bitumen-bearing complexes: Lower Permian (P_1) carbonate, Ufa (P_1u) terrigenous, Kazan (P_2kz) terrigenous-carbonate and carbonate-terrigenous [3]. In the Republic of Tatarstan, the development of natural bitumen deposits is associated with thermal methods of formation effects [3]. Thermal development methods are intra-layer combustion, steam-thermal treatment of the well bottom-hole zone, injection of heat-transfer fluids into the formation — steam or hot water. However, the use of thermal methods is significantly limited by the conditions of stability of the most important element of the bitumen deposit – fluid seals.

2. Methodology

We carried out a laboratory study of the stability of fluid seals of bitumen deposits in Cheremshano-Yamashinskaya structural zone (Figure 1). They are represented by a dense clay formation of "lingul clays". The object of study is the core material of the fluid seals from wells of natural bitumens. Special attention was paid to the study of the rocks of the fluid seals of the Ashalchinskoye field. At Ashalchinskoye field, the development of natural bitumen is carried out using thermal methods of extraction.

The main research methods are macro-structural-texture analysis, optical methods on a Leica microscope. The method of X-ray diffractometry was used to study the phase composition of clay minerals of the initial rocks and subjected to thermal effects (diffractometer DRON-3M, Cu Kα radiation, shooting modes: 20kv x 20ma, slits 0.5= =0.5, preparation shooting range 2-36°2Θ). Oriented preparations of clay minerals were made by centrifugation method, studied in air-dry state and after thermal exposure. Heat treatment of initial samples of rocks of fluid seals was carried out in the muffle furnace SUOL in the temperature range from 200 °C to 600 °C.

The study of the features of the distribution of manganese ions was carried out by electron paramagnetic analysis (EPR) on a spectrometer CMS8400 3 cm range. The samples were heated for 30 minutes in a hydrogen atmosphere at temperatures of 350 °C and 600 °C in a horizontal electric furnace of the SUOL type. Morphological transformations take place at temperature effects of fluid seals samples, because for their study was used the method of scanning electron microscopy. Samples of fresh rock chips, decorated with a thin layer of gold in the Agar sputtering system, were studied. The investigations were carried out with a Philips XL-30 scanning electron microscope in the Hi-vac mode, detector of backward scattered electrons SE, voltage of 20 kv, spot 4, magnification from 500x to 2500x.

3. Results and Discussions

The rocks of the fluid seals are composed of a clay-aleuritic series with complex mineral composition. The bulk of the rock is a complex of clay minerals: illite, mixed-layer minerals such as mica-smectite with a ratio of 80:20 layers, chlorite [4, 5]. The fine substance consists of quartz, feldspar, and pyrite framboids (Figure 2). Carbonaceous excrements and fragments of organic residues are also present in the samples.

In order to determine the stability of the rocks of the fluid seals deposits during thermal effects on the reservoir, differential thermal analysis (DTA) was performed in the temperature range from 20 ° to 950 °C. According to the data of DTA, a number of thermal effects are observed in the studied samples of fluid seals. In the range of 110-250 °C, the loss of interlayer water molecules in the mixed-layer phase mica-smectite occurs. Then there is a loss of hydroxyl groups of the tetrahedral and octahedral layers of the crystal lattice. The most active losses occur in the temperature range 650-750 °C. According to X-ray phase analysis, the final degradation of the illite phase occurs with its transition to a stable mica phase (Figure 2). Thus there is a restructuring of the clay component of the fluid seals samples. Rocks almost completely lose their plastic properties, which is accompanied by deformation processes: the emergence of micro-fracturing, secondary porosity, etc. (Figure 3).

IOP Conf. Series: Earth and Environmental Science **282** (2019) 012001 doi:10.1088/1755-1315/282/1/012001

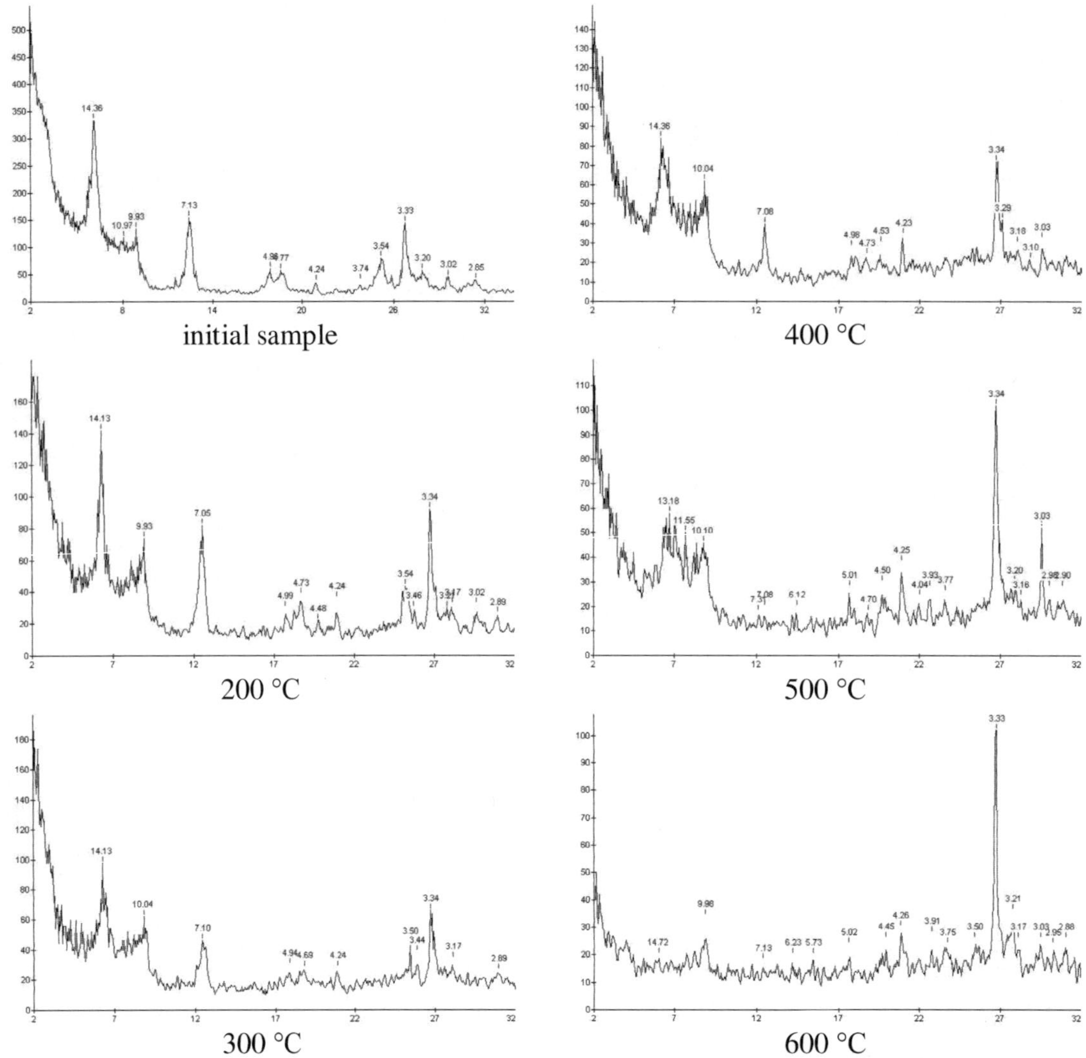

Figure 2. Degradation of the phase composition of clay components (illite, mixed-layer minerals such as mica-smectite, chlorite) and fine minerals (quartz, calcite, orthoclase) at thermal processing of samples of fluid seals rocks.

The heterogeneity of the mineral composition of fluid seals samples is also confirmed by the results of the EPR analysis [6]. The survey spectrum of the EPR analysis (360 mT scan) shows structural iron lines (g= 4.3), six characteristic lines of the impurity radical Mn^{2+} in the carbonate structure, and lines corresponding to the organic radical C_{org} (Figure 4).

The change in the intensity of the EPR signal of manganese ion in the carbonate part of the rock before and after thermal (350 °C) exposure is shown in the diagram (Figure 5).

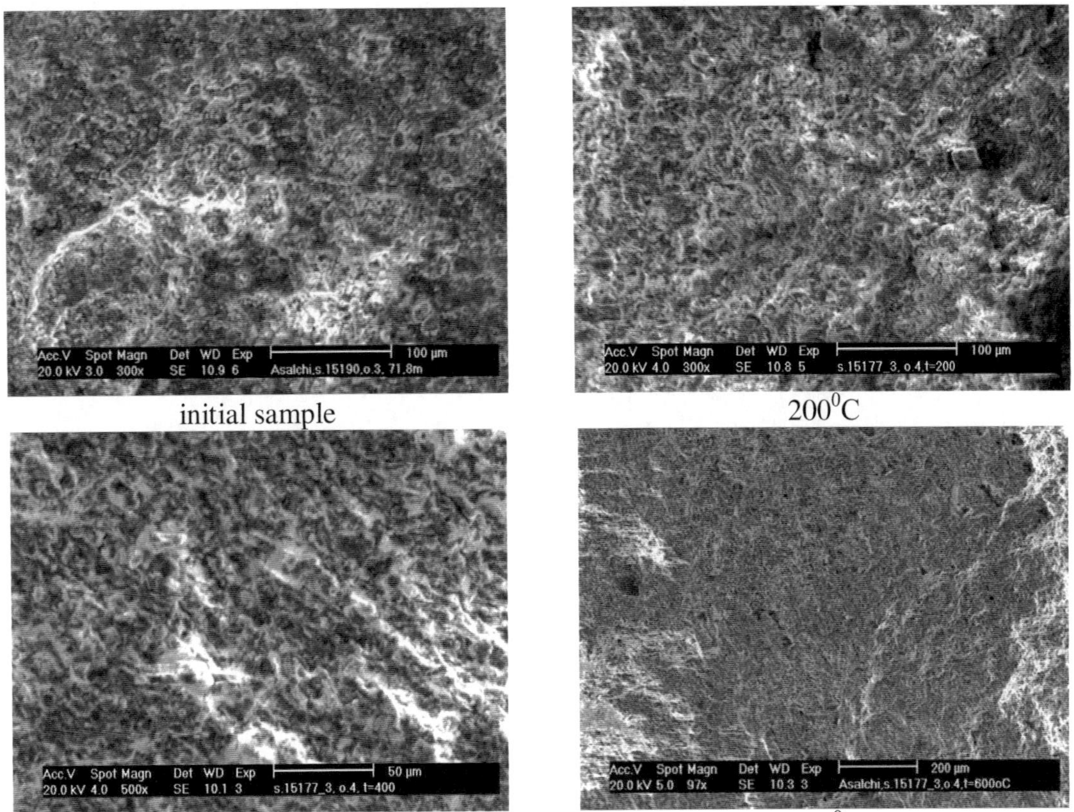

initial sample　　200^0C

400^0C　　600^0C

Figure 3. Changes in the morphological properties of samples during thermal effects on the rocks of fluid seals.

The variations in the intensity values of the EPR signal under thermal effects even within a few meters of thickness are established. At a depth of 83.64 m and 84.14 m (samples 1 and 3), the manganese signal decreases as a result of its possible leaching from the rock. As a result, the pore space of the sample increases. Minor changes in the intensity of the manganese impurity ion at depths of 83.94 m and 84.44 m (samples 2 and 4) are only indicative of a change in post-sedimentary conditions and the preservation of pore space.

Also, according to the results of the EPR analysis, it was noted that at the depths of 84.64 m and 85.4 m (samples 5 and 6) there was a sharp change in the material composition of the rocks from substantially carbonate to substantially clay.

Thermal effects up to temperatures of 600 °C, unlike exposure at 350 °C, are accompanied by intensive leaching and oxidation of iron ions with the formation of supramagnetic fractions, which contributes to the formation of spin density and favorable conditions for the generation of hydrocarbon gas fractions [7-9].

IOP Conf. Series: Earth and Environmental Science **282** (2019) 012001 doi:10.1088/1755-1315/282/1/012001

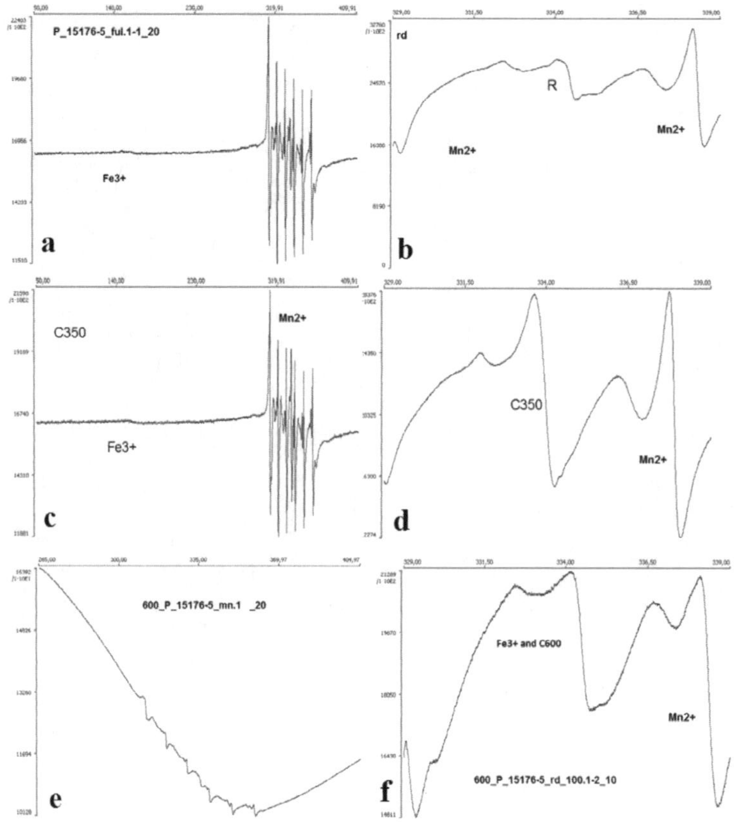

Figure 4. Overview EPR spectra of samples: a, b - initial samples, c, d - after thermal exposure at temperatures of 350 °C, e, f - after thermal exposure at temperatures of 600 °C.

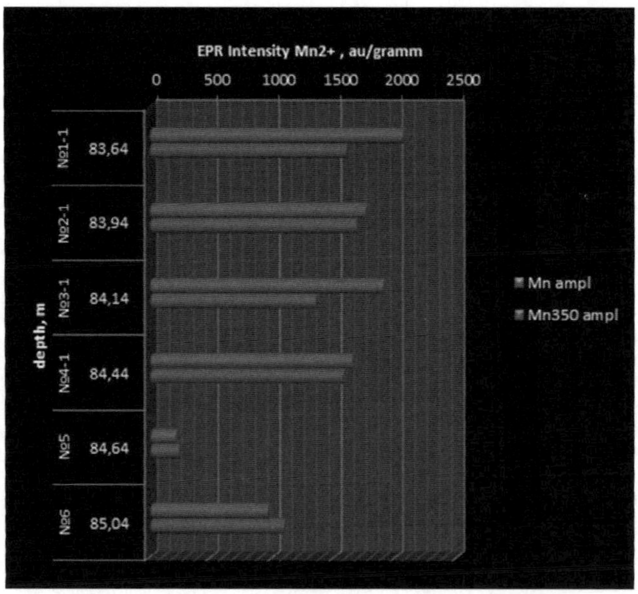

Figure 5. The change in the intensity of the EPR signal of manganese ion in the carbonate part of the rock before and after thermal (350 °C) effects.

4. Conclusions

The deposits of natural bitumen of the Republic of Tatarstan are distinguished by a high lithological heterogeneity, uneven distribution of bitumen. Deposits of natural bitumen in Ufa sediments (P_1u) are being actively developed. For the preservation of deposits of bitumen, the quality of fluid seals horizons becomes relevant. It is especially important to take into account the state of the fluid seals rocks in the thermal methods of oil field development.

Fluid seals horizons of bitumen and bitumen-bearing rocks are clay rocks of Kazan age (P_1kz), the so-called pack of "lingul clays". Illite, mixed-layer complexes of illite-smectite, chlorite and finely dispersed minerals predominate in the mineral composition of fluid seals rocks.

According to the results of laboratory tests of fluid seals samples, a number of thermal effects can be stated. At temperatures up to 200-250 °C, loss of interlayer water molecules of the swelling phase of clay minerals occurs. At higher temperatures, the restructuring of their crystal lattice begins with the loss of hydroxyl groups of tetrahedral and octahedral layers. The most active losses of hydroxyl groups at temperatures of 650–750 °C are associated with the processes of degradation of the illite phase with its transition to stable mica.

Temperature transformations of rocks cause morphological transformations that change the overall appearance of the rock. Initial homogeneous microporous samples at temperatures above 400 °C experience the appearance of microcracks and interconnected pores. There is a microstructural restructuring of the matrix of fluid seals rocks with the loss of their plastic properties.

Increasing porosity values in reservoir conditions can lead to breakthroughs of the coolant, liquid and gaseous components of bitumen in the overlying rocks, leading to negative environmental consequences. It is necessary to take into account the thickness of fluid seals horizons, to conduct a detailed study and control of the mineral composition of the fluid seals rocks and the physical condition of the rocks during active thermal methods of development.

Acknowledgements

The work is performed according to the Russian Government Program of Competitive Growth of Kazan Federal University.

References

[1] Troepolskiy V I and Ellern S S 1978 On the issue of the conditions for the formation and destruction of hydrocarbon accumulations in the Volga-Ural basin *Modern problems of oil and gas geology* 172

[2] Muslimov R Kh, Romanov G V, Kayukova G P et al 2012 Complex development of heavy oils and natural bitumen from the Permian system of the Republic of Tatarstan 396 p

[3] Khisamov R S, Gatiyatullin N S, Shargorodsky I E et al 2007 Geology and development of natural bitumen deposits of the Republic of Tatarstan 295 p

[4] Sitdikova L M, Khasanova N M, Talipov I P et al 2012 Residual organic matter the Jurassic deposits West Siberia *Georesources* **1** 4

[5] Havkin A Y, Izotov V G., Sitdikova L M. et al 2013 *Georesources* **4** 68
[6] Sitdikova L M, Khasanova N M, Izotov V G et al 2017 *International Multidisciplinary Scientific GeoConference Surveying Geology and Mining Ecology Managemen* **17** 565
[7] Vakhin A V, Onishchenko, Y V, Chemodanov A E et al 2016 *Neftyanoe Khozyaystvo - Oil Industry* **10** 32
[8] Ivanova A G, Vakhin A V, Voronina E V et al 2017 *Bul. Rus. Academy Sci. Physics* **81** 817
[9] Sitnov S A, Mukhamatdinov I I, Vakhin A V et al 2018 *J. Petrol. Sci. Eng.* **169** 44

The study of the selectivity of the catalyst based on Pt-Re/γ-Al_2O_3 in the cracking reactions of n-hexane

S Petrov[1,2], A Valieva[2]

[1]Kazan Federal University, 18 Kremlevskaya Str., Kazan, Russia
[2]Kazan National Research Technological University, 68 Karl Marx St., Kazan, Russia

E-mail: psergeim@rambler.ru

Abstract. The actual task in the fuel and energy sector is the conversation of alkanes, which are a part of the light hydrocarbon fractions of petroleum, into the top requested chemical feedstock. Perspective technologies are the technologies ofther olefinic and aromatic hydrocarbons production which are used as feedstock for the synthesis of valuable chemical products with a high discount of the market price from the cost of production. A study of the n-hexane conversion over a stationary layer of a catalytic system based on γ-Al_2O_3 containing Pt, Re, In, Ti at 400 and 450 °C and a rate of volume flow of 4 h^{-1} was performed. The specific surface of the catalytic system is more than 190 m^2/g. It was revealed that an increase in the catalysis temperature by 50 °C leads to an increase in both the conversion of n-hexane by a factor of 2 and the methyl-substituted benzenes selectivity to 33.2 wt %. The catalytic system did not change before and after the experiments according to X-ray fluorescence and X-ray spectral analyzes, the elemental and phase composition.

1. Introduction

Catalytic processes are widely used in oil refining [1-5]. Aluminium oxide supporting structure of various phase compositions are mainly used in the main catalytic processes in petrochemical synthesis. The improvement of catalysts based on γ-Al_2O_3 is carried out by combinations of bi-and trimetallic catalysts. Thus, the addition of In to the Pt–Sn/γ–Al_2O_3 system promotes an increase in the reduction of the active phase in the dehydrogenation reaction. A decrease in the number of acid centres favours a decrease in cracking side reactions. The presence of In also contributes to the decline in coke formation, thereby improving the stability of the catalyst [6-8].

Re, which was linked to the supporting structure, prevents the RuOx molecules migration on the catalyst surface. Thereby stabilization of the Ru phase occurs. The content of the Re

in the catalyst to 3 wt % increases the catalytic activity of Ru-Re/γ-Al_2O_3, if it is compared to ruthenium catalysts. A further increase results in a decrease in catalyst activity [9-11].
The introduction of 5 % of molybdenum exercise influence over the acidity of γ-Al_2O_3, because new medium-acid centres are formed, which improve the catalytic activity. The conversion of fatty acid methyl esters and the yield of alkanes are 98.4 wt % and 94.2 wt % [12–13].

In [14], the effect of phosphorus on NiW/γ-Al_2O_3 obtained by the impregnation method was studied. It was found that the addition of 1.0-1.5 wt % increases the pore diameter and dispersiveness of Ni and W. More Ni-W-O phase was formed and the proportion of weak acid was increased. Nickel bind to tetracoordinated unsaturated aluminumsites of the γ-Al_2O_3supporting structure. An increase in NiO loading lead to the appearance of layers of MoS_2 clusters in sulfided NiMo /γ-Al_2O_3 catalysts due to the average length. Increasing the stacking of the layers increases the hydrogenation reaction activity, however, the probability of the hydrogenation reaction is reduced slightly [15-16].

The addition of 0.6-1.5 wt % of calcium compounds to the PtSnIn/xCa-Al catalyst resulted in improved stability of the catalytic system. The conversion of propane on the PtSnIn/1,5Ca-Al catalyst is above 58.0 wt % and remains to 25 hours at this level [17–18].

In turn, studies on the influence of the catalytic system supporting structure show that the particles of Pd, Re, Ir metals deposited on the Al_2O_3 surface led to a homogeneous distribution and dispersiveness over the supporting structure surface compared to silicon-based supporting structure. Thus, the activity of catalysts studied in the reactions of methane conversion with n-pentane was higher for catalysts based on Al_2O_3 [19–20].

The authors of article [21] carried out a study on the effect of catalyst supports Al_2O_3, SiO_2 and Al_2O_3-SiO_2 on the properties of mono- and bimetallic catalysts Rh-Pb in the selective opening of methylcyclohexane and decalin. The total and Bronsted acidity corresponded to the following order: SIRAL 40>Al_2O_3>SiO_2. Catalyst supports on SiO_2 and Al_2O_3, prepared by traditional methods, have a monomodal pore size distribution, which leads to low diffusion. Bimodal supports are distinguished by the presence of large and small pores, thereby the availability of reagents to the active centres of the catalytic agent increases. The Al-Si support shows the highest reaction rate due to a larger specific surface area of 177 m^2/g and a pore volume of 1.15 cm^3/g and a bimodal distribution of pores compared with the samples under study [17].

With an inert support, aerogel [18-19], the catalytic system is a monofunctional catalyst. The conversion of n-hexane was studied by changing the temperature, atmosphere, and contact time in the atmosphere, and the catalytic behavior was associated with Pt loading and porosity of the samples.

Catalytic activity of a bifunctional Pt-Re / γ-Al_2O_3 catalyst containing Cl, In, Ti and having a specific surface of 192 m^2/g was investigated in n-hexane cracking reactions at 400 and 450 °C.

2. Methodology

The object of the study was a single-phase catalytic system, which was obtained by impregnating of the γ-Al_2O_3 with solutions of compounds of active components. The microelement composition of the catalyst was presented according to spectral analysis, shown in Figure 1.

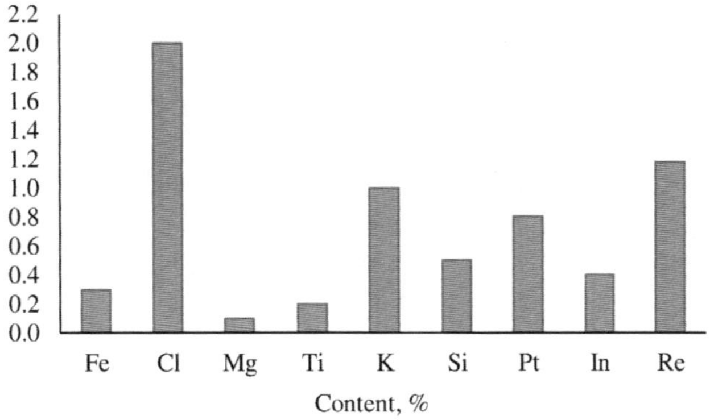

Figure 1. The elemental composition of the catalytic system according to spectral analysis.

The textural characteristics of the catalytic system were investigated by the method of nitrogen thermal desorption on the NOVA 2200e of the Quantachrom company. The specific surface was calculated by the Brunauer–Emmet–Teller method. The catalytic system is characterized by a mesopores structure, a high specific surface area of 193 m^2/g and a pore volume of 0.52 cm^3/g, and has thin pores with an average diameter of 5-20 nm.The pore volume distribution is shown in Figure 2.

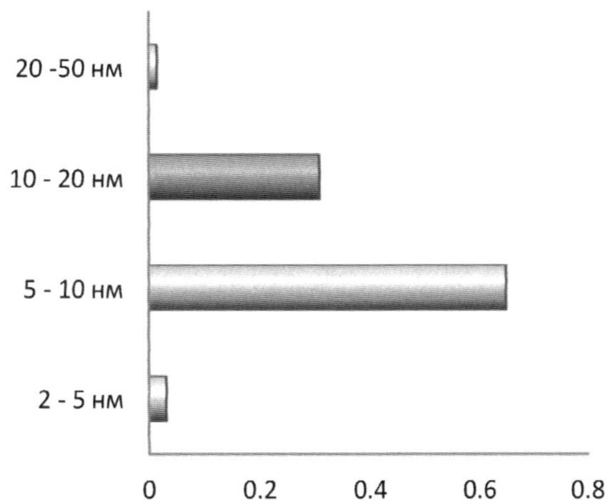

Figure 2. Pore volume distribution (wt %) in the area of mesopore diameters, nm.

The study of the Al-Cl-Re (Pt) activity in the high-temperature transformation of n-hexane was carried out in a laboratory setup at atmospheric pressure in a flow tubular-type reactor. The temperature in the reactor was 400 °C during the first experiment, 450 °C during the second experiment. Exit to the operating mode of the setup was carried out in a stream of nitrogen. Consumption of n-hexane was 1.2 ml/min, the ratio of the volume of n-hexane to the volume of the catalyst was 4: 1 h^{-1}, the catalyst bulk density was 0.76 g/ml.

The phase composition of the catalytic system before and after the experiments was investigated on a BrukerAxs X-ray diffractometer with Bragg-Brentano geometry using the DIFFRAC.EVA software module and the PDF-2 Release 2013 diffraction data base.

The elemental composition was obtained using an energy dispersive fluorescence X-ray spectroscopy EDX-800HS2 of the company "Shimadzu".

Analysis of the liquid conversion products composition was performed by chromatography-mass spectrometry using a PerkinElmerTurboMassGold GS/MS gas-liquid chromatography-mass spectrometer. The separation was carried out on a capillary column with a length of 30 m, a diameter of 0.32 mm with a phase of PE-XLB, a carrier gas was helium, flow rate equaled 2 ml/min.

The trace element composition was determined by the method of emission spectral analysis on a DFS-458 diffraction spectrograph.

To study the changes occurring on thecatalytic system surface, we used the method of scanning electron microscopy using a Hitachi TM 1000 scanning electron microscope.

3. Results and Discussions

Increasing the temperature of n-hexane conversion to 450 °C does not lead to an increase in the content of coaly matter, however, the yield of gaseous products increases significantly from 13.9 to 17.1 wt % as shown in Table 1.

Table 1. Material balance of n-hexane conversion.

Feedstock	wt %	Products	Composition wt % at the temperature of the experiment: 400°C	450°C
N-hexane	100	Gas	13.95	17.08
		Catalyzate	81.29	77.98
		Coaly matter	4.76	4.94
Total amount	100	Total amount	100	100

The gases include methane, propane, butane, up to 35 wt % iso-butane, pentane and hydrogen. This distribution is shown in Figure 3.It is known that increasing the cracking temperature of linear alkanes above 450-500 °C while maintaining the atmospheric pressure in the reaction medium leads to an increase in the yield of gaseous products with high selectivity for ethylene and propylene.

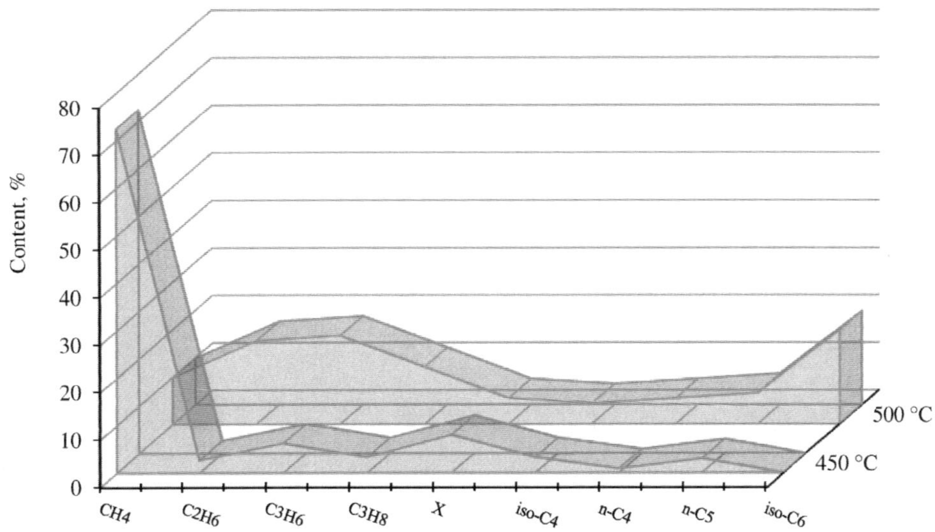

Figure 3. The gases composition of the catalytic conversion n-hexane.

According to the X-ray fluorescence analysis after catalysis, the content of chlorine and aluminum in the catalytic system did not change and amounted to 5.2 and 91.6 wt %. You can see it from Figure 4. The ratio of other metals varies within 0.2 wt %, which is due to the presence of coaly matter (coke) in the spent catalyst. According to X-ray analysis of the catalytic system, it contains gamma aluminium oxide with crystallite sizes of 4.8 nm and lattice parameters (a) 7.92 Å. No other independent phases were observed before and after catalysis on the sample of the catalytic system.

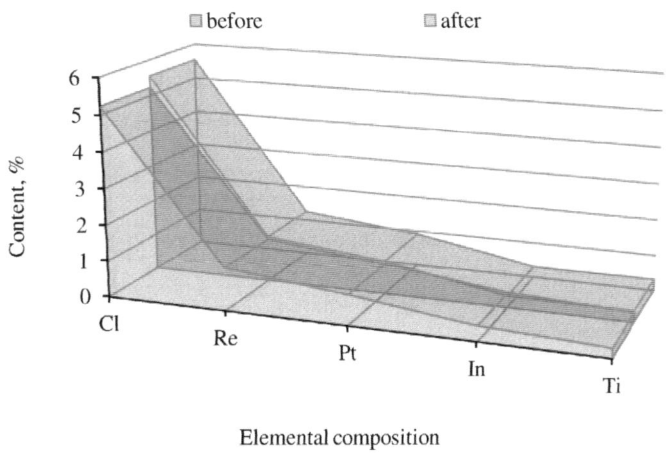

Figure 4. The elemental composition of the catalytic system according to x-ray fluorescence analysis.

Thus, according to X-ray fluorescence and X-ray spectral analysis, the elemental and phase composition of the catalytic system does not change before and after the experiments. Analyzing the morphology of the catalytic system surface before and after catalysis, which was obtained by scanning electron microscopy, we can distinguish the following features of the surface potential image. The surface is represented by grains of rounded shape. The grains have sizes of less than 200 nm and predominantly fused with each other. There are no obvious changes in the surface of the catalytic system after the catalytic tests, what can be observed in Figure 5.

a *b*

Figure 5. The surface of the catalytic system: a) initial surface; b) after catalysis.

Analysis of the catalyzate composition testifies that the highest degree of n-hexane conversion is observed during the second experiment at a temperature of 450 °C and is 15.5 wt %. In the first experiment at a temperature of 400 °C and n-hexane conversion of 8 wt %, the selectivity for aromatic compounds is 45.3 wt %, including the selectivity for methyl-substituted benzenes reaches 23 wt %. In the second experiment, the selectivity for aromatic compounds is 45.6 wt %, including the selectivity for methyl-substituted benzenes increases to 33.2 wt %.The composition of the liquid products of the experiments is presented in Figure 6 in more detail.

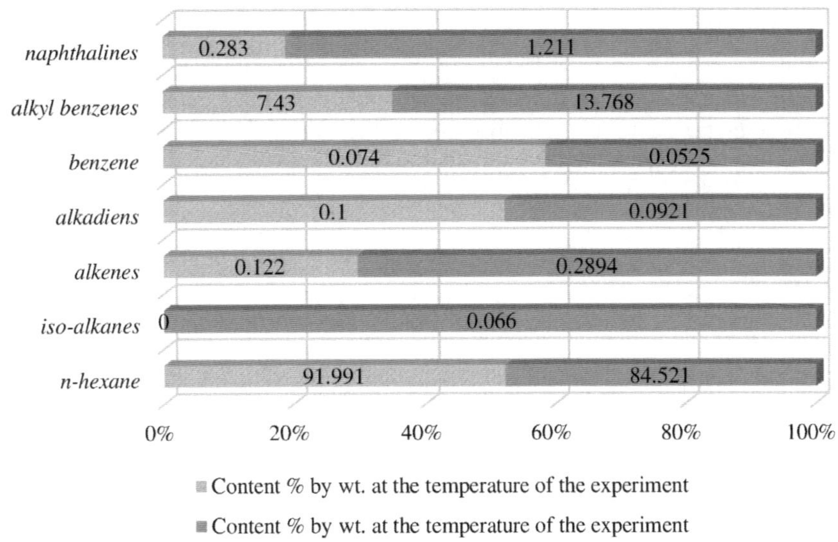

Figure 6. The composition of the liquid products of experiments.

Along with the course of the aromatization reactions of n-hexane, side reactions of cracking and disproportionation occur, leading to the formation of olefins and naphthalenes which are the precursors of coke formation. The presence of platinum in the catalytic system in the reaction conditions for the n-hexane conversion ensures its dehydrogenation with the formation of olefins. Their selectivity in the first and second experiments is 2.8 and 2.4 wt %, respectively. The highest yield of dienes is observed at a conversion temperature of 400 °C. The catalytic system in the absence of hydrogen has low resistance to coking by reaction byproducts, resulting in a reduced selectivity of the process.

The scheme for the n-hexane conversion can be represented as follows. At the initial stage, the dehydrogenation of n-hexane on the metal centres of the catalyst occurs with the formation of hexadiene-1.4. The resulting diene is converted to a carbonium ion, which is isomerized with the participation of γ-Al_2O_3, followed by stabilization to 4-methylpentene-2 at the acid centre. Isomeric cyclization of the carbocation is also possible. The isomeric carbonium ion, returning the proton to the acid centre of the catalyst, is converted into the corresponding olefin, methyl cyclopentene, which is then hydrogenated on the metal centre of the catalyst, turning into methyl cyclopentane. The carbocation resulting from cyclization redistributes the charge to the radical. Skeletal isomerization occurs with the participation of γ-Al_2O_3. The isomeric carbonium ion returns the proton to the acid centre of the catalyst, turning into cyclohexene. The resulting cyclohexene is hydrogenated on the metal centre of the catalyst to form cyclohexane.

Thus, under the action of the catalyst, thermal decomposition of hexane occurs with the formation of a mixture of propane and propylene. Hexane is dehydrated at the metal centre to form methylcyclopentane. The resulting cyclohexene is also dehydrated on the metal centre of the catalyst, forming benzene. In turn, benzene interacts with propylene, forming possible products under the action of γ-Al_2O_3: propyl benzene and isopropyl benzene. Benzene, interacting with two propylene molecules, with the participation of γ-Al_2O_3 turns into 1-propyl, 4-isopropylbenzene.

4. Conclusions

During catalysis, the elemental and phase composition of the catalytic system does not change, coaly matters are formed on the surface in the amount of 4.8 wt %. When the n-hexane conversion temperature increases from 400 to 450 °C, the yield of gaseous hydrocarbons increases by 22 wt %, the conversion increases by 2 times, the total selectivity for olefinic and aromatic compounds remains constant at 45 wt %, while the selectivity for methyl-substituted benzenes increases by 44 wt %.

Acknowledgments

The study was carried out by a grant from the Russian Science Foundation project No. 18-77-10023.

References

[1] Kayukova G P, Mikhailova A N, Kosachev I P et al 2018 *Energy and Fuels* **32** 6488
[2] Vakhin A V, Mukhamatdinov I I, Aliev, F A et al 2018 *Petrol. Sci. Technol.* **36** 1850
[3] Sitnov S A, Vakhin A V, Mukhamatdinov I I et al 2019 *Petrol. Sci. Technol.* **37** 687
[4] Mukhamatdinov I I, Vakhin A V, Sitnov S A et al 2018 *Chem. Technol. Fuels and Oils* **54** 574
[5] Feoktistov D A, Kayukova G P, Vakhin A V et al 2018 *Chem. Technol. Fuels Oils* **53** 905
[6] Gianotti E et al 2014 *Applied Catalysis B: Environmental* **160** 574
[7] Liu X et al 2014 *Chem. Eng. J.* **247** 183
[8] Xia K et al 2016 *J. Catalysis* **338** 104
[9] Baranowska K and Okal J 2015 *Applied Catalysis A: General* **499** 158
[10] Okal J and Zawadzki M 2016 *Applied Catalysis B: Environmental* **194** 22
[11] Baranowska K et al 2016 *Applied Catalysis A: General* **511** 117
[12] Jing Z et al 2018 *J. Fuel Chem. Technol.* **464** 427
[13] Shi C, Zhang P 2015 *Applied Catalysis B: Environmental* **170** 43
[14] Tong R et al 2015 *J. Fuel Chem. Technol.* **4312** 1461
[15] Zhao R et al 2016 *J. Fuel Chem. Technol.* **44 5** 564
[16] Ayodele et al 2014 *Energy Conversion and Management* **88** 1111
[17] Long L et al 2014 *Chem. Eng. J.* **257** 209
[18] Shan Y et al 2015 *Chem. Eng. J.* **278** 240
[19] Vinichenko N et al 2016 *Procedia Eng.* **152** 101
[20] Ziaei-Azad H and Sayari A 2016 *J. Catalysis* **344** 729
[21] D'Ippolito S et al 2015 *J. Molecular Catalysis A: Chemical* **398** 203
[22] Rahmati M et al 2018 *J. Catalysis* **359** 92
[23] Maldonado-Hódar F 2012 *Catalysis Communications* **17** 89
[24] Maldonado-Hódar F 2013 *Catalysis today* **218** 43

IOP Conf. Series: Earth and Environmental Science **282** (2019) 012003 doi:10.1088/1755-1315/282/1/012003

Conversion of n-alkanes C_{11}-C_{22} using a catalyst based on Ni-Al mixed oxides

S Petrov[1,2], A Valieva[1]

[1]Kazan National Research Technological University, 68 Karl Marx St., Kazan, Russia
[2]Kazan Federal University, 18 Kremlevskaya Str., Kazan, Russia

E-mail: psergeim@rambler.ru

Abstract. During the work composition and properties of Ni-Al mixed oxides samples were investigated. These oxides were obtained from Ni-Al layered double hydroxide using thermal and hydrothermal methods. Catalytic cracking of a model mixture of alkanes was carried out in the presence of Ni-Al mixed oxides samples. Gas chromatography, chromatography-mass spectrometry, and IR-spectroscopy were used for the analysis of gaseous and liquid products. X-ray structural, X-ray fluorescence, IR-Fourier spectrometry, the specific surface analyzer, particle size analyzer, microphotographic analysis were used for the analysis of the composition and properties of catalytic systems. According to X-ray structural and X-ray fluorescence analyzes, the composition of the supporting structure of the catalyst Ni-Al mixed oxide did not change after catalysis. The content of their low-boiling homologs increases, alkanes of a branched structure is formed in the products of n-alkanes cracking in the presence of the obtained mixed Ni-Al oxide.

1. Introduction

To date, the latest research in the field of catalysts set a goal to increase the catalytic activity and the specific surface area of catalytic systems for secondary refining processes.

In recent decades, layered double hydroxides are of particular interest in the global scientific communities due to the developed specific surface, high concentration of active centres, the possibility of varying metal cations, thereby changing the acid-base properties, as well as ease of preparation. They are used widely in the field of heterogeneous catalysis, mainly as precursors for the synthesis of catalysts based on mixed oxides. Through the use of various transition metals, it is possible to control compatibility and interconnection of metal-substrate. All these features extend the catalytic scope of layered hydroxides [1].

When the LDH is heated at about 250 °C, the water in the interlayer space is released with following dehydroxylation of hydroxide layers and non-stoichiometric mixed metal oxides with characteristic properties form. They are used as active catalysts for many catalytically important reactions [2]. The advantages of the basic and / or redox properties of layered double hydroxides are used in catalysis, since the dispersion of cations in hydroxylated layers is preserved after calcination, which leads to the formation of well-stabilized metal particles [3].

In the literature [4], various forms of layered double hydroxides such as calcined, rehydrated, functionalized and as supporting structure, have been studied in various catalytic processes, such as partial oxidation and steam reforming of methane to produce hydrogen, one-step liquid-phase heterogeneous synthesis, combined oxidation and dry reforming of methane to synthesis gas, alkylation, isomerization, aromatization of olefins, selective oxidation of phenol to catechin and hydroquinone, benzene to phenol, benzyl hydroxylation, liquid phase oxidation of aromatic hydrocarbons and etc.

The catalytic activity of the catalyst supporting structure based on Ni – Al mixed oxides in the process of higher alkanes cracking is determined in this work.

2. Materials and Methods

The layered hydroxide was obtained by the coprecipitation method [5]. Samples of mixed oxides were obtained in two ways. Ni-Al mixed oxide (sample No. 1) obtained by calcination at 450-500 °C for 4 hours in a nitrogenous medium. Ni-Al mixed oxide (Sample No. 2) was obtained in the reaction medium by heating the reactor 450-500 °C for one hour.

The phase composition of the catalytic system was investigated on a BrukerAxs X-ray diffractometer with Bragg-Brentano geometry using the DIFFRAC.EVA software module and the PDF-2 Release 2013 diffraction data base.

The textural characteristics of the catalytic system were investigated by the method of nitrogen thermal desorption on the NOVA 2200e of the Quantachrom company. The specific surface was calculated by the Brunauer–Emmet–Teller method.

Spectrometric analysis of the samples was carried out on a IR-Fourier spectrometer of the Frontier PerkinElmer.

The elemental composition was obtained using an energy dispersive fluorescence X-ray spectroscopy EDX-800HS2 of the company "Shimadzu".

The size of the aggregates of the samples was determined by scattering and detection of reflected / refracted laser light, red and blue spectra (650 and 405 nm), in accordance with ISO 13320-1 on a laser analyzer of particle size distribution HORIBA LA-950. Measurement range: from 0.01 μm to 3000 μm, measurement frequency 5000 per second. Micrographs of the studied materials were obtained on a Hitachi TM 1000 microscope.

A model mixture of $C_{11}H_{24}$, $C_{12}H_{26}$, $C_{15}H_{32}$, $C_{16}H_{34}$, $C_{17}H_{36}$, $C_{20}H_{42}$, $C_{22}H_{46}$ was selected as the feedstock. The ratios of n-alkanes illustrated in Figure 1. Catalytic cracking of n-alkanes of C_{11}-C_{22} using 1 and 2 samples were carried out on the laboratory "REXO Engineering" catalytic cracking unit with automated reactor system. The process was carried out at a

IOP Conf. Series: Earth and Environmental Science **282** (2019) 012003 doi:10.1088/1755-1315/282/1/012003

temperature of 500 °C, a pressure of 5 atm and a consumption of feedstock of 0.5 ml/min. The multiplicity of feedstock to the catalyst was 9 in experiment No. 1, in experiment No. 2 this ratio equalled 22.

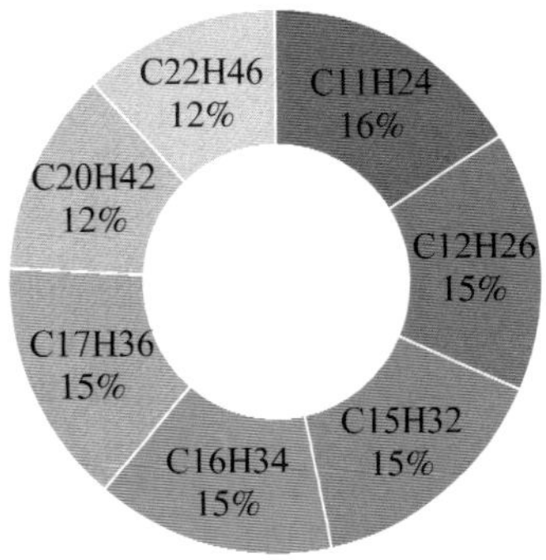

Figure 1. The composition of the model mixture.

The products of conversion were analyzed by gas-liquid chromatography on Crystal-2000M devices of the company Hromatek with a flame ionization detector, a DB-1 capillary column is 15 m long, internal diameter of 0.32 mm, in the linear programming mode temperature from 50 to 320 ° C at a rate of 20 °C/min The sample volume is 1 ml, the dilution is 1:20 in CCl4, PerkinElmerClarus 680 with a SQ8 MS mass spectrometric detector, a capillary column, in the temperature programming mode from 70 to 270 °C. Gas analysis was investigated on a GC 2010 Plus Shimadzu gas chromatograph.

3. Results and Discussions

The resulting catalyst supporting structure based on layered double hydroxide correspond to the structure of hydrotalcite, wich is a natural layered double hydroxide. The coherent scattering region was 7 nm. The diffractogram of the obtained system based on Ni-Al layered double oxide (sample No. 1) contains nickel oxide with a reduced lattice parameter value, which is due to the replacement of a part of Ni^{2+} cations with Al^{3+}. The substitution of Ni^{2+} for Al^{3+} in nickel positions is 16 wt %. The crystallite size of the oxide phases is 4 nm. The Ni-Al oxide phase consists of bunsenite-like octahedral layers with a certain number of spinel layers. The specific surface area of sample No. 1 was 169.9 m^2/g. The diffractogram of the obtained systems is shown in Figure 2.

According to X-ray fluorescence and X-ray diffraction analyzes, the composition of the samples studied remained unchanged. The nickel content in the resulting layered double hydroxides was 86 wt %, aluminium – 14 wt %. This ratio remains the same after catalysis. Layered double hydroxides were preliminarily deposited on ceramic wool, which explains its presence on diffractograms and in elemental composition. The appearance of iron and chromium in the catalytic system after the reaction is explained by the fact that the reactor is made of alloyed steel, the particles of which were adsorbed on the catalyst. Traces of

IOP Conf. Series: Earth and Environmental Science **282** (2019) 012003 doi:10.1088/1755-1315/282/1/012003

elements such as sulfur, zirconium, calcium, potassium are explained by the degree of purity of the raw materials used in the preparation of layered double hydroxides. All these data can see from Figures 3 and 4.

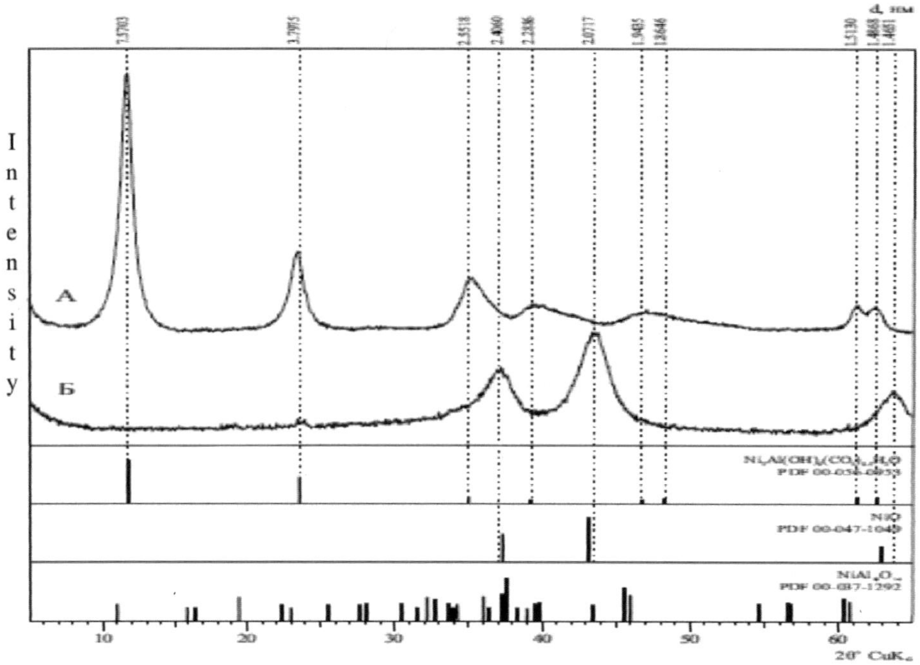

Figure 2. Diffractograms a) Ni-Al layered double hydroxide,
b) Ni-Al mixed oxide (sample No.1).

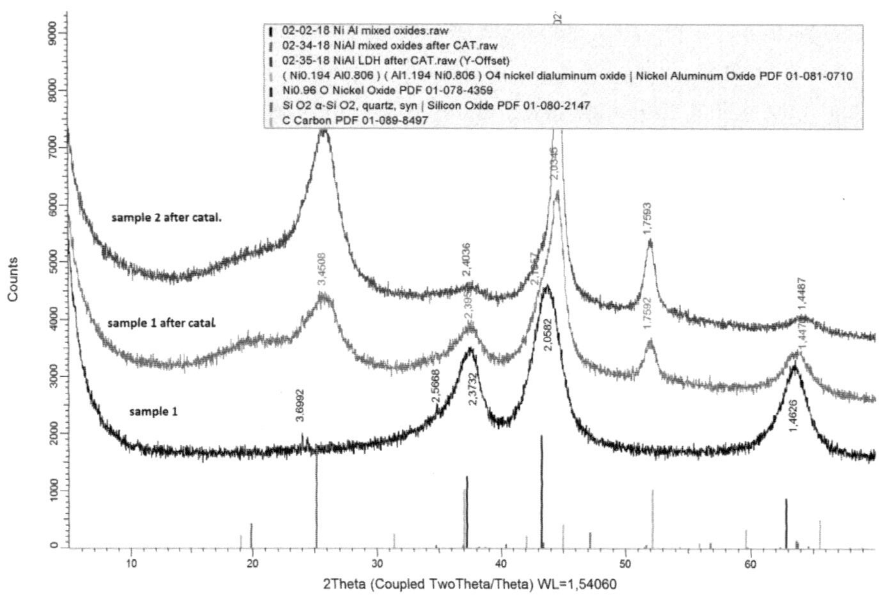

Figure 3. Diffractograms of mixed oxides before and after catalysis.

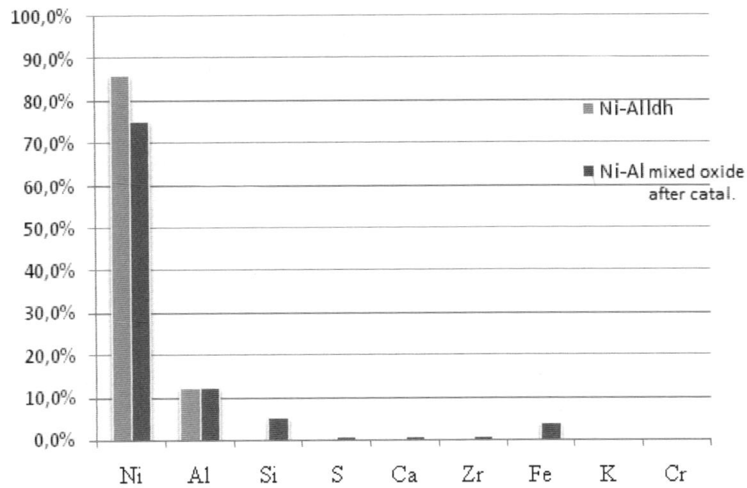

Figure 4. X-ray fluorescence analysis of layered binaries hydroxides before and after catalysis.

The IR spectra of Ni-Al layered double hydroxide, which is shown in Figure 5, have bands characteristic of layered double hydroxides. Reflexes with wavelengths less than 800 cm^{-1} (770.06 cm^{-1} and 561.15 cm^{-1}) are attributed to oscillations in brucite-like layers, the corresponding oscillations of metal – oxygen and the band with a maximum of about 3450 cm^{-1} are oscillations of hydroxyl groups. The interval 3000–3150 cm^{-1} corresponds to the stretching oscillations of water molecules in the interlayer space. The band with a maximum of 1360.17 cm^{-1} indicates the presence of O–C–O carbonate anions in the interlayer space, and the band of 1650 cm^{-1} reflects the deformational oscillations of water in the interlayer space [6]. In the IR spectra of samples 1 and 2, an absorption band of 1050 cm^{-1} is observed, which is characteristic of the ether –C–C–C bond. The formation of this compound is associated with the oxidation process, accompanied by the release of heat, which is confirmed by the data of thermal analysis.

Analysis of microphotographs, which are presented in Figure 6, allows us to conclude that the surface of the LDH is composed of rod-like aggregates with dimensions of 245 nm and diameters of 50-80 nm. In turn, the aggregates form agglomerates larger than 1 micron. During calcination, the relief is represented by pronounced rod-like forms of nickel oxide particles with a length of 600 nm, and the mixed oxide phase is represented by spherical particles with a diameter of 131 nm. When a layered double hydroxide is converted to a mixed oxide under the cracking conditions of the initial mixture of n-alkanes at a temperature of 500 °C, there are no obvious changes in the grain size in the composition of the aggregates, however, the boundaries of their contact degenerate. The probable cause of the observed effects may be coke formation products, which are deposited on the supporting structure surface in the amount of 5 wt %.

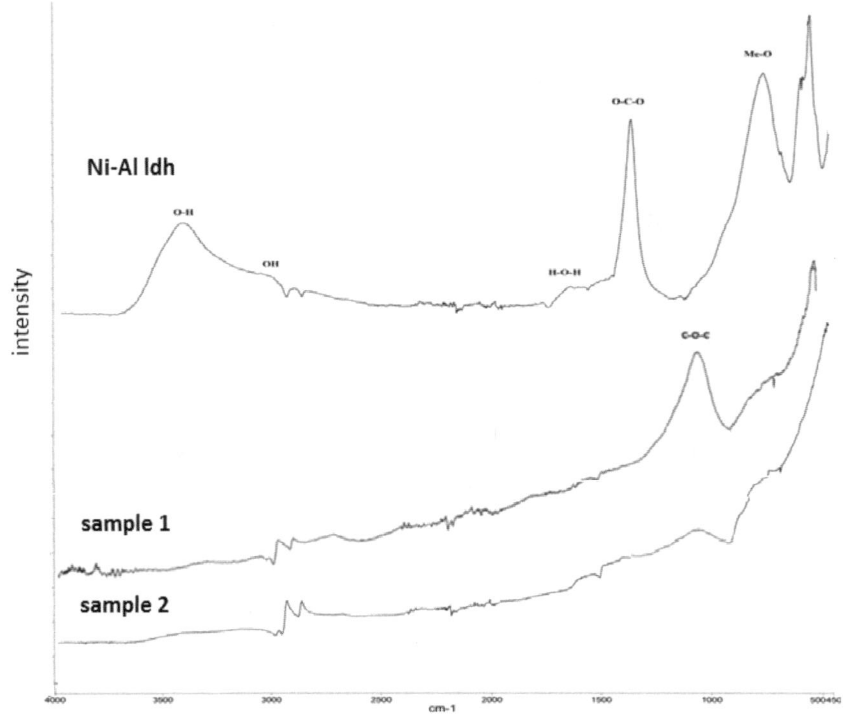

Figure 5. IR spectra of layered double hydroxide and mixed oxides after catalysis.

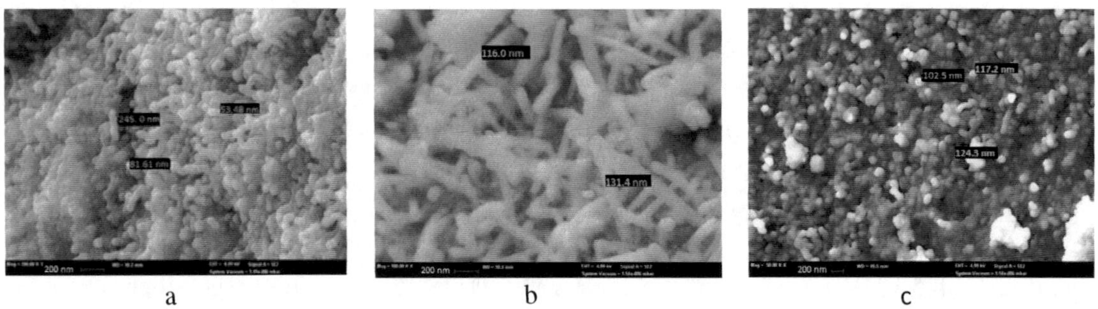

Figure 6. Microphotographs of a) NiAll layered double hydroxides, b) NiAl mixed oxides of oxides (sample No. 1) c) NiAl mixed oxides of oxides after catalysis (sample No. 2).

The composition of the gases produced is shown in the diagram in Figure 7.

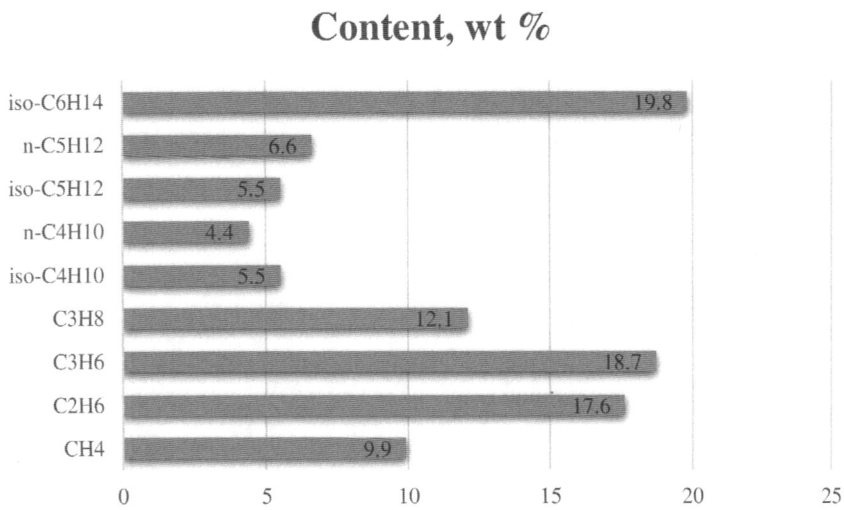

Figure 7. The composition of the resulting gases.

According to the Figure 8, the highest yield of newly formed hydrocarbons is 16.32 wt % of sample No. 2. The yield of hydrocarbons for sample No. 1 was 11.13 wt %. For the studied samples, the selectivity for isoalkanes is observed and amounts to 7.3 wt % (sample No. 2) and 5.5 wt % (sample No. 1). Conversion of sample No. 1 was 39.66 wt %, sample No. 2 was 28.37 wt %. The high conversion for sample No. 1 is associated with a lower feedstock / catalyst ratio. The increase in yield of newly formed hydrocarbons in the presence of sample No. 2 is associated with the calcination mode. According to [7], the maximum concentration of the basic centres is observed at a calcination temperature of 450 °C. The share of the basic centres decreases at 500 °C. The basic centres of the catalytic system have isomerizing functions, which were confirmed by the experiments.

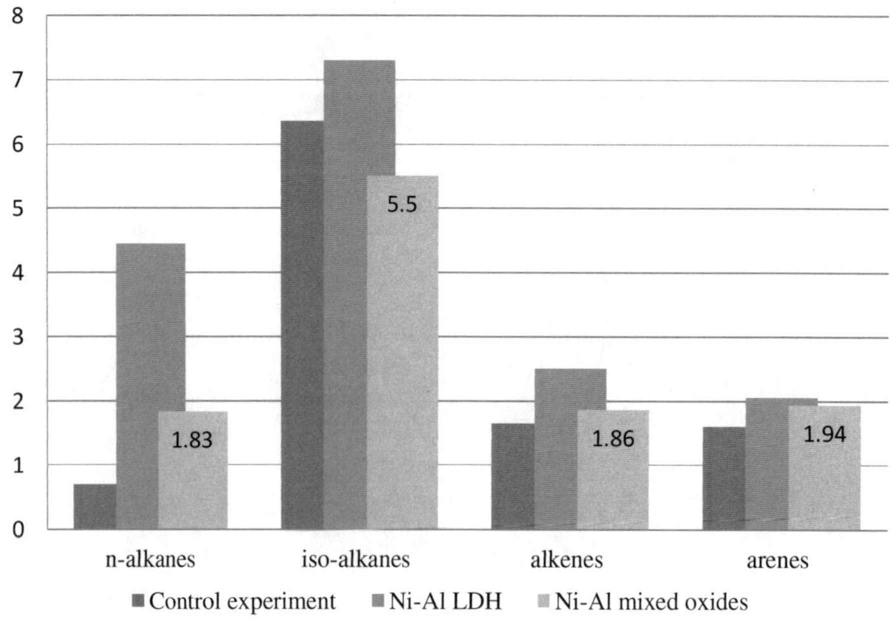

Figure 8. The composition of the resulting hydrocarbons.

4. Conclusions

The composition of Ni-Al mixed oxides samples, which was obtained from Ni-Al layered double hydroxide, was studied. Catalytic cracking of a model mixture of alkanes was carried out in the presence of Ni-Al mixed oxides samples. According to X-ray structural and X-ray fluorescence analyzes, the composition of the supporting structure of the catalyst Ni-Al mixed oxide did not change after catalysis. According to the resulting data, sample No. 2 has the highest yield of iso-alkanes, which is associated with a high concentration of the basic centres on its surface. Conversion for sample No. 1 was 39.66 wt %, and for sample No. 2 equalled 28.37 wt %.

Acknowledgments

The study was carried out by a grant from the Russian Science Foundation (Project No. 18-77-10023).

References

[1] Braterman P et al, *Handbook of layered materials* **1** 373
[2] Duan X and Evans D 2006 *Springer Science & Business Media* **119** 1
[3] Ötvös S et al 2015 *Applied Catalysis A: General* **501** 63
[4] Kannan S 2006 *Catalysis Surveys from Asia* **103-4** 117
[5] Valieva G et al, 2017 *Vestnik of Kazan technological University* **2017** 20
[6] Valente J et al 2009 *Applied Catalysis B: Environmental* **903** 330

[7] Kooli F et al 1997 *Clays and Clay Minerals*vol **451** 92
[8] Xu Z 1999 *Chem. Mater* **111** 67

Features of heavy oil spraying with single evaporation

S Petrov[1,2], A Nosova[1], N Bashkirtseva[1], R Fakhrutdinov[1]

[1]Kazan National Research Technological University, 68 Karl Marx St., Kazan, Russia
[2]Kazan Federal University, 18 Kremlevskaya Str., Kazan, Russia

E-mail: nostya_98@mail.ru

Abstract. Development of resources of heavy hydrocarbons is possible due to the evolution of technologies of their processing, including through the intensification of technological methods implemented in the field of their production at oil treatment plants. The research is devoted to the creation of a method of refining heavy hydrocarbon raw materials based on multiple increases in the working surface of the liquid phase during its intensive evaporation. The intensification of the single evaporation process is based on the fact that the heated raw material is sprayed in a hollow apparatus with the division of the final product into a distillate and a bottomless residue. Due to the spraying of raw materials, the interface of the phases is greatly increased, the supramolecular convective mass transfer becomes so large that it is comparable to the diffusion transfer. The aim of the study is to obtain the data necessary to create energy-saving technologies for the separation of heavy hydrocarbon resources. The paper discusses the possibility of improving the process of single evaporation of heavy oil due to the development of the working surface of the liquid phase, which is realized by spraying the latter. The process of crushing oil into droplets is considered in unity and interrelation with all phenomena occurring in the apparatus. The phase boundary increases several times, which makes supramolecular convective mass transfer comparable to diffusive transfer, which allowed for the separation of heavy oil to increase the distillate selection more than 3 times, with the weighting of the residual product.

1. Introduction

Oil-producing companies are increasingly paying attention to the development of heavy oil and natural bitumen deposits. Consideration of super-viscous oil as a source of hydrocarbon feedstock is associated with large volumes, on the one hand, and with a significant depletion of reserves of oils traditionally used for these purposes, on the other. Their reserves are among the main alternative sources of energy raw materials of the future [1, 2]. However, their processing with the help of known technologies is very difficult, energy-consuming and sometimes unprofitable, because by their physicochemical properties, heavy oils differ significantly from standard oil in low content of fuel fractions, high content of sulfur compounds, resins, asphaltenes, which leads to their anomalously high density , viscosity and lower thermal stability [2, 3, 4]. To ensure an acceptable refining depth for such oil, significant investments are required, as well as high percentage rates of operating costs and working capital. In some cases, the use of features of the flow of gas-liquid media from the nozzles allows to intensify hydrodynamic and heat exchange processes in chemical technology and to reconstruct existing installations for the preparation and processing of oil without attracting large investments. According to most specialists of oil producing companies, the development of heavy oil can become cost-effective due to the evolution of technologies of primary oil refining that can be realized at the production sites at crude oil treatment plants [3, 4, 5].

The work is devoted to the study of the intensification of the process of single evaporation of heavy oil through a multiple increase in the working surface of the liquid phase, due to its dispersion.

2. Experimental section

One of the promising fields of super-viscous oil in the territory of Tatarstan is Ashalchinskoye, which is currently in pilot development. The highly viscous oil of the Ashalchinskoye field is characterized by high viscosity 10000 mPa·s , density 0.9715 g/cm³, content of a significant amount of resins 25-28 wt %, asphaltenes 4.8-5.5 wt %, sulfur 4.0-4.56 wt % and a low concentration of paraffins 0.9-1.6 wt %. The initial thermodynamic parameters of the single evaporation process were: volume flow 1.0 m³/h, temperature 300 °C, nozzle pressure 1.0 MPa. To determine the specific surface area, it is necessary to make a sequential calculation, including. Calculate the median droplet diameter: $d_m = \delta_c * 15 * (\frac{\delta_c * W_c}{v} * 10^{-3})^{-0,4}$, δ_c – film thickness of the fluid in the outlet section of the nozzle, $\frac{\delta_c * W_c}{v} = Re$ – Reynolds number. Calculation of surface linear diameter: $d_o = \sqrt[3]{\frac{6*m}{\pi*\rho}}$, m - distribution coefficient of droplets by fractions; ρ – the density of heavy oil at t = 300 °C. The calculation of the increase in surface: $\frac{n*\pi*d_m^2}{\pi*d_o^2} = \frac{d_o}{d_m}$. The diameter of the nozzle of the nozzle is calculated by the following formula: $d_c = \sqrt{\frac{4*\dot{m}}{\pi*\mu 1*3600*\sqrt{2*\rho*\Delta P}}}$, $\dot{m}$ – mass flow through the nozzle kg / s, $\dot{m} = \varrho^{20} * V^{20}$, ϱ^{20} - the density of heavy oil at t = 20 °C, kg/m³, V^{20} – volumetric flow rate through the nozzle at t = 20 °C, P = 0.1 MPa, 1 nm / h, $\rho^{300} = \rho^{20} - \beta * (t - 20)$ – the density of heavy oil at t = 300 °C, β – expansion coefficient, $V^{300} = \frac{\dot{m}}{\rho^{300}}$ – volumetric flow rate through the nozzle at t = 300 °C, m³/h, μ - consumption coefficient. Film thickness in nozzle exit section: $\delta = 0,5 * d_c * (1 - \sqrt{1 - \varphi})$, $\bar{\delta} = \frac{\delta}{d_c} = 0,5 * (1 - \sqrt{1 - \varphi})$, ϕ – live stream coefficient. The formula for calculating the warm-up time: $\tau_\theta = \frac{(\delta/2)^2}{3*Bi*a_ж} * ln\frac{\theta_o}{\theta}$, δ – film thickness; Bi – criterion "Bio"; a_g – thermal diffusivity.

The experimental part of the work was carried out in a laboratory unit operating at atmospheric pressure, equipped with lines for feeding raw materials, distilling the distillate vapor-liquid mixture, and discharging residual fractions (Figure 1). The installation includes a heated R-102 single-evaporation reactor with a volume of 1 l. Raw material enters R-102, is sprayed through a nozzle located at the bottom of the reactor. The evaporated part of the raw material is discharged through the top of the vessel, enters the heat exchanger-condenser C-101, then through the back pressure regulator BPR-101 enters the separator S-101. From the separator S-101 gaseous reaction products are removed through the upper part, liquid - the drain valve at the bottom of the vessel. VAT residue is pre-cooled in a heat exchanger-condenser C-102, then pumped R-103 in the heated tank T-102. Installation is intended for work in the flowing mode.

3. Results and discussion

In the first approximation, the process was regulated by the aero-thermodynamic parameters in the reactor, depending on the dispersion of the ensemble of drops of their speed, spray angle, contact time of the vapor and liquid phases. With a decrease in the diameter of the centrifugal nozzle and an increase in pressure, the average droplet size decreases (Figure 2), the specific surface area is predicted to be 100 m^2/kg. The result obtained is quite consistent with the information on the increase in the surface when 1 l is disintegrated into droplets with a diameter of 100 nm: ≈ 1200 times and 1240 times. However, bearing in mind the presence of droplets of both large and smaller median diameters, the increase in the surface will be ≈6400 times, since the initial sphere of 0.052 m^2 turns into a collection of drops with a total surface of 330 m^2.

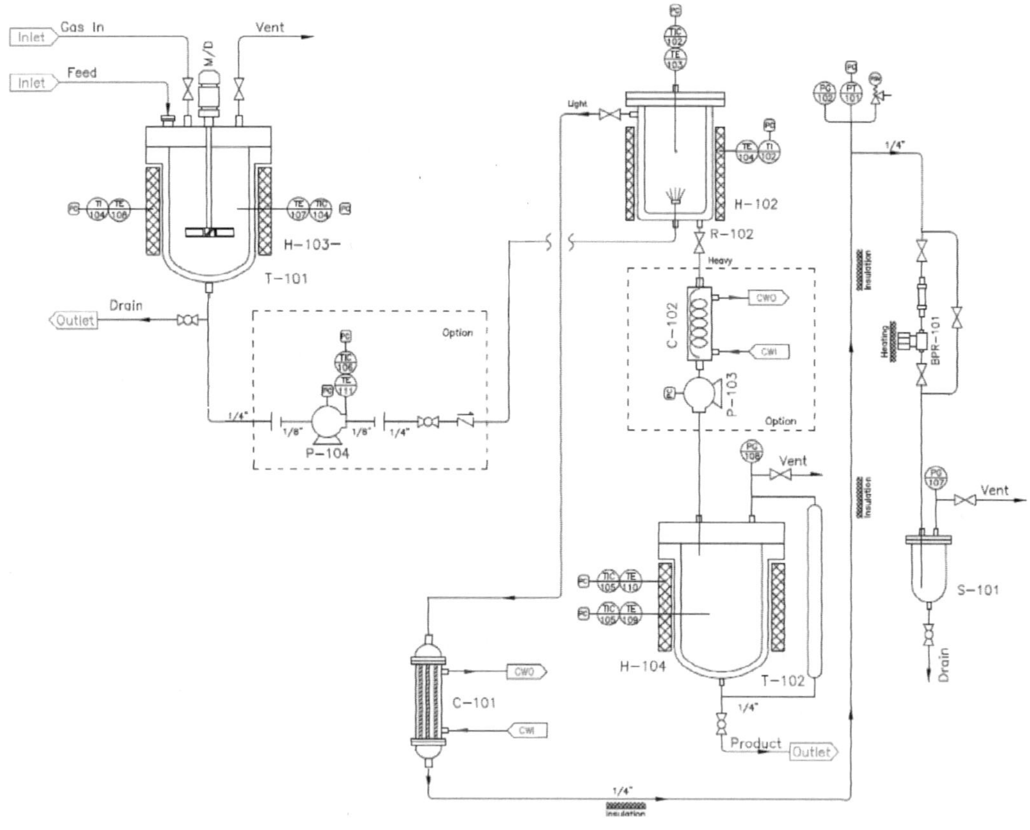

Figure 1. Schematic diagram of the installation.

On the contrary, an increase in the nozzle diameter leads to an increase in the thickness of the liquid film and, accordingly, to an increase in the median droplet diameter and a decrease in the total surface of the droplets (Figure 3). The pressure in front of the nozzles of about 5 MPa prevents heavy oil feedstock from boiling in the pipeline and delays the process of vaporization inside the screwing chamber of the nozzle, while the dispersion of the torch improves, the total surface of the droplets increases.

The residence time of the liquid phase in the atomized state after exiting the nozzle provides evaporation of the droplets, for large droplets the evaporation process is delayed due to poor heating. With an increase in the size of the droplets, the heating time also increases, the evaporation process of the droplets deteriorates, and the range and diameter of the torch increase. Preference should be given to torches with a moderate root-spray pattern. Opened to a large angle, the torch increases the diameter of the apparatus, on the contrary "narrow" torches are better assembled, but the height of the apparatus increases markedly.

The warm-up times increase very rapidly with increasing size of the droplets. For heating large drops, $\theta \approx 1$ s are needed. However, it is difficult to implement such regimes. Therefore, warming up the small drops in a dispersed state, when they form an aggregate torch, you need to take care of the heating of large drops. Since the process of heating large droplets is delayed in time, it is necessary to organize the flow of the liquid film vertically downwards, and the thickness of the film should be minimal and comparable, for example, with a median diameter. The residence time of the liquid in this state should be an order of magnitude longer than the warm-up time for large droplets. Small film thickness will not prevent steam bubbles from escaping.

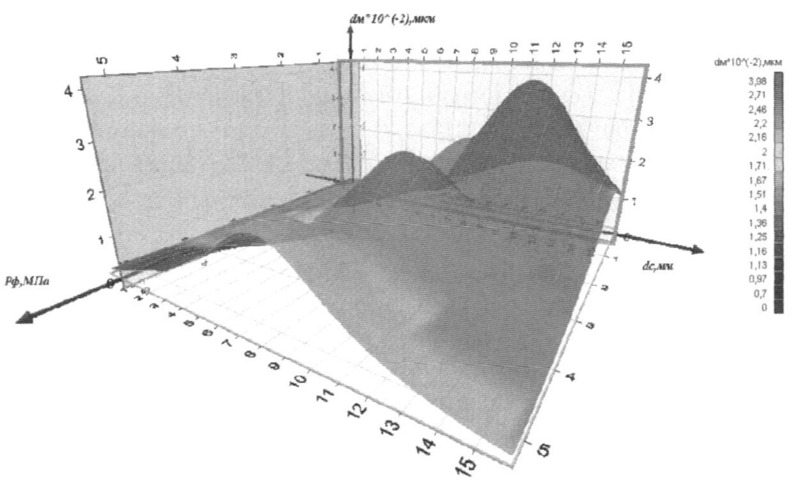

Figure 2. A plot of the nozzle diameter of the centrifugal nozzle, the pressure and the average size of the droplets.

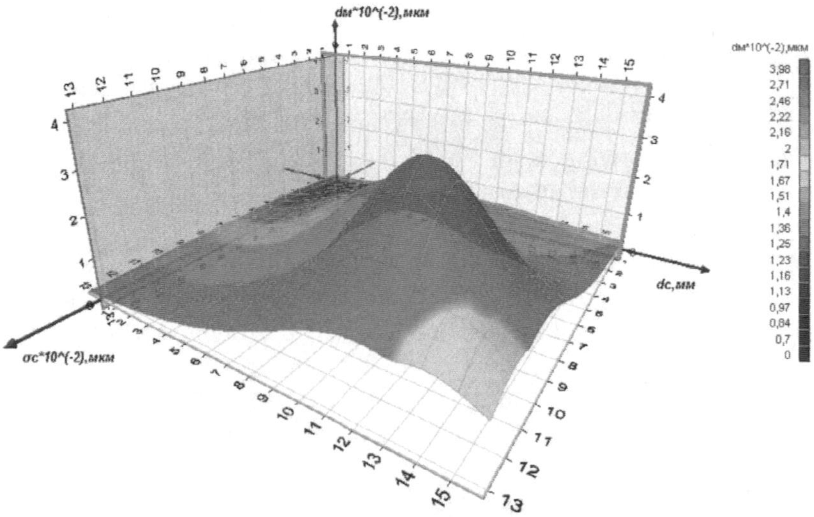

Figure 3. A plot of the nozzle diameter of the centrifugal nozzle, the thickness of the liquid film and the median size of the droplets.

The necessary pressure for the high-quality spraying of heavy oil heated to a temperature of 200-350 °C is 3-5 MPa. To increase the dispersion, it is possible to use acute water vapor or nozzles with rotating elements, thus it is possible to achieve a polydisperse plume of sprayed oil with an average size of droplets from 30 μm to 40 μm. Also, the supply of raw materials in the form of a water-oil emulsion with a water content of up to 10 wt % leads to an improvement in the dispersion of the spray of crude oil due to the additional crushing of droplets due to the instantaneous evaporation of moisture and rupture of the droplets (microexplosions) the diameter of the droplets of liquid fuel sharply decreases to 2 microns.

Under established thermophysical conditions, the heavy oil even before the nozzle will be in a superheated or at least metastable state when it enters the laboratory unit, where the pressure is an order of magnitude lower and the temperature is higher than in the history of the stream, the sprayed oil boils condenses in the refrigerator and is discharged as a distillate. The liquid phase, which is the remainder of the process of single evaporation, is also cooled and collected in a collection. Spraying the raw material in the bottom-up direction allows the light-boiling components enclosed in the drops to diffuse to its surface and evaporate. The evaporation from the surface of the droplet also contributes to their blowing in of the lighter fraction rising from the bottom upwards, while drip entrainment of more high-boiling components occurs, which remain in the residue during classical evaporation (Figure 4). Along with molecular diffusion, convective mass transfer across the interface is initiated.

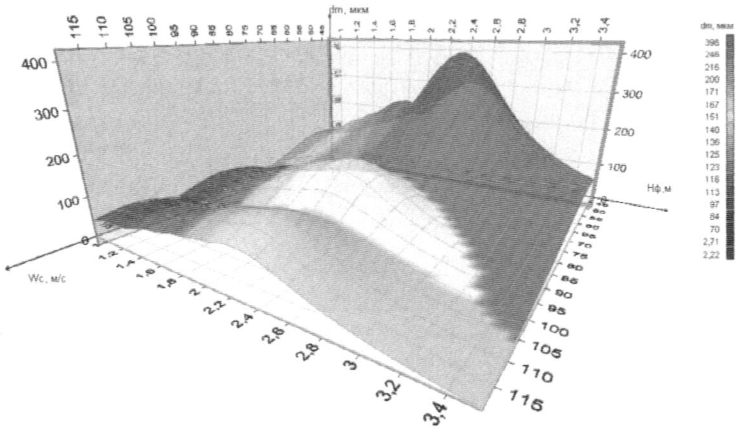

Figure 4. Graph of droplet diameter versus speed and ablation distance.

A significant effect is achieved with increasing process temperature. Comparison of the results of single evaporation of heavy oil without spraying and with spraying through nozzles shows an increase in the distillate yield with raising the process temperature from 15 to 44 wt %.

The spraying of a heavy hydrocarbon feedstock, which is a highly viscous multicomponent liquid, was achieved through the use of a hydraulic centrifugal nozzle. A compact region of suspended droplets of boiling liquid was obtained using a tangential feed of raw materials to the nozzle. Structurally, the nozzle consists of two main parts: a twist chamber and a nozzle (Figure 5, 6).

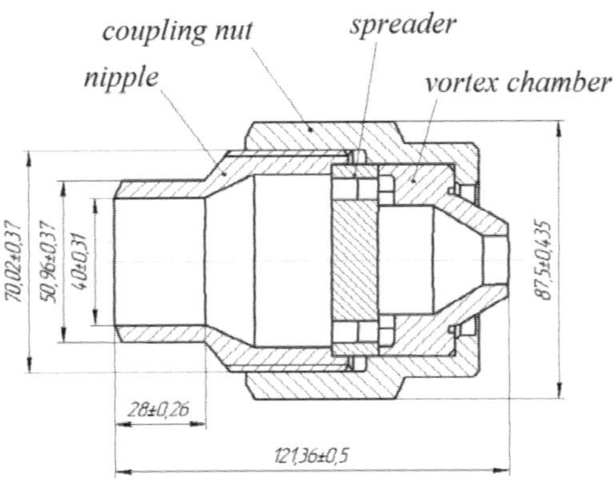

Figure 5. Sketchnozzle.

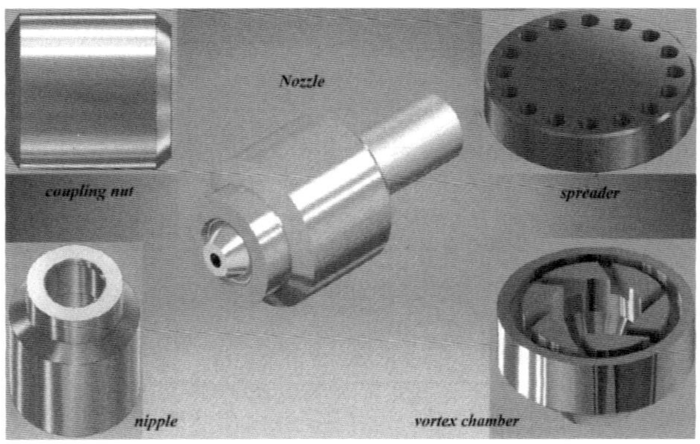

Figure 6. 3D nozzle model.

In the swirling chamber, the oil makes a complex rotational-translational motion and is pressed by the centrifugal force to the walls of the chamber and the nozzle, a space not occupied by oil is formed around the axis of the nozzle - a gas vortex cord that expands when moving towards the nozzle.

Thus, the outflow of oil from the nozzle occurs through the annular section of the nozzle ,minus the radius of the gas vortex. The pressure in the cavity of the vortex cord is atmospheric and equal to the pressure in the apparatus, in which the separation of heavy oil.

An increase in temperature leads to a significant ablation of high-boiling compounds. From this point of view, the small residence time of the droplets in the column and their large sizes should be avoided, which results in an incomplete selection of the distillate part from the potential content in the feedstock. So, for example, the yield of the light part when separated in the apparatus of intensive evaporation at 300 °C was 43 wt %, with the potential content of light fraction boiling up to 300 °C in oil 29 wt %, in turn the yield of the heavy part was 57 wt % (Table 1).

Table 1. Fractional composition.

Single Evaporation	Process temperature	Yield, wt %	
		Distillate	Remainder
Classical	200	12	88
Spray	200	22	78
Classical	250	19	81
Spray	250	33	67
Classical	300	25	75
Spray	300	43	57

At the same time, the content of the gasoline fraction boiling up to 200 °C in the light part decreased 1.5 times to 32% compared with the classical version of the process. From the data obtained it follows that with an increase in the distillate yield at process temperatures of 250 °C and 300 °C to 35 and 43 wt %, respectively, their viscosity at 20 °C does not exceed 20 mm^2 / s, that is, is similar to traditional light oils (Table 2). As a result of intensive single evaporation, sulfur compounds are concentrated in the residual part (Table 3). The sulfur content in the distillate decreases by 2 times, and increases with increasing process temperature.

Table 2. Distillate output.

Fractional composition, i.b.p.-e.b.p., °C	The output of distillate, wt %					
	Classical 200 °C	Spray 200 °C	Classical 250 °C	Spray 250 °C	Classical 300 °C	Spray, 300 °C
i.b.p.-200	91	63	72	49	57	37
200-250	8	15	16	11	14	11
250-300	1	16	9	21	17	17
300-350	-	6	3	19	12	35

Table 3. Residue.

Indicators	Heavy oil	200°C	250°C	300°C	Brent	WTI
		Distillate				
Density at 20 °C, g/cm³	925.1	765.9	791.5	819.9	0.825-0.828	0.817-0.822
Output fractions up to 350 °C, wt %	32	100	100	76	>62	>72
Sulfur content, wt %	3.47	0.95	1.44	1.9	0.37	0.24

The heavy residue resulting from the process includes significant amounts of asphalt-resinous substances with an asphaltene content of up to 15 wt %, and boiling fractions in the temperature range 350-500 °C, ~ 30 wt %.

The separation efficiency, at a constant temperature and pressure gradient, is determined by the equilibrium state between the liquid and vapor phases, which depends on the residence time of oil in the apparatus and the specific evaporation surface area that the raw material will have when sprayed through the nozzle. The optimum is the state of hydrodynamic equilibrium, at which at each moment in time the number of molecules moving from the liquid phase to the vapor phase and back increases. The composition of the distillate can be set over a wide range by adjusting the droplet entrainment of the higher-boiling oil components with a vapor phase, by changing the aero-thermodynamic conditions, the design of the nozzle and its location in the apparatus. Thus, it is possible to ensure the maximum possible yield of distillate with given characteristics.

The proposed process flow diagram for the installation of this process is shown in Figure 7, where heavy oil, pump *H*, is taken from the heated tank *E*-1 to the furnace *P*, where it is heated to the required temperature, then enters the EV evaporator, the vapor phase is then condensed in a heat exchanger *T*-1 and discharged in the form of distillate. The liquid phase, which is the remainder of the process of single evaporation, is also cooled by the heat exchanger *T*-2 and is collected in tank *E*-3.

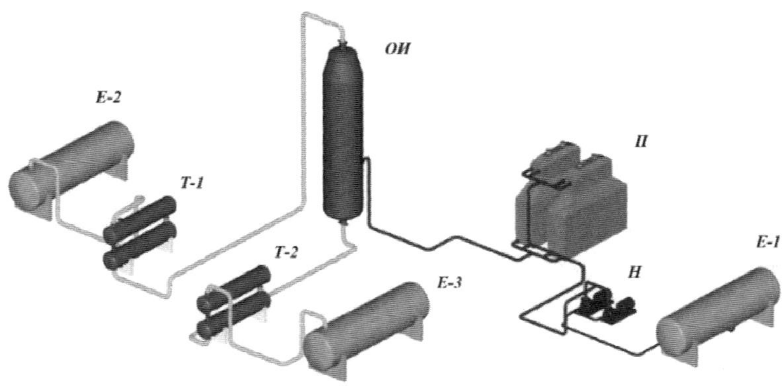

Figure 7. 3D model of technological installation.

4. Conclusions

As a result of the research, the following conclusion can be made: due to spraying of raw materials, the interface of the phases increased significantly, supramolecular convective mass transfer became significant and comparable to diffusion transfer, the shift from phase equilibrium occurs towards increasing the yield of distillate, enriching it with heavy components.

The process of intensive single evaporation can be successfully applied to refining heavy oil. As a result of the process, oil is divided into two components: the distillate is a high-quality light oil enriched in fuel fractions, and the remainder may be suitable for the production of residual bitumens.

Acknowledgments
The work is performed according to the Russian Government Program of Competitive Growth of Kazan Federal University.

References

[1] Petrov S M et al 2015 *Chem. Technol. Fuels Oils* **1** 80
[2] Kayukova G P. Mikhailova A M, Feoktistov D A et al 2017 *Energy Fuels* **31** 7789
[3] Kayukova G P, Mikhailova A N, Kosachev I P et al 2018 *Energy Fuels* **32** 6488
[4] Pat. 2013120452 (RF)
[5] Pat. 2566775 (RF)

IOP Conf. Series: Earth and Environmental Science **282** (2019) 012005 doi:10.1088/1755-1315/282/1/012005

Temperature influence on the composition of high-carbonic Domanic rocks organic matter during hydrothermal treatment in CO_2 atmosphere

G Kayukova[1,2], A Mikhailova[2], Z Nasurova[1], O Sotnikov[3], E Nazimov[3]

[1]Kazan Federal University, 18 Kremlevskaya Str., Kazan, Russia
[2]A.E. Arbuzov Institute of Organic and Physical Chemistry, Kazan Scientific Center, Russian Academy of Sciences, Kazan, Russia
[3]Institute TatNIPIneft PJSC "Tatneft", 75 Lenin St., Almetyevsk, Russia

E-mail: galina.kayukova@rambler.ru

Abstract. Depositions of the Domanic type in the territory of Tatarstan are stratigraphically associated with carbonate-siliceous deposits of the Upper Devonian part of the section. These deposits, enriched with organic matter, are characterized by the ability to generate liquid and gaseous hydrocarbons and accumulate them in the form of deposits of unconventional type. In this paper, the results of geochemical studies of rock samples from the depth interval 1705-1728 m of the Semi-Bug-Buregian deposits of the Berezovskaya area of the Romashkino oil field are presented. It is shown that the rocks, in spite of the narrow core sampling interval, differ in the content of quartz and calcite, as well as in the admixture of minerals such as muscovite, pyrite and albite. The content of organic matter according to thermal analysis data varies from 1.97 to 35.48 %, a significant part of which falls on an insoluble organic matter - kerogen. The influence of hydrothermal processes on the production capacity of the organic matter of these rocks is shown to generate hydrocarbons at temperatures of 200, 250, 300 and 350°C in a carbon dioxide medium. The yield and quality of the products obtained is estimated. The results of the research give grounds to believe that: firstly, the rocks of the Berezovskaya area of the Romashkino oil field have not fully realized their oil potential and the hydrothermal effect on them contributes to the thermal maturation of organic matter in the rock; second, at these temperatures, more intensive recovery of free hydrocarbons from the rock occurs. Thus, the hydrocarbon potential of the house rock is determined by the composition and content of liquid bituminous components that are in a free state in the rock, as well as the products of destruction of tar-asphaltene substances and macromolecules of insoluble kerogen and, consequently, it can be successfully realized with further optimization of hydrothermal processes directed to increase the quantity and improve the quality of the extracted fluids.

1. Introduction

Domanic deposits in the territory of Tatarstan are stratigraphically associated with carbonate-siliceous deposits of the Upper Devonian part of the section. These deposits, enriched with organic matter, are characterized by the ability to generate liquid and gaseous hydrocarbons and accumulate them in the form of unconventional reservoirs. There is a big interest in high-carbon containing rocks, which are enriched with organic matter (OM), because of its reserves and perspective of shale oil production [1-7]. The share of shale oil in world energy balance is significantly increasing. The uniqueness of discovering hydrocarbon reservoirs in shales is not only about new rock types containing huge hydrocarbon accumulation, but also they are the main oil and gas pay zones in their location region and bedded around their formation. Domanic deposits of Volga-Ural petroleum province of Russia are presently considered as an analogue of shale formations [7]. The main part of Domanic formation section takes bituminous rocks with high content of OM from Semikul and Mendim deposits of upper Devonian, which are considered as underexplored oil bearing horizons in territory of Tatarstan and currently are not explored [2-3, 6]. A large part of Domanic OM take bituminous resin-asphaltene components and kerogen. According to modern concepts kerogens are natural geopolymers with irregular structures and are regarded as potential source of petroleum generation [8-9]. Kerogen has solid state, stable chemical structure, specific composition and generally is not dissolved in mineral and organic acids. The Investigation structure of kerogen can be done in laboratory or industrial scale by modeling hydrocarbon formation processes [10]. Thereby, studying Domanic shale is very important, initially for general understanding of oil generation, migration and accumulation issues. Secondly, for solving problems regarding discovery of industrial reservoirs directly in Domanic formations. Thirdly, to reveal the production perspectives of Dominic hydrocarbon resources by modern secondary production techniques.

2. Materials and Methods

The object of geochemical study were samples from Romashkino oil field of Semiluksko-Mendim (Domanic) deposits, Berezev area. The interval of coring sample 1705-1728 m. The data, gained by Simultaneous Thermogravimetry and Differential Scanning Calorimetry allows to evaluate the content of organic matter in rocks and its thermal stability in the range of 20-1000 °C (Table 1). For the modeling of hydrothermal experiments Domanic rocks are used, constituting clay and limestone of the same interval mentioned before. However, the content of OM in the given rock sample is very high – 35.48 %, from this 25.43 % stands for kerogen. The hydrothermal conversion experiments of OM of Domanic rocks are carried in laboratory autoclaves with a volume of 1 L (Parr Instruments, USA) at temperatures 200, 250, 300 and 350 °C during 5 hours in CO_2 environment. The water content in system is 30 wt % of rock sample taken for the experiments. The initial pressure of the system was 2 MPa. During the process, depending from temperature, the vapor pressure of mixtures rises up to 17 MPa. The organic matter content in given samples was evaluated by loss of mass in 200-600 °C temperature range. For rock samples more than 650 °C, the mass loss is explained by thermal cracking of carbonates. And the mass losses for temperatures 20-150 °C are related with evaporation of light fractions and cleaning free and adsorbed water in clay-carbonic minerals [11], [12].

3. Results and Discussions

Table 1. The thermal analysis data of Domanic (Berezev area) rock samples from different coring intervals.

Sample	Coring interval, m	Mass loss of rocks in temperature ranges of 20-1000 °C						
		20-200	200–400	400–600	600–800	800-1000	∑ OB	F_{OB}**
1	1705.0-1712.5 (3.56-3.59)	0.34	1.26	4.48	9.04	7.74	5.74	0.28
2	1712.5-1718.5 (2.67-2.73)	0.27	2.17	9.56	11.77	3.12	11.73	0.23
2*		0.50	1.43	8.76	11.04	5.44	10.19	0.16
3	1712.5-1718.5 (4.84-5.00)	0.14	0.50	1.47	7.48	32.04	1.97	0.34
3*		0.15	0.16	0.99	10.16	29.30	1.15	0.16
4	1718.5-1724.5 (4.00-4.09)	0.34	2.68	11.12	12.53	10.47	13.80	0.24
5	1718.5-1724.5 (5.19-5.28)	0.48	2.65	8.01	10.85	1.79	10.66	0.33
6	1724.5-1728.0 (2.77-2.85)	0.25	0.82	4.18	10.35	20.86	5.00	0.20
6*		0.19	0.32	3.62	9.05	24.01	3.94	0.09

*Rocks after extraction; **$F_{OB}=\Delta m_1(200\text{-}400°C)/\Delta m_2(200\text{-}600°C)$

Analyzing samples from Romashkino oil field of Berezev area of Semilukskoe horizon, Domanic deposits revealed heterogenic rock composition in small coring interval. The content of OM in rock samples varies from 1.97 to 13.80 %. The minimal OM amount is observed in samples 3 and 6. The former constitutes for dark gray clays and the latter for transition of dark gray clays to limestone. Oppositely, samples 2 and 4 are rich in OM. The former is also constitutes for dark gray clay, but sample 4 is recrystallized limestone transiting from gray to dark gray color. The given samples are characterized by enough high content of kerogen. This is explained by mass loss in temperature range of 400-600 °C from 1.47 to 11.12 %. Based on fractional value $FOM=\Delta m_1(200\text{-}400\ °C)/\Delta m_2(200\text{-}600\ °C)$ after extraction, the OM content in rocks decreases, which is mostly realized in samples 3 and 6, which shows free hydrocarbon content is high enough. Table 2 provides the thermal analysis data before and after hydrothermal treatment at various temperatures.

Table 2. The thermal analysis data before and after hydrothermal treatment at various temperatures.

samples	The mass loss of rocks in temperature range of 20-1000 °C						
	20-200	200–400	400–600	600–800	800-1000	∑ OB	F_{OB} **
Initial rock							
1	0.96	10.05	25.43	11.32	0.45	35.48	0.39
1*	0.76	9.33	24.99	12.94	0.36	34.32	0..37
After thermal treatment T 200 °C, P 7.6 MPa							
1.1	0.76	13.04	18.41	18.85	0.18	31.45	0.71
1.1*	0.61	11.35	18.21	18.80	0.41	29.56	0.62
After thermal treatment T 250 °C, P 8.3 MPa							
1.2	0.49	12.40	16.14	13.43	0.21	28.54	0.77
1.2*	0.28	10.19	15.06	10.54	0.45	25.25	0.68
	After thermal treatment T 300 °C, P 10.5 MPa						
1.3	0.89	11.17	15.28	13.50	0.32	26.45	0.73
1.3*	0.35	9.30	12.45	12.54	0.35	21.95	0.75
After thermal treatment T 350 °C, P 17.0 MPa							
1.4	1.02	8.34	16.36	13.36	0.13	24.69	0.73
1.4*	0.46	8.31	10.58	12.57	0.38	18.89	0.79

*rock sample after extraction; **$F_{OB} = \Delta m_1(200\text{-}400\ °C)/\Delta m_2(200\text{-}600\ °C)$

According to thermal analysis of initial rock samples, the low value of fractional composition FOM = 0.39 is observed, that proves the small amount of mobile hydrocarbon and large content of kerogen are in rocks. Increasing the given value for rock samples after hydrothermal treating indicates about formation of light hydrocarbons due to destruction of high molecules in free bituminous components, which are in both rocks and insoluble kerogens. After experiments and extraction, the OM content decreases, because of new formed hydrocarbons. The kerogen content doesn't change much and stays actually the same as in experiment at 350 °C. However, the mass loss of rock samples in 200-400 °C range is less, which indicates less destruction of OM with free hydrocarbon formation. The hydrothermal treatment of rock samples comparatively at lower temperatures, i.e. 250, 200 °C results to intensive extraction of free hydrocarbons existing in the structure of kerogen as liquid phase [13]. It is important to mention that kerogen from Domanic rocks of Berezev area are not destructed totally even at 350°C (11.35 vis. 25.43 %). The extraction yield from Domanic rock samples after hydrothermal treatment at all investigated temperature ranges is increasing according to initial rocks (Table 3). The saturate content is high in the products of group composition of experiments. The aromatic compounds, as well as resins are decreasing. The products of asphaltenes consist of two fractions, containing normal asphaltenes (Fraction A), which are dissolved in aromatic solvents and their modification products – not dissolving carbene-carboid types (Fraction B).

Table 3. The group composition of extracts from rock samples of Berezev area before and after hydrothermal experiments at various temperature ranges in CO medium.

Sample	Content of bitumoid, wt %	*Group composition, wt %				
		SH	AC	Resins	Asphaltenes	Carbene-Carboids
Initial rock						
1	0.25	18.58	38.94	28.32	14.16	-
Rock after experiment T 200°C, P 7.6 MPa						
1.1	1.90	29.22	25.80	24.89	18.72	1.37
Rock after experiment T 250°C, P 8.3 MPa						
1.2	2.28	32.46	28.24	22.98	14.21	2.11
Rock after experiment T 300°C, P 10.5 MPa						
1.3	3.85	31.86	24.11	24.05	13.84	6.14
Rock after experiment T 350°C, P 17.0 MPa						
1.4	4.56	35.69	23.45	24.48	12.07	4.31

SH – saturate hydrocarbons, AC – aromatic compounds

In the work [14] from rock samples of Domaic deposits (Sarmanov area, Romashkino Oil Field) we have determined the main relations, characterizing hydrothermal treatment processes to OM and kerogens. These dependencies show the transformation of structural elements in kerogen, by aliphatic bonding with further transition to free micro oil of large fragments, containing naphthenic and aromatic hydrocarbons. Hydrothermal treatment breaks molecular aggregates of asphaltenes as well. This is explained by dealkylation reaction, which brings compaction of carbon skeleton, increasing the degree of structurization and carbonization. The existence of two solid disperse phases that varies by their structure and dissolving in organic solvents, in the products of hydrothermal experiments is also a result of dealkylation reaction. The element analysis of asphaltenes (Fraction A) and carbene-carboids (Fraction B) after hydrothermal treatment indicates a significance increase in sulfur content, most part of which is concentrated in carbene-carboid compositions. In this fraction a hydrogen is decreasing noticeably, but carbon content is increasing. This reflects to declining the proportion value H/C_{at}. The nitrogen content also increases in asphaltenes of initial rocks. Besides the most part takes place in asphaltenes of fraction A. In experimental products of asphaltenes (Fraction A) the free radical concentration relative to vanadium (V^{+4}) concentration, which is in vanadyl-porphyrin complex composition, is increased as temperature rises [15]. This indicates to more carbonization of its structure. This reflects to increasing value of R^*/V^{+4} (from 0.35 to 0.55, 0.51, 0.74 and 0.8), as temperature rises. The abnormal high values (2.42, 3.2, 2.39 and 3) belong to carbene and carboid type compounds, which are differed from asphaltenes by low content of alkyl radicals, vanadyl-porphyrin complexes and other functional groups.

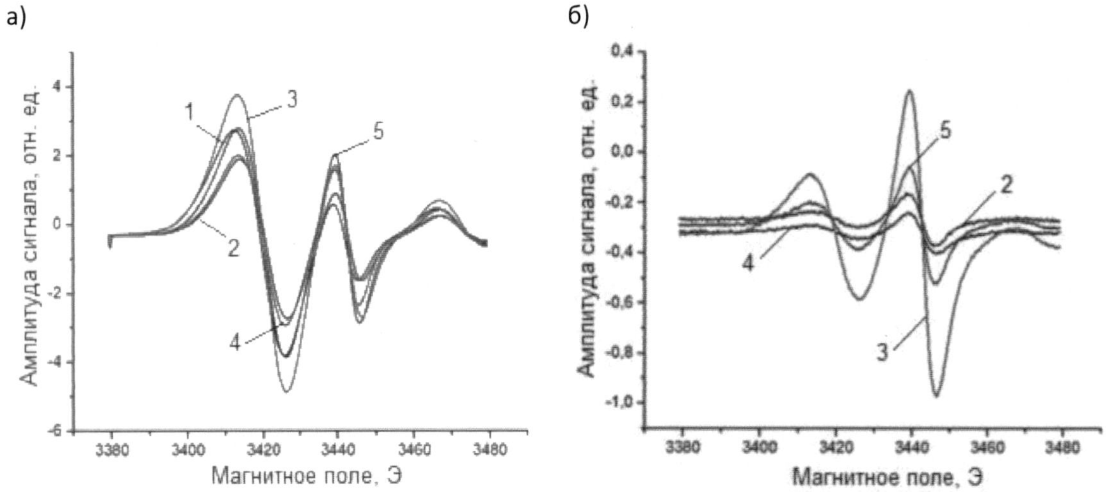

Figure 1. EPR spectrum of asphaltenes (left) and carbene-carboids (right) from Domanic (Berezev area) deposits: 1 - initial rock; 2 - product of experiments at 200 °C, 3 -products at 250 °C , 4 - products at 300 °C, 5 – products at 350 °C

Thus, the kerogen destruction after hydrothermal experiment and increasing high molecular component in the products indicate Domanic rocks from Romashkinskoe Oil Field are not suffered by high temperature effects in the genesis process and consequently their oil generating potential may be practiced by application of hydrothermal technologies.

4. Conclusion

Depositions of the Domanic type in the territory of Tatarstan are stratigraphically associated with carbonate-siliceous deposits of the Upper Devonian part of the section. These deposits, enriched with organic matter, are characterized by the ability to generate liquid and gaseous hydrocarbons and accumulate them in the form of deposits of unconventional type.

The effect of hydrothermal processes to yield-producing ability of OM to generate carbohydrates at temperatures 200, 250, 300 and 350 °C in CO surrounding is revealed. The results of experiment indicate that oil generating potential of rocks from Domanic horizon, Berezev area, Romashkino Oil Field are not totally implemented. The hydrothermal influences on these rocks result to thermal maturity of OM. In the given temperature ranges free hydrocarbons may be extracted more intensively from rocks. The hydrocarbon potential of Domanic rocks are defined by composition and liquid bituminous content. As well as destruction products of resin-asphaltene compounds and macromolecule of kerogens. Hence, oil generating potential of Domanic rocks may be successfully realized by further optimizing hydrothermal processes, directed to enhancing hydrocarbon recovery.

Acknowledgments
The work is performed according to the Russian Government Program of Competitive Growth of Kazan Federal University.

References

[1] Speirs J, McGlade C, Slade R 2015 *Energy Policy* **87** 654
[2] Khisamov R S, Bazarevskaya V G, Timirov D A et al 2014 SPE Russian Oil and Gas Exploration and Production Technical Conference and Exhibition, Paper No. 112320
[3] Burnham A K. 2017 *Fuel* **203** 208
[4] Ivanova A G, Vakhin A V, Voronina E V et al 2017 *Bul. Rus. Academy Sci. Physics* **81** 817
[5] Khisamov R S, Bazarevskaya V G, Tarasova T I et al 2016 *Neftyanoe Khozyaystvo - Oil Industry* **7** 10
[6] Kayukova G P. Mikhailova A M, Feoktistov D A et al 2017 *Energy Fuels* **31** 7789
[7] Muslimov R Kh, Plotnikova I N 2014 *Neftyanoe Khozyaistvo - Oil Ind.* **1** 12
[8] Bushnev D.A., Burdel'Naya N.S., Mokeev M.V. et al 2010 *Doklady Earth Sc.* **430** 210
[9] Burdelnaya N S, Bushnev D A, Mokeev M V 2013 *Geochemistry* **9** 819
[10] Patent RU 2263774 C2. Passi K.R., Thomas M.M., Bohaks K.M. 2015
[11] Kök M V 2008 J. *Thermal Analysis and Calorimetry* **91** 763
[12] Almyashev V.I., Gusarov V.V., Thermal analysis methods: Teaching experience. Allowance. SPb.: Publishing and printing center SPbGETU (LETI), 1999. - 40 p
[13] Batalin O Yu., Wafina N G 2013 *Int. J. Appl. Fund. Res.* **10** 418
[14] Kayukova G P., Kiyamova A M., Mikhailova A. N et al 2016 *Chem. Technol. Fuels Oils* **52** 149
[15] Gilinskaya L G 2008 *J. Struc. Chem.* **49** 245

Investigating the effect of iron oxide (II, III) nanoparticles on aquathermolysis of heavy oil

S Sitnov[1], I Mukhamatdinov[1], Y Tubman[1], O Slavkina[2], K Bugaev[2]

[1] Kazan Federal University, 18 Kremlevskaya Str., Kazan, Russia
[2] OJSC RITEK, 3 Bolshaya Ordynka, Moscow, Russia

E-mail: sers11@mail.ru

Abstract. For many years, enhancing heavy crude oil recovery has been widely considered as an important challenge for many researchers who, as a result, proposed the use of catalytic systems as an alternative way to solve it. Among a wide range of catalytic systems, nano-sized catalysts based on transition metals in several studies showed an effective catalytic activity and high diffusion through reservoir rock porous media during the process of oil aquathermolysis. This paper aims to widen our current knowledge about catalytic oil aquathermolysis by studying the ability of using catalysts based on iron oxide (II, III) nanoparticles combined with hydrogen donor in order to improve heavy oil hydrocarbon content and to reduce its viscosity. Therefore, we performed oil aquathermolysis at 200 and 300°C for 24 hours. In the case of the catalytic process, catalyst content was 0.3 wt % and hydrogen donor content was 3 wt %. As result, SARA analysis of upgraded oil showed a decrease in the resins content by 25 % and an increase in aromatic fractions due to the destruction of C-S bonds compared to the initial oil. Despite the fact that asphaltenes content remained practically constant, their molecular weight decreased. The evidence from this study suggests that heavy oil viscosity decreases by more than 67 % in the presence of nanoparticles based on iron oxide (II, III).

1. Introduction

Heavy oil extraction has been always associated with quite complex and costly technical solutions due to its high density and viscosity. Thermal methods particularly, steam injection technology, from one hand, are among the most widely applied approaches for the deposits of such resources. From another hand, their coupling with catalysts provides an increase in oil recovery factor by reducing the content of high-molecular compounds, and hence result an irreversible decrease in viscosity [1-3]. In the literature, there are various types of catalysts as oil-soluble, water-soluble metal compounds, and in form of dispersed nanoparticles [4].

Two major advantages of using nano-catalysts: the first one is related to their high surface area, which ensures a high activity during oil aquathermolysis while the second advantage is associated with their ability to penetrate deep into the pore space of the reservoir rock [5, 6]. In addition, nanocatalysts enhance oil recovery and facilitate its separation from oil after thermal treatment [7].

Metal oxides nanoparticles especially iron oxides have been receiving much attention from many scientists due to their high catalytic activity in cracking heavy oil [8–10]. Experiments on heavy oil aquathermolysis in the presence of iron oxide nanoparticles were conducted by of researchers [11, 12] who showed a high efficiency of various concentrations of Fe_2O_3 nanoparticles on heavy oil aquathermolysis at different temperatures. Experimental data led to a reduction in oil viscosity by more than 50 % as a result of activating some reactions by the catalyst. It has been established that injection of α-Fe_2O_3 nanoparticles together with steam increases the flow rate of heavy oil due to cracking reactions, which destroy the C–S, C=C, and C≡C bonds of the heavy components of heavy oil and change its composition by converting them into light components [13].

In this work, catalytic efficiency of synthesized iron oxide (II, III) nanoparticles for improving steam-heat treatment efficiency of heavy oil was studied in combination with a hydrogen donor.

2. Methodology

In order to synthesize iron oxide (magnetite) we have mixed two salts in an aqueous medium according to the precipitation method.

Aquathermolysis process laboratory modeling was carried out in a high-pressure reactor (300 ml volume) manufactured by Parr Instruments, USA. Physical modeling of the catalytic and non-catalytic aquathermolysis were performed using a sample of high-viscous oil from Ashalcha field of the Republic of Tatarstan at 200 and 300 °C for 24 hours. The catalyst and the hydrogen donor amounts to oil were 0.3 wt % and 3.0 wt % respectively. The aqueous phase/oil ratio was fixed at 30:70.

In accordance to ASTM D4124-09 and GOST 32269-2013 in several stages, we applied SARA analysis method in order to estimate oil fractions. This method aims to separate bitumen into four groups of fractions: saturated hydrocarbons, aromatic compounds, resins and asphaltens (SARA) depending on their solubility and polarity.

Dynamic viscosity (at 10 °C) was determined on a Fungilab Alpha L rotational viscometer (Spain) in combination with a Huber K-6 cryostat thermostat (Germany).

MALDI experiments were carried out in a Bruker Daltonics Reflex IV MALDI-TOF mass spectrometer.

At the end of oil catalytic aquathermolysis process, solid catalyst particles were extracted from oil using toluene. Surface morphology analysis of the sample was provided by Merlin high-resolution field-emission scanning electron microscope from Carl Zeiss (Germany). Analysis showed spherical particles of iron oxide with 70 nm size (Figure 1).

IOP Conf. Series: Earth and Environmental Science **282** (2019) 012006 doi:10.1088/1755-1315/282/1/012006

Figure 1. SEM image of the catalyst particles surface.

3. Results and Discussions

SARA (Saturates, Aromatic hydrocarbons, Resins, Asphaltenes) analysis data of oil are presented in Table 1.

Table 1. SARA analysis data of oil.

Objects	SARA fractions, wt %			
	Saturates	Aromatics	Resins	Asphaltenes
Initial oil	34.31	34.91	24.68	6.10
Oil after non - catalytic aquathermolysis at 200 °C	28.62	30.84	35.09	5.44
Oil after catalytic aquathermolysis at 200 °C	30.89	40.44	22.85	5.82
Oil after non - catalytic aquathermolysis at 300 °C	33.46	38.47	22.50	5.58
Oil after catalytic aquathermolysis at 300 °C	35.32	41.04	18.43	5.21

According to SARA analysis, the non-catalytic process at 200 °C is includes a high content of resinous compounds and a low content of total light fractions due to the compaction of the system during the recombination of radicals resulting from the steam-heat effect. However, the thermocatalytic effect and the destruction of C–S bonds in the high molecular weight components of heavy oil lead to decrease the amount of resins and the transfer of detached fragments to the aromatic fraction.

IOP Conf. Series: Earth and Environmental Science **282** (2019) 012006 doi:10.1088/1755-1315/282/1/012006

Since the studied oil is biodegraded [3], the content of saturated hydrocarbons and asphaltenes practically does not change due to the absence of the process of alkyl substituents separation from the latter molecules. Nevertheless, the conversion of asphaltenes occurs by heteroatomic bonds, as shown by the results of determining the molecular weight (Figure 2). Thermocatalytic destruction ensures the destruction of dense packs of asphaltenes and, as a result, a decrease in their molecular weight by almost 1.5 times at a temperature of 300 °C.

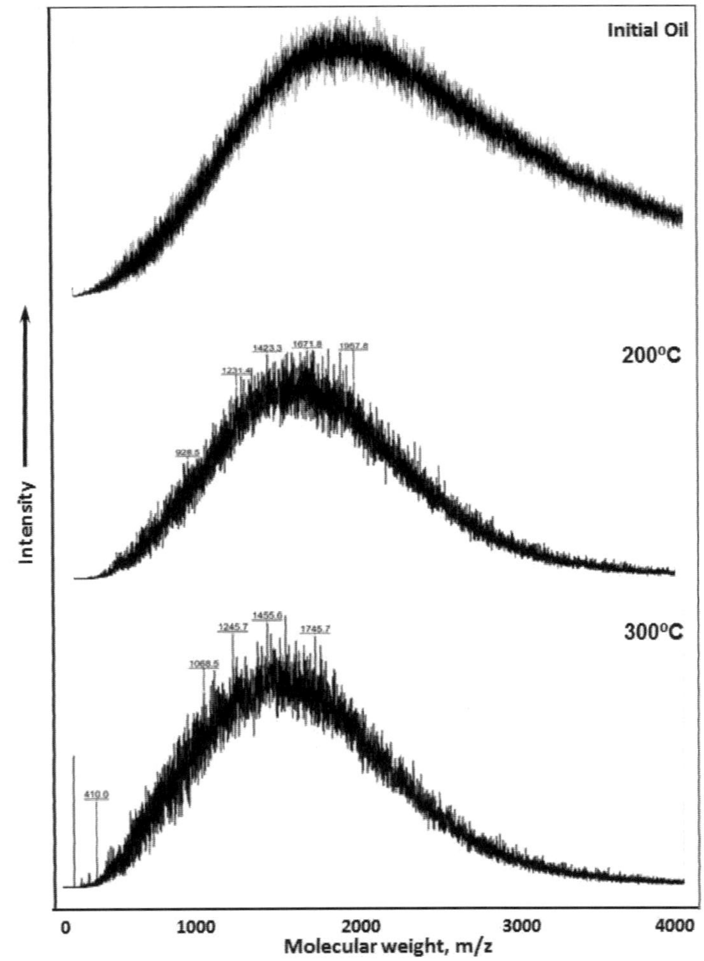

Figure 2. MALDI mass spectra of the original oil asphaltenes and the products of catalytic aquathermolysis at different temperatures during 24 hours in the autoclave.

All this is reflected in the values of the dynamic viscosity of the selected objects for study, which correlate with SARA-analysis data (Figure 3).

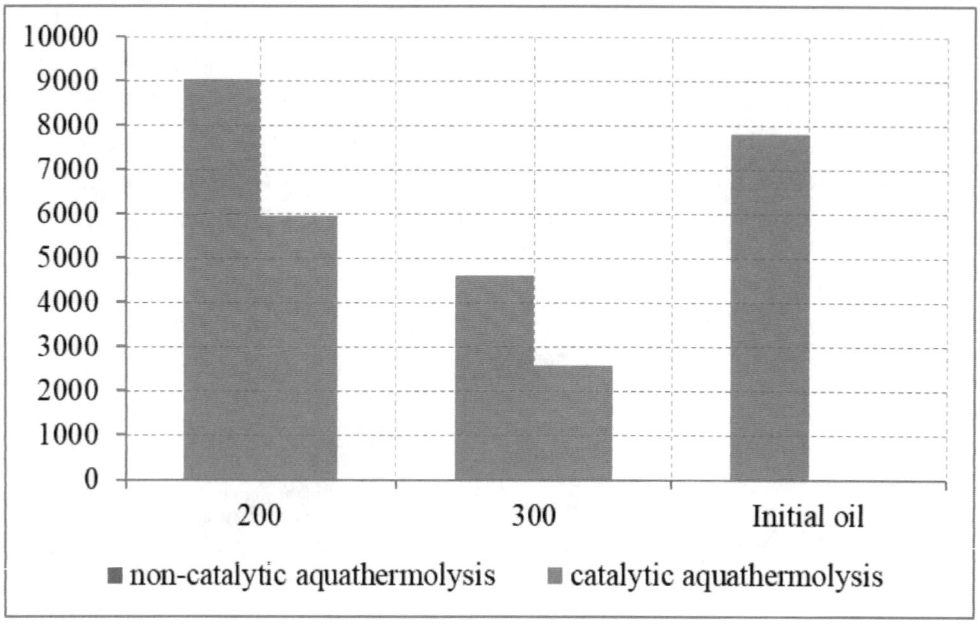

Figure 3. Dynamic viscosity dependence (measurement at 10 °C) on the temperature of the steam-thermal effect at 24 hours of exposure.

As shown on the histogram, oil mobility in all cases becomes higher after the catalytic process comparing to experiments without catalyst and to initial oil, which highlights the high effectiveness of introducing the catalyst in the process of heavy oil steam-heat treatment. Therefore, the greatest effect obtained from the experiment in the presence of the catalyst was at a temperature of 300 °C where viscosity reduced to more than 67 % compared to the initial oil.

4. Conclusion

This paper has investigated the effect of iron oxide nanoparticles on heavy oil fields of the Republic of Tatarstan during steam-heat treatment at 200 and 300 °C for 24 hours. In general, analysis results combined with dynamic viscosity determination point toward the high efficiency of iron oxide as a catalyst of oil heavy components cracking, mainly resins, which content decreased by almost 25 %. We have found that the molecular weight of asphaltenes decreased by (almost 1.5 times), even if their content practically does not change when increasing the temperature of aquathermolysis. Our investigations into this area are in progress and seem likely to confirm our hypothesis.

Acknowledgments
The work is performed according to the grant of the President of the Russian Federation for young scientists-candidates of sciences 075-02-2018-110. The work is performed according to the Russian Government Program of Competitive Growth of Kazan Federal University.

References

[1] Sitnov S A, Mukhamatdinov I I, Vakhin A V et al V 2018 *J. Petrol. Sci. Eng.* **169** 44
[2] Wu B, Zhu J, and Li X 2014 *Petrol. Sci.* **11** 590
[3] Vakhin A V, Mukhamatdinov I I, Aliev F A et al 2018 *Petroleum Sci. Technol.* **36** 1850
[4] Maity S K, Ancheyta J, and Marroquın G 2010 *Energy Fuels* **24** 2809
[5] Rivera Olvera J N, Gutierrez G J, Romero Serrano J A, Medina Ovando A, Garibay Febles V, and Barriga Arceo L D 2014 *Catalysis Communications* **43** 131
[6] Liu X, Yang Z, Li X et al 2015 *Micro and Nano Letters* **10** 167
[7] Junaid A S M, Rahman M M, Rocha G et al 2014 *Energy Fuels* **28** 3367
[8] Tarboush B J A, and Husein M M 2015 *Fuel Processing Technol.* **5** 133
[9] Nassar N N, Nedal Marei, and Vitale G 2015 *World Heavy Oil Congress* **2015** 1015
[10] Liu X, Yang Z, Li X, and Zhang Z 2015 *Micro & Nano Letters* **10** 167
[11] Kayukova G P, Foss L E, Feoktistov D A et al 2017 *Petrol. Chem.* **57** 657
[12] Mukhamatdinov I I, Vakhin A V, Sitnov S A et al 2018 *Chem. Technol. Fuels Oils* **54** 574
[13] Ivanova A G, Vakhin A V, Voronina E V et al 2017 *Bul. Rus. Academy Sci. Physics* **81** 817

Acid compositions for enhanced oil recovery from carbonate reservoir

D Ibragimova[1,2], I Ivanova[2], D Ivanov, R Akhmetzanova

[1]Kazan Federal University, 18 Kremlevskaya Str., Kazan, Russia
[2]Kazan National Research Technological University, 68 Karl Marx St., Kazan, Russia

E-mail: khalidina@mail.ru

Abstract. This work is devoted to selection of acidic compositions for use in the process of acid fracturing of carbonate oilfields. The selection is based on the stability of the compositions and their compatibility with oil and corrosion influence on oilfield equipment. The object of research was highly viscous bituminous oil and carbonate rock of Tatarstan Republic (Russian Federation). The object of research was oil and carbonate rock samples from the Vereian horizon of the Vishnevo-Polyanskoe oil field, which is characterized by oil containing in carbonate rocks. The chosen oil can be pertained to highly viscous bituminous oil according to its density and viscosity, what makes the process of its extraction from oil formations complicated. As the result there were selected two acid compositions which contained hydrochloric acid 15 wt %, demulsifier, dispersant, iron converter (for Fe^{3+} 2000 and 5000 ppm concentration) and corrosion inhibitor. The effectiveness degree of selected compositions was also demonstrated.

1. Introduction

Despite the fact that the International Energy Agency (IEA) affirms that traditional oil reserves will still provide up to 90 % of the global industry up to 2030 [1], the century of light liquid fuel production has passed, and now the expansion of the resource base can be carried out only by development of hard-to-recover oil and gas reserves.

Global recoverable non-traditional oil reserves (heavy oil, natural bitumen and shale, etc.) are about 4.495×10^{8} tons, which almost coincides with the quantity of traditional oil resources; Non-regular extractable gas reserves (coal strata methane, dense sand gas, shale gas and gas hydrate, etc.) are about 4.000×10^{12} m^3, which is approximately eight times higher than regular gas reserves [2]. The share of recoverable oil and gas reserves of the whole world, contained in carbonate reservoirs, is evaluated by experts at a rate of 50-60 %. A large number of huge fields of carbonate type has a prospect of development for the next 50 years. There is nothing surprising that scientists all over the world are interested in the development of carbonate reservoirs. Specialists in the sphere of oil and gas extraction emphasize the complexity of developing carbonate deposits in comparison with siliciclastic deposits. Despite the large number of complex siliceous-clastic reservoirs, carbonate deposits require the development of a new fundamental base for their effective mastering [3].

The main characteristics of carbonate rocks are porosity and permeability. When the rock contains different types of grains and is crushed by stress and deformation, the pore structure become very complex. Often for convenience, porous space is presented as porous bodies that connect with other pores with the help of the necks [4].

The authors of articles [5-7] give several effective and promising ways of developing carbonate reservoirs. Among them, flooding with slightly saline solutions, water and gas impact, pressurizing foam into the reservoir, pumping carbon dioxide into the reservoir, saturation with surfactants, equalizing the rate of intake capacity, increasing the degree of oil extraction by microbiological methods, gravity drainage during steam injection, "smart flooding" [8], concomitant [9] and fracturing.

The meaning of the acid fracturing is to create a pressure at the bottom of the oil well that is higher than the geostatic pressure. The volume of productive stratum is torn along the planes of the minimum stresses of the rock pressure when the liquid is pumped into the reservoir and is accompanied by the fracture splits. After creating a non-natural fracture, the acid is pumped into the reservoir under the pressure, which is greater than the opening pressure of the crack. The acid interacts with the rock on the surface of the split, what leads to appearance of a heterogeneous rough surface. Therefore, after the excess pressure is removed, the interconnected gaps remain in the crack. The result of the acid fracturing is a significant increase of the permeability of the bottomhole stratum zone, which in turn causes an increase the productivity of the oil well, as well as the oil recovery coefficient as a result of the increased drainage zone.

Another important direction of the development of carbonate reservoirs is the selection of acid compositions for treatment, which are an important part of the acid fracturing. The research [10] was also carried out on carbonate rock samples with a permeability of 50-120 mD. For surfactant concentrations selection, hydrochloric acid with a concentration of 15 % by weight was used as the basis compound of the composition. As a result, it was found that the injection speed of acid is the main factor for characterizing the zone being impacted by the composition. It is also known to use emulsified liquids containing acids for treating underground formations. Emulsified acids result a reduced or slowed reaction rate of the acid under the influence of the external oil phase.

2. Methodology

The object of research was oil and carbonate rock samples from the Vereian horizon of the Vishnevo-Polyanskoe oil field. Oil characteristics: density at 20 °C - 0.9058 g/cm^3, kinematic viscosity at 20 °C - 155.7 mm^2/s. Fractional yield (IBP - 200 °C) - 12.4 %vol. The analyzed oil can be pertained to highly viscous bituminous oil, which makes the process of its extraction from oil formations complicated.

According to the data given in [11], the carbonate rock of the Vereian horizon is characterized by a porosity of 15.5 % and a permeability of 0.055 mm^2.

The following substances were chosen as additives: corrosion inhibitor, iron converter, demulsifier, dispersant. For the preparation of acidic compositions, the substances were matched in various ratios. Fe^{3+} ions in concentrations of 2000 ppm and 5000 ppm were introduced into the test flasks.

The selection of acid compositions is made according to the appendix to the technological instruction of Rosneft Oil Company in four stages.

Checking the stability of the acid composition. Flasks 1-2 and 4-5 are filled in accordance with Table 1 and placed in a thermostat for 30 minutes to preheat to the bottomhole temperature. This way the stability of the acid composition under conditions of reservoir temperature is determined. Flasks 3, 6 - check flasks, maintained at 20 °C for comparison with four other tests.

Table 1. Samples for stability test.

№ of flask	Content	Fe^{3+}, ppm
1	50 ml of composition (HCl 15 %+ all additions)	0
2		5000
3		0
4		0
5		2000
6		0

For compatibility test of active acid composition and oil, the cylinders were filled with required compound in accordance with Table 2.

Table 2. Samples for stability test of active acid composition and oil.

№ of cylinder	Content of test-tube		Content of oil in test-tube by volume, %
1	100 ml (acid composition + oil)	5000 ppm Fe^{3+}	75
2			50
3			25
5		2000 ppm Fe^{3+}	75
6			50
7			25

All test-tubes were placed in thermostat, heated to the bottomhole temperature, for statistic sedimentation no more than 30 minutes. Decantation time being over, samples are checked for phase separation.

If there is no complete separation of the emulsions into two phases or the interface is blurred, then the composition should be changed with selecting the appropriate additive. Content of every test-tube is filtrated through a sieve (100 mesh) to check the presence of clots or sediment.

Compatibility test of partly neutralized acid composition and oil. Acid and calcium carbonate are mixed in quantity according to the Table 3. As chemistry reaction finished, content should be filtrated .The cylinders are filled with oil accordingly data from Table 3, mixture is stirred for 30 s.

Table 3. Samples for compatibility test of partly neutralized acid composition and oil

№ of cylinder	Content of acid composition		Ratio neutralized acid comp.:oil, ml:ml
4	50 ml acid composition + 11.05 g $CaCO_3$	5000 ppm Fe^{3+}	44:44
8		2000 ppm Fe^{3+}	45:45

Test-flasks 4 and 8 are placed in thermostat for 4 hours, heated to the static bottomhole temperature. Content of every test-tube is filtrated through a sieve (100 mesh) to check the presence of clots or sediment. Test is passed if there is no sediment, emulsion or sedimentation of impurities.

The check for the corrosion activity of the acid composition was estimated by the propagation speed of corrosion through the influence of the acid composition on the sample of pump-compressor tube within 12 hours. The corrosion rate was determined by the change in mass of the sample per unit surface, per unit time:

$$v = \frac{m_1 - m_2}{S \times t}, \quad (1)$$

where v - corrosion rate, g/(cm^2×h); m_1, m_2- mass of steel sample before and after, g; S - surface of the sample, cm^2; t - time of the test, h.

Corrosion rate mustn't be higher than values from the Table 4.

Table 4. Maximum values of corrosion rate.

Downhole temperature, °C	Corrosion rate, g/cm^3×h
< 93	0.007638
> 93	0.002010

Effectiveness of acid composition is proved by increasing the coefficient of reservoir permeability using this method: 10 ml of acid composition with the selected concentration were poured into three measuring cylinders and rock samples were placed in for a time (Sample 1 - for 12 hours; Sample 2 - for 18 hours; Sample 3 - for 24 hours).

Porosity after acid influence is calculated with following formula:

$$\varnothing_2 = \frac{V_i - V_f}{V_i}, \quad (2)$$

where ø2 – porosity of sample after use of acid; Vi – full volume of the rock sample before use of acid; Vf- full volume of the rock sample after acid impact.

In order to estimate the changes in the permeability of the reservoir, the following formula is used:

$$\frac{K_{per1}}{\varnothing_1} = \frac{K_{per2}}{\varnothing_2}, \quad (3)$$

where $ø_1$ – porosity of carbonate rock sample before use of acid; K_{per1} – permeability before acid impact; K_{per2} – permeability after acid impact.

3. Results and Discussions

During the stability test of acid compound in cylinders *1 – 6* separation of mixtures or sedimentation weren't observed, that indicates the suitability of composition for use according to the method [14]. This suggests that the composition during injection into the well will not cause additional difficulties associated with phase separation and loss of any sediment that can not only settle on the equipment, but also clog the pores of the formation, which in turn is not permissible. Results of oil compatibility test represents us sufficient separation of acid from oil emulsionon (Figure 1), absence of sedimentation after 30 min. heating at the bottomhole temperature, and free pass of substance through the sieve.

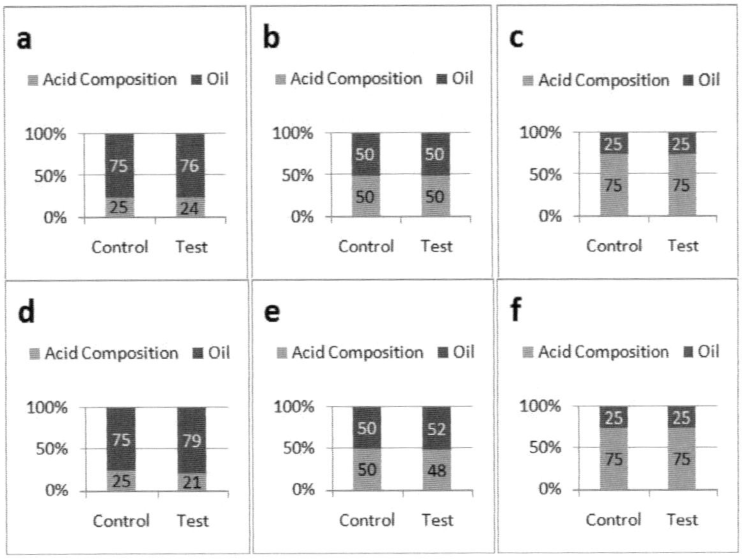

Figure 1. Results of separation "Acid comp.-oil":a, b, c – Fe^{3+} conc. 5000 ppm; d, e, f – Fe^{3+} conc. 2000 ppm.

Thus, the selected acid composition is completely separated from the oil produced after removing the emulsion from the reservoir. Such kind of composition (1) does not affect crude oil characteristics, what is also confirmed by the free flow of the experimental emulsions through a sieve (without the formation of clots), and (2) does not requireadditional energy or reagents for the emulsion separation.

Relating the compatibility test for partly neutralized acid and crude oil, it is also possible to note that required degree of separation of test samples 4 and 8 under experimental conditions is achieved (Figure 2).

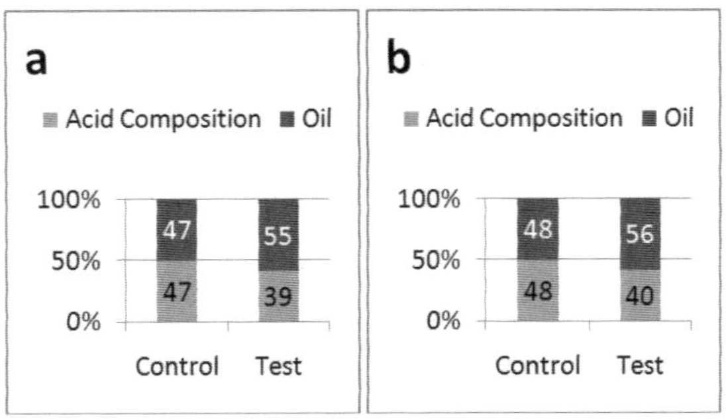

Figure 2. Results of emulsion separation «neut. Acid Comp. - Oil»:a – Fe^{3+} conc. 5000 ppm; b – Fe^{3+} conc. 2000 ppm.

It can be concluded that an increase in pH as the result of acid interaction with carbonate rocks does not affect the stability of the emulsion and still allows separatingit by settling.

The tests also demonstrated the absence of clots and sediment in experimental mixtures, what eliminates the possibility of blocking the channels of an oil-bearing formation and pumping equipmentand does not create complications with the transportation of recovered oil.

Results of corrosion rate test has demonstrated that mass of the sample decreased from 29.30680 g to 29.2893 g. Full square of sample - is 20.0875 cm^2. Following calculations use formula (1):

$$v = \frac{29.30680-29.2893}{20.0875\times 12} = 4.36\times 10^{-4}\ g/(cm^2\times h)$$

As the bottomhole temperature is lower than 93 °C, then as a norm of corrosion rate we use maximum corrosion speed value 0.09765 g/cm^2 for 12 hours, which is greater than received value, therefore inhibitor of corrosion was chosen correctly.

As result of efficiency test of acid compound for its dissolving properties were estimated following changes (Table 5) in volume of the samples.

Table 5. Volume change of rock samples.

№ ofsample	Initial volume, V_i, ml	Duration of impact, h	Final volume, ml
1	3.0	12	1.8
2	3.0	18	1.8
3	3.0	24	1.5

Use formula (2) for approximated calculations of porosity:

a) Sample *1*: $\varnothing_2=\frac{3.0-1.8}{3.0}\times 100\ \% = 40\ \%$;

b) Sample *2*: $\varnothing_2=\frac{3.0-1.8}{3.0}\times 100\ \% = 40\ \%$;

c) Sample *3*: $\varnothing_2=\frac{3.0-1.8}{3.0}\times 100\ \% = 50\ \%$

Thus, average value of porosity (40+40+50) / 3 = 43.3 %.

Applying formula (3), to measure, how much the permeability of reservoir is increased:

$$K_{per2}=\frac{K_{per1}\times\varnothing_2}{\varnothing_1}=\frac{0.055\times 43.3}{15.5}=0.154\ mcm^2$$

Based on obtained data we make a conclusion about efficiency of acid compound, made of HCl 15 %, for acid fracturing and development of Verian reservoir at Vishnevo-Polyanskoe oil field.

In result of all tests two effective for the impact on Vereian reservoir compositions, satisfying the oil compatibility requirements, were chosen for given conc. of Fe^{3+} ions (Table 6).

Table 6. Content of acid composition.

Component	5000 ppm Fe^{3+}	2000 ppm Fe^{3+}
	Content, vol. %	
HCl aq.solution15 %	98.18	98.84
Iron converter	1.32	0.66
Demulsifier	0.20	0.20
Dispersant	0.20	0.20
Corrosion inhibitor	0.10	0.10

4. Conclusions

In the process of the study was carried out the selection of acidic compositions for use in the process of acid fracturing. The selection was made out through a series of tests aimed at testing the stability of the compositions and their compatibility with oil and corrosion influence on oilfield equipment, the analysis of which was made in advance.

The object of research was oil and carbonate rock samples from the Vereian horizon of the Vishnevo-Polyanskoe oil field, which is characterized by oil containing in carbonate rocks.The chosen oil can be pertained to highly viscous bituminous oil according to its density and viscosity, what makes the process of its extraction from oil formations complicated.

In addition, the choice of the basis of acidic compositions (hydrochloric acid with a concentration of 15 wt %) was justified and its effectiveness was proved for use in the development of a carbonate reservoir. All tasks were performed.

As a result, two hydrochloric acid compositions were selected for use at the stage of acid treatment in the acid fracturing process of the carbonate reservoir of the Vishnevo-Polyanskoe field at a concentration of Fe (III) ions of 2000 and 5000 ppm.

Competent selection of compounds for the acid fracturing simplifies the time-consuming process of carbonate reservoirs development, increases the degree of oil recovery from productive formations, reduces outgoings and increases the economic efficiency of oil production, while enriching the global resource base.

Acknowledgments

The work is performed according to the Russian Government Program of Competitive Growth of Kazan Federal University.

IOP Conf. Series: Earth and Environmental Science **282** (2019) 012007 doi:10.1088/1755-1315/282/1/012007

References

[1] Sorrell S *et al* 2010 *Energy policy* **38** 4990
[2] Caineng Z *et al* 2010 *Petrol. Exploration and Development* **37** 129
[3] Burchete T 2012 *The Geological Society of London* **370** 17
[4] Xu S and Payne M *Rock physics* **28** 66
[5] Aladasani A and Bai B 2010 *SPE* **130726** 1
[6] Alvarado V and Manrique E 2010 *Energies* **3** 1529
[7] Shah A *et al* 2010 *Energy & Environmental Science* **6** 700
[8] Yousef A *et al* 2011 *SPE Reservoir Evaluation & Engineering* 578
[9] Qi W*et al* 2012 *Petroleum Exploration and Development* **39** 377
[10] Al-Gramdi A *et al* 2009 *SPE* **121713** 1
[11] Muslimov R *et al* 2007 *Geology and development of oil fields* **2** 524

IOP Conf. Series: Earth and Environmental Science **282** (2019) 012008 doi:10.1088/1755-1315/282/1/012008

Distribution of vanadyl complexes and free radicals in asphaltenes fractions from electron paramagnetic resonance

A Rodionov, I Mukhamatdinov, G Mamin, M Gafurov, S Orlinskii, I Salih, and A Vakhin

Kazan Federal University, 18 Kremlevskaya Str., Kazan, Russia

E-mail: IIMuhamatdinov@gmail.com

Abstract. Structural characterization of asphaltenes in complex systems is in the focus of scientific and industrial interests since many years. Asphaltenes isolated from the high-viscous Ashal'cha oil (2500 mPa·s) were studied. Five asphaltene fractions were obtained by stepwise extraction with an alcohol-benzene solvent with different alcohol contents after precipitating the asphaltenes from the oil with hot isooctane. Redistribution of paramagnetic centers between the fractions of asphaltenes is revealed by electron paramagentic resonance (EPR) while no significant changes in EPR spectral parameters are found. The results can be used for tracking the asphatenes formation as well as following the changes of their supramolecular structure.

1. Introduction

Paramagnetic properties and electron paramagnetic resonance (EPR) spectra in complex hydrocarbon systems (like oil rocks, crude oil, bitumen, their fractions, etc. which we will denote in the text as PDS – petroleum disperse system) are mainly defined by the presence of complexes of *d*-metals (first of all V, Ni) and stable "free" radicals (FR). It is usually assumed that FR are mainly concentrated in the heaviest PDS fraction - asphaltenes and arise due to the delocalized π-electrons of the aromatic rings and stable organic radicals of the side chains. Vanadyl based porphiryns in heavy oils were first discovered in 1930s by Treibs [3, 4], who postulated the biological origin of these compounds. Vanadyl (VO) porphyrin is one of the most abundant forms of metalloporphyrin, and is of particular significance as a biomarker in the context of oil formation research [5]. PDS porphyrins exist in homologous manifolds of several structural classes and manifest great structural diversity [6, 7]. The molecule can contain different types of substitutions, including alkyl, cycloalkane, and aromatic groups and possesses near planar anisotropic structure [8-12].

Because of its high boiling point, minimal vanadium is contained in the distillates. Additionally, the low solubility of several vanadium compounds in most solvents, specifically metalloporphyrins, facilitates their partitioning between the asphaltenes and resins [1, 2]. It is self-evident that different asphaltene fractions can also have various amount of paramagnetic centres. Dickie and Yen were the first to suggest that, having many of the same interactive groups as asphaltenes, metalloporphyrins could also play a role in asphaltene self-assembly, adsorption, and phase separation [13]. Although the exact location and structure of FR and VO in PDS and asphaltenes are still undefined, some recent results based on the advanced EPR techniques show that VO and FR are bound in asphaltenes and their EPR parameters can serve at least as complimentary measures to follow thermal, catalytic, organizational changes in asphaltenes' structures [14-20]. To the knowledge of authors, there are no detailed studies of the asphaltenes fractions by EPR and of redistribution of paramagnetic centres in asphaltenes fractions. It can be important for the understanding the mechanisms of asphatenes formation and precipitation. In this paper, we present the results of conventional EPR measurements of asphaltenes fraction for Ashal'cha oil to evaluate the capabilities of conventional EPR techniques to characterize asphaltenes fractions.

2. Materials and Methods

Crude oil with the viscosity of 2500 mPa·s at room temperature was extracted from the Ashal'cha oilfield (Volga-Ural basin, Republic of Tatarstan, Russia). Compositional analysis of the oil sample (SARA analysis) was conducted according to the common technique describes in [21], for example, which includes asphaltenes precipitation from hexane and further chromatographic separation of the non-asphaltic oil components through column filled with alumina. SARA (saturated, aromatic, resins, asphaltenes) analysis data of the studied crude oil are presented in Table 1. Besides, of the presented fractions volatile organic compounds (VOC, 5.6 %) with the boiling point of less than 110 °C are found.

Table 1. Results of SARA analysis for the crude oil sample.

SARA fractions, wt %			
Saturated	Aromatic	Resins	Asphaltenes
24.9	37.4	25.9	6.4

The oil samples (5-10 g) were treated by hot isooctane (200 ml). Then by using the mixtures of isopropanol and benzene containing 20, 40, 60, 80, 100 % of benzene in filter paper in a Soxhlet apparatus, asphaltenes fractions A_1-A_5, correspondingly, were obtained. Their molecular mass was determined with the Matrix-Activated Laser Desorption/Ionization (MALDI) spectrometer with time-of-flight analyzer (TOF) UltraFlex III TOF/TOF (Bruker) on the 2.5-dihydroxybenzoic acid matrix. It gives the values of 786, 1190, 1233, 1202, and 1280 a.u. for them.

Elemental analysis of crude oil asphaltene and asphaltene fractions was carried out on Perkin Elmer 2400 Series II elemental analyzer [21]. No significant difference in nitrogen (1.5-1.7 wt %) or sulfur (3.6-3.8 wt %) content with the sample number was observed. Data for carbon and hydrogen content along with the NMR analysis are published in the complimentary paper. No vanadium was detected by using the mentioned equipment.

From literature [22] it is known that the total amount of vanadium in Ashal'cha oil is estimated to be about 0.023 wt % distributed mainly in asphaltenes (0.186 wt. %) and resins (0.042 wt %). Presence of S allows to suggest that vanadyl in Ashal'cha oil has (apart of $VO(N_4)$), $VO(NS_3)$, $VO(N_2S_2)$ and $VO(N_3S_2)$ as possible environments but increase in sulfur

IOP Conf. Series: Earth and Environmental Science **282** (2019) 012008 doi:10.1088/1755-1315/282/1/012008

content could be also explained by incorporation of thiophenes, i.e. various porphyrins in oil have the same $VO(N_4)$ skeleton [23, 24]. Several series of porphyrins containing a core structure of N_4VO have been identified in fossil fuels like etioporphyrins (ETIO), deoxophylloerythroetio porphyrins (DPEP), dicyclic-deoxophylloerythroetio porphyrins (Di-DPEP), rhodo-etioporphyrins (rhodo-ETIO), rhodo-deoxophylloerythroetio porphyrins (rhodo-DPEP), and rhodo-dicyclic-deoxophylloerythroetio (rhodo-Di-DPEP). According to the studies of ref. [22], apart from etio- and phylloporphyrins, there are also rhodo- and DPEP found in asphaltenes of Ashal'cha oil and about 20-60 % of the are parts of the paramagnetic complexes.

EPR spectra of the samples in continuous wave (cw) mode were obtained using an ESP-300 and ElexSys 680 (Bruker) spectrometers operating at 9.4 – 9.9 GHz (X-band). FR and VO concentration was estimated at room temperature in the double cavity ER4105DR by comparing the integrated intensities of the simulated with EasySpin utility [25] the integrated intensities of the spectra of a test sample and a reference sample (powder of DPPH, Mn^{2+} in MgO, and a series of Cu-DETC solutions). The value of the microwave power was chosen to be 2 μW to avoid EPR saturation effects.

3. Results and Discussion.

EPR spectra detected at X-band (Figure 1) for all fractions A_1-A_5 are very similar to each other and are defined by the presence of stable "free" radical (FR, single line, electron spin $S = 1/2$, $g \approx 2.003$, linewidth of 6.4 G) and paramagnetic vanadyl complexes. Taking into account a nearly planar skeleton structure of single VO molecule (VO^{2+}, $S = ½$), the EPR spectra could be satisfactory described by the g-tensor of axial symmetry with $g_{\|} \approx 1.96$, $g_{\perp} \approx 1.98$ and anisotropic hyperfine interaction with a magnetic moment $I = 7/2$ (therefore, 8 projections of I with the quantum numbers $m_I = \pm 7/2; \pm 5/2; \pm 3/2$ and $\pm 1/2$ are possible) for ^{51}V nuclei (natural abundance of 99.75 %) with hyperfine structure constants $A_{\|} \approx 480$ MHz, $A_{\perp} \approx 157$ MHz (Table 2). In our notations the values of $(g, A)_{\|}$ correspond to the orientation perpendicular to VO^{2+} plane (out of plane), along the direction c and $(g, A)_{\perp}$ correspond to the orientation in the VO^{2+} plane, ab plane. As for the most VO^{2+} related investigations, we suppose that the direction of the main axes coincide for g- and A- tensors. Comprehensive description of VO spectra, their spectra simulations and spectra changes with external treatment are given in papers [11,15,17].

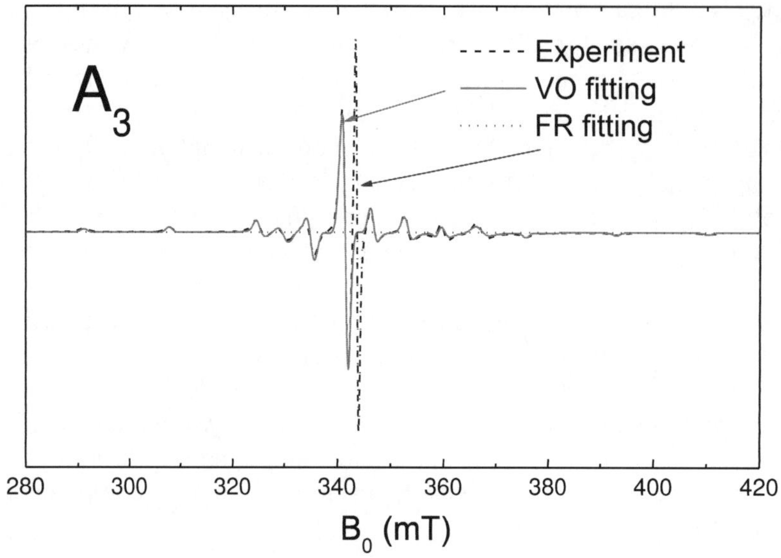

Figure 1. Conventional X-band EPR spectra of A_3 fraction and its simulation in the self-written environment based on the EasySpin package for Matlab [25] with the parameters given in Table 2.

Table 2. Values of of g- and A- components for VO and FR paramagnetic centers for asphaltenes.

VO				*FR*	
$g_{\parallel}$	$g_{\perp}$	$A_{\parallel}$, MHz	$A_{\perp}$, MHz	g	ΔHpp, G
1.9610(8)	1.9825(5)	475(15)	155(5)	2.0030(3)	6.4(1)

Within the accuracy of our experiments, no changes with the number of fraction in the presented spectroscopic parameters are found. It is often suggested that a difference in g-factors can or should be revealed [26]. It means that no great structural changes of single VO containing molecule or FR with fractioning in our experiments is observed. We hope to address this issue in details by using the more elaborated EPR techniques like electron-nuclear double resonance (ENDOR) [15,17].

Results for the concentration measurements in the investigated powders are presented in Table 3 and Figure 2. It follows that generally concentration of VO reduces with fraction number while the intensity of FR is growing till A_3 and then slightly drops. A very similar behaviour was reported in [15] in which precipitation of asphaltens on alumina surface by EPR technique was investigated. It was found that the reducing of VO/FR ratio occurs with the lowering amount of the precipitated asphaltenes. It was suggested that asphaltenes with a more condensed structure and lower VO content were adsorbed on alumina. It agrees with the previously revealed trend of inverse dependence of the concentration of FR on the content of vanadyl complexes [27] and shows the potential ability (correctness) of investigation of asphaltenes' fractioning and formation by their precipitation on various supports [15,17]. The results became even more interesting in the light of a new wave of interest to the unraveling of features of archipelago-island asphaltene structures [28].

Table 3. Absolute and relative concentrations of FR and VO in the investigated species.

	Mass from MALDI-TOF, a.u.	VO concentration, spin/g	FR concentration, spin/g	Relative concentration [VO/FR]
Initial asphaltene				9.1 [15]
A_1	786	7.5(7) 10^{18}	0.81(7) 10^{18}	9.1
A_2	1190	6.0(6) 10^{18}	1.5(1) 10^{18}	3.7
A_3	1233	6.0(6) 10^{18}	1.8(2) 10^{18}	3.45
A_4	1202	6.7(7) 10^{18}	1.8(2) 10^{18}	3.68
A_5	1280	4.0(5) 10^{18}	1.42(8) 10^{18}	2.77

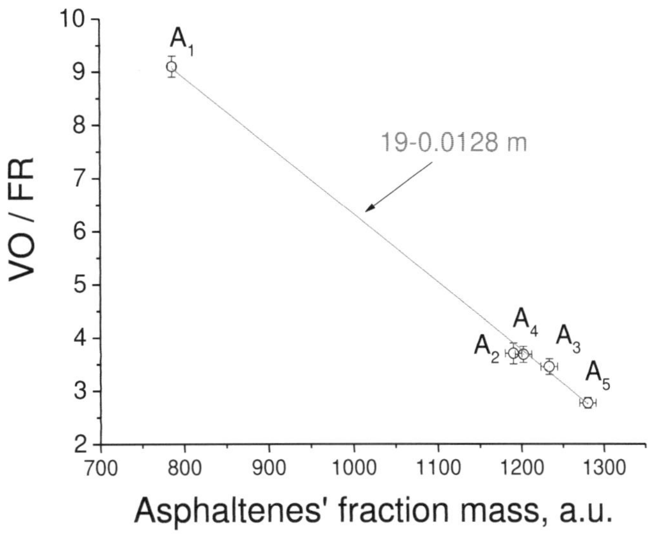

Figure 2. Dependence of the relative concentration VO/FR on asphaltenes mass and its linear approximation.

4. Conclusions

While no significant changes in EPR spectroscopic parameters of paramagnetic vanadyl complexes and "free" radical in Ashal'cha asphaltenes' fractions was observed, perceptible lowering of their relative concentrations which is found to be nearly inverse proportional to the extracted atomic masses was found. The obtained feature can be used for the elucidation of the asphaltenes formation and their precipitation.

Acknowledgments
The work is performed according to the Russian Government Program of Competitive Growth of Kazan Federal University.

References

[1] Yen T F and Chilingarian G V 2000 *Asphaltenes and asphalts.* Vol. 2 *(Developments in petroleum science)* 40B (Amsterdam: Elsiever Science) p 622
[2] Yen T F and Chilingarian G V 1994 *Asphaltenes and Asphalts.* Vol. 1 (*Developments in Petroleum Science*) 40A (Amsterdam: Elsiever Science) p 459
[3] Treibs A 1934 *Ann. Chem.* **510** 42
[4] Treibs A 1936 *Angew. Chem.* **49** 682
[5] Baker E and Louda J 1986 in *Biological Markers in the sedimentary record* ed. JohnR B (Amsterdam: Elsiever) 125-225
[6] Zhao X, Liu Y, Xu C et al 2013 *Energy Fuels* **27** 2874
[7] Zhao X, Shi Q, Gray M, R and Xu C 2014 *Scientific Reports* **4** 5373
[8] Dechaine G P and Gray M R 2010 *Energy Fuels* **24** 2795
[9] Gracheva I N, Gafurov M R, Mamin G V et al 2016 *Magn. Reson. Solids* **18** 16102
[10] Biktagirov T, Gafurov M, Mamin G et al 2017 *Energy Fuels* **31** 1243
[11] Biktagirov T B, Gafurov M R, Volodin M A et al 2014 *Energy Fuels* **28** 6683
[12] Gafurov M R, Volodin M A, Rodionov A A, et al. *J. Petrol. Sci. Eng.* 2018 **166** 363
[13] Dickie J P, Yen T F 1967 *Anal. Chem.* **39** 1847
[14] Mamin G V, Gafurov M R, Yusupov R V et al 2016 *Energy Fuels* **30** 6942
[15] Gafurov M R, Gracheva I N, Mamin G V, et al. 2018 *Russ. J. Gen. Chem.* **88** 2374
[16] Alexandrov A S, Archipov R V, Ivanov A A et al 2014 *Appl. Magn. Reson.* **45** 1275
[17] Gafurov M R, Galukhin A V, Osin Y N et al 2019 *Magn. Reson. Solids* **21** 19102
[18] Mehrabi-Kalajahi S S, Varfolomeev M A, Yuan C et al 2018 *J. Petrol. Sci. Eng.* **169** 673
[19] Galukhin A, Khelkhal M A, Gerasimov A et al 2016 *Energy Fuels* **30** 7731
[20] Mukhamatdinov I I, Gafurov M R, Kemalov A F et al 2017 *IOP Conference Series: Earth and Environmental Science* **155** 012004
[21] Vakhin A V, Mukhamatdinov I I, Aliev F A et al 2018 *Petrol. Sci. Technol.* **36** 1850
[22] Yakubov M R, Abilova G R, Sinyashin K O et al 2017 *Phthalocyanines and Some Current Applications*, InTechOpen: Zagreb pp. 155–169
[23] Zaragoza, I.P., Santamaria, R., Salcedo, R 2009 *J. Mol. Catalysis A: Chemical* 307 64
[24] Qian K, Fredriksen T R, Mennito A S et al 2019 *Fuel*, **239** 1258
[25] Stoll S and Schweiger A 2006 *J. Magn. Reson.* **178** 42
[26] Cui Q, Nakabayashi K, Ma X et al 2017 *RSC Advances* **7** 37908
[27] Yakubov M R, Milordov D V, Yakubova S G et al 2017 *Petrol. Sci. Technol.* **35** 1468
[28] Lin Y J, Cao T, Chacón-Patiño M L et al 2019 *Energy Fuels* doi:10.1021/acs.energyfuels.8b03835 (in press)

Study of catalytic activity of modified montmorillonitis sustainable in supercritical water

S Petrov[1,2], D Ibragimva[1,2]

[1]Kazan Federal University, 18 Kremlevskaya Str., Kazan, Russia
[2]Kazan National Research Technological University, 68 Karl Marx St., Kazan, Russia

E-mail: khalidina@mail.ru

Abstract. Currently, the world is intensively exploring the possibility of obtaining highly active catalysts for hydrothermal processing of high-viscosity oils through the synthesis of new carriers, optimization of the composition, structural parameters in order to increase the degree of conversion of high-molecular hydrocarbons. The work is devoted to the study of the composition and properties of a montmorillonite intercalated with γ-$Al_{13}O_4(OH)_{24}(OH_2)_{11}(OH)_6^+$ and containing catalytically active pillars agents γ-Al_2O_3 in the reactions of hydrothermal transformation of high-viscosity oils. Experiments conducted in a catalytic laboratory setup in a flow-down reactor with a stationary layer of pillared montmorillonite at a temperature above 450 °C and a pressure of not more than 0.5 MPa showed an increase in the conversion of high molecular weight n-alkanes 1.3 times, with the formation of low-boiling n-alkanes and alkanes branched structure.

1. Introduction

To date, Russia sharply raised the issue of the production of highly active catalysts of deep processing of high-viscosity oil, corresponding to world level [1-4]. Poor quality or complete lack of modern domestic highly active catalysts, together with the negative trend of the foreign producers focused on the import is not the most modern brands catalysts, constrained by high pace of development of high-viscosity oil in the Republic of Tatarstan and the Russian Federation. Leading manufacturers improve technology of highly active catalysts due to the synthesis of new carriers, optimization of the structural parameters and the stability of the catalyst to deactivation in order to increase the degree of conversion of high-molecular hydrocarbons of petroleum.

The work is dedicated to study of composition, properties and the catalytic activity is stable in supercritical water, modified montmorillonite in the hydrothermal conversion reactions of high-viscosity oil.

Layered silicates have a two-dimensional channel system zeolites and comparable to increase the availability of the hydrocarbon molecules in the interlayer space. The introduction of the silicate layers of certain cations Keggin, may increase the channel, making them wider than zeolites, that makes pillar clays convenient carrier for the active components of catalysts cracking large hydrocarbon molecules highly viscous oils, which cannot penetrate the pore system of zeolites. The acidic nature of the layered silicate is caused by the presence of Johannes Nicolaus Bronsted centers, which are proton donors, and Lewis is an electron acceptor pair. The presence of Lewis Center provides pillar agents. Due to the acidic properties of silicate can be used in the cracking reactions, hydroisomerization, dehydration, dehydrogenation, hydrogenation, aromatization, disproportionation of hydrocarbons. The most common metals for the intercalation into the interlayer space of the clay materials are Al, Zr, Cr, Fe and Ti [5, 6]. The object of investigation is selected selected from bentonite intercalated montmorillonite pillar agent is a Keggin (Al_{13}) – γ-$Al_{13}O_4(OH)_{24}(OH_2)_{11}(OH)_6^+$. This complex is composed of 12 octahedral Al^{3+} ions and a central tetrahedral Al^{3+}.

2. Methodology

The modified layered silicate is prepared in two stages. First, polyhydroxocation intercalation into the interlayer space of montmorillonite aluminosilicate with negatively charged layers, the second: heat treatment intercalated silicate. As a result of the heat treatment hydroxocations Al_{13}, embedded into the interlayer space of the silicate layers are transformed into γ-Al_2O_3, which create columnar periodic system and stitching silicate layers form a nano porous structure. The elemental composition of the obtained using energy dispersive X-ray fluorescence spectrometer, at the final stage after calcination montmorillonite contained more than 18 % γ-Al_2O_3. Textural characteristics were investigated by thermal desorption of nitrogen, the specific surface area calculations performed according to Brunauer-Emmett-Teller effect. The intercalated montmorillonite is characterized by the nanoporous structure, high specific surface area greater than 0.8 cm^2/g and a pore volume of 5-20 nm ~ 0.42 cm^3/g. Phase composition of montmorillonite before and after implementation has been investigated by a powder X-ray diffractometer, the basal spacing after calcination was ~ 1.3 nm. To study the changes occurring on the surface of montmorillonite used scanning electron microscopy.

Pre-catalytic activity of the modified montmorillonite during hydrothermal exposure was studied using a model mixture of high n-alkanes with different geometric characteristics in a laboratory setup catalyst flow type. Installation equipped with a pre-heated raw material container with a stirrer, feed lines pumps the mixture of the evaporator, gas lines nitrogen supply installation, an evaporator, a tubular reactor with a fixed bed of the modified montmorillonite exchanger-condenser for the final reaction products, a backpressure regulator, a separator for separating products reaction [7]. The reactor temperature was 450 °C, in the evaporator 400 °C, pressure 0.5 MPa. The yield of the operating temperature was carried out in nitrogen flow at identical temperatures in the evaporator and reactor. The temperature in the heat exchanger-condenser ~ 10 °C. Yield on predetermined setting mode carried out in a nitrogen stream, and then nitrogen feed was stopped, and feeding of the starting mixture of n-alkanes at a rate of 0.5 ml / min, the volume ratio of n-alkanes to the volume of the modified montmorillonite oxide was 9:1 $hour^{-1}$. According to X-ray fluorescence analysis after catalysis content of γ-Al_2O_3 in pillars silicate has changed. On a sample of the modified montmorillonite before and after catalysis of any other independent phases was observed. It should be noted that the catalysis of higher n-alkanes in the presence of the modified montmorillonite at 450 °C does not lead to an increase in the content of carbonaceous substances, but significantly increases yield gaseous products from 14 (control) to 17 wt %. Analysis of the gaseous products of the cracking of n-alkanes was conducted by gas chromatography. The composition gases are methane and propane to 35 wt %, as well as butane, iso-butane, pentane. From the literature it is known that an increase in cracking temperature linear alkanes ~ 500 °C while maintaining the atmospheric pressure increases the yield of the product gas with high selectivity to ethylene and propylene.

3. Results and Discussions

Analysis of the micrographs shown in Figure 1, leads to the conclusion that the surface is characterized by a pronounced relief formed aggregates of rounded grains no larger than ~200 nm. After calcination, the surface of the relief is smoothed, and the obvious changes in the composition of the grain size of the aggregates is not observed, but the degeneration of their contact boundaries occurs.

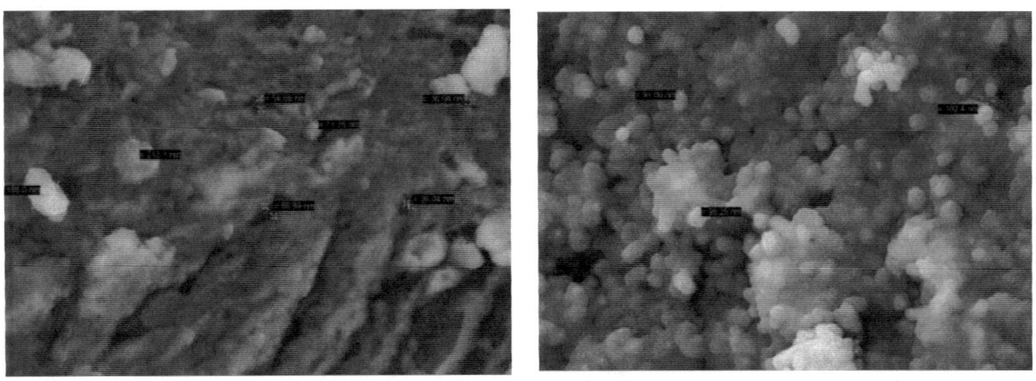

Figure 1. Microstructure of the surface.

Analysis of the composition of liquid conversion products was carried out by chromatography and mass spectrometry using gas-liquid chromatography-mass spectrometer PerkinElmer Turbo Mass Gold GS/MS. Separation was conducted on a capillary column 30 m long, 0.32 mm diameter with PE-XLB phase, carrier gas - helium flow rate of - 2 ml/min. Impact modified montmorillonite in the composition of the liquid product illustrated data table. Note that, in the presence of the modified montmorillonite degree of conversion of n-

IOP Conf. Series: Earth and Environmental Science **282** (2019) 012009 doi:10.1088/1755-1315/282/1/012009

alkanes is increased by more than twice. Parallel isomeric cyclization may go carbocation, a structure which is desorbed to form methylcyclopentyl (Table 1). Cycloolefin further adds hydrogen to the metal center of the catalyst, turning to methyl cyclopentane. The resulting olefin further adds hydrogen to form cyclohexane or cleaves hydrogen to benzene. In turn, benzene is reacted with alkenes formed during disproportionation of n-alkanes. This is supported by the presence in the reaction mixture of alkylbenzenes.

Table 1. The difference in the composition of newly formed hydrocarbons in the products of the control experiment and cracking of n-alkanes in the presence of modified montmorillonite.

Control experiment		Product experiment	
Component	content, wt %	Component	content, wt %
hexane	0.1764	hexane	0.1052
heptane	0.3386	heptane	0.1762
3-methyl-2-ethylhexane	0.1186	3- ethylhexane	0.3081
3-methylheptane	0.4981	octane	0.2824
3- ethylhexane	0.3842	cis-1,2-ethylmethylcyclopentane	0.1249
octane	0.1415	o-xylene	0.6329
2,4-dimethylheptan	0.1099	nonane	0.5649
3-methyloctane	0.7527	2,5-dimethyloctane	0.1266
o-xylene	0.2379	1-methyl-3- isopropylcyclohexane	0.7867
nonane	0.4581	1-methyl-4- isopropylcyclohexane	0.2586
2,5- dimethyloctane	0.1348	decane	0.5899
1-methyl-2- butylcyclopentane	0.8236	1,2,3- trimethylbenzene	0.1298
1-methyl-3- isopropylcyclohexane	0.1944	1-ethyl-2-i- propylbenzene	0.8945
decane	0.4694	4- undecene (e)	0.1357
1,2,3-trimethylbenzene	0.1322	6-dodecene (z)	0.2765
1-ethyl-2-i- propyl benzene	0.9113	2-methyl decane	0.5464
3-methyl undecane	0.1030	1-butyl -2-pentyl cis-cyclopropane	0.8626
6- dodecene (z)	0.2230	3-dodecane (e)	0.1218
1-butyl -2-pentyl cyclopropane	0.8124	Tridecane	0.4116
1-ethyl-4,2- methylpropylbenzene	0.1324	1-tetradecene	0.3970
2- methyldecane	0.4863	tetradecene	0.3641
tridecene	0.2987	cis-2-methyl-7-octadecane	0.3287
1- tetradecene	0.3607	7-hexadecene	0.1120
tetradecane	0.2600	octadecane	0.1048
cis-2-methyl-7-octadecene	0.4767	nanodecane	0.1079
3-ethyl5-(2-ethylbutyl)octadecene	0.1134	2-methyl -eicosane	0.2986
2-methyl-7-hexadecene	0.1062	2-methyl-7-nondecane	0.6633
2-methyl-eicosane,	0.1436	4-methyldocosane	0.2030
2-methyl-7-nondecane	0.3569		
10 methyl-eicosane	0.2068		

Symbatically the conversion increases alkyl-substituted benzene selectivity (from 23 to 33 wt %), while the cumulative selectivity to aromatics practically unchanged (45.3 and 45.6 wt %) (Figure 2). Along with the progress of the aromatization reactions in the system tested side cracking and disproportionation reaction resulting in the formation of mono- and di-olefins and naphthalene derivatives, which subsequently take part in the coking reactions in the absence of excess hydrogen in the reaction medium.

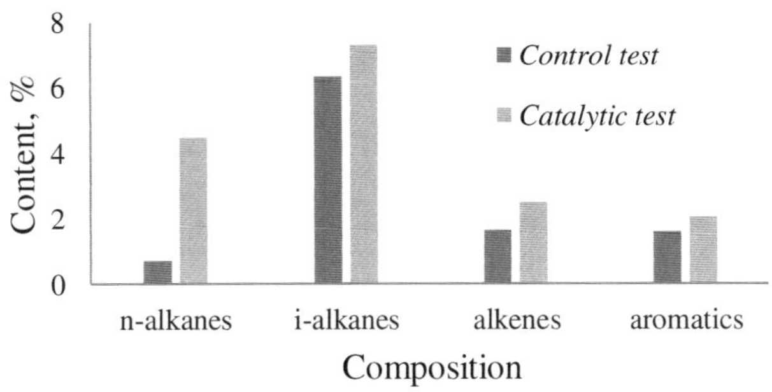

Figure 2. Group composition of new formed hydrocarbons in the final experimental products.

This is supported by, in particular, the reduction of the integral output diene. For the intended transformation circuit of n-alkanes in the first step occurs at dehydrogenation γ-Al_2O_3 centers of montmorillonite to form dienes. In chemisorption dienes in a protic acidic center formed carbonium ion is subsequently isomerized to a stable structure alkanes branched structure.

Conclusions

Studies have shown that the silicate intercalated with hydroxocations γ-$Al_{13}O_4(OH)_{24}(OH_2)_{11}(OH)^{6+}$, embedded in the interlayer space, after calcination contains 18 % γ-Al_2O_3 and has a nanoporous structure with a volume of ~0.42 cm^2/g, as well as a specific surface area of more than 0.8 m^2/g. The experiments carried out on a catalytic unit in a flow-down reactor with a stationary layer of modified layered silicate at a temperature above 450 °C and a pressure of not more than 0.5 MPa showed an increase in the degree of conversion of high-molecular n-alkanes of high-viscosity oils by 1.3 times, with the formation of low-boiling n-alkanes and alkanes of a branched structure. According to x-ray fluorescence and x-ray spectral analyses, the elemental and phase composition of intercalated layered silicate does not change before and after catalysis.

Acknowledgments

The work was carried out with the financial support of RFBR and the government of the Republic of Tatarstan in the framework of scientific project № 18-48-160053.

References

[1] Parida K M et al 1999 Applied Clay Science **15** 463
[2] Sitnov S A, Vakhin A V, Mukhamatdinov I I et al 2019 *Petrol. Sci. Technol.* **37** 687
[3] Vakhin A V, Mukhamatdinov I I, Aliev, F A et al 2018 *Petrol. Sci. Technol.* **36** 1850
[4] Kayukova G P, Mikhailova A N, Kosachev I P et al 2018 *Energy Fuels* 32 6488
[5] Rightor E G, Tzou M S, Pinnavaia TJ 1991 *Journal of Catalysis* **130** 29

IOP Conf. Series: Earth and Environmental Science **282** (2019) 012009 doi:10.1088/1755-1315/282/1/012009

[6] Abeysinghe S et al 2013 *Langmuir* **29** 14124
[7] Bineesh K V 2011D*alton Trans* **40** 3938

Comparison of reaction media of Aquathermolysis: water in different physical states

R Soldatova[1], S Petrov[1,2], N Bashkirceva[1], A Lakhova[1], E Vasilev[1]

[1]Kazan National Research Technological University, 68 Karl Marx St., Kazan, Russia
[2]Kazan Federal University, 18 Kremlevskaya Str., Kazan, Russia

E-mail: psergeim@gmail.com

Abstract. The experimental results and the character of the heavy oil transformation in a vapor-air medium, at temperatures and pressures providing the water availability in superheated steam state, in addition in subcritical and supercritical states are shown. Due to various thermodynamic conditions, there was the detection that water reactivity to the transformation of heavy oil hydrocarbon components is different. It was proved that supercritical water ensures the vastest yield of light fractions, boiling point of which is above 350 °C

1. Introduction

Due to the deterioration in the structure of hydrocarbon reserves and the growing consumption of energy resources, there is a necessity in the restoration of non-conventional heavy oils, which account for more than 50 % of the world's oil reserves [1-6]. The most general methods for upgrading heavy oils are thermo-destructive methods, in particular hydrothermal methods using water in various thermodynamic states. In this case, the use of water is an environmentally friendly alternative method for upgrading heavy oil and meets the principles of "green chemistry". The change of oil composition during water vapour exposure at high temperatures are considered to be "aquathermolysis" [7-12]. In 1982, this definition was firstly suggested by J. B. Hyne.

The study of the hydrothermal conversion of hydrocarbons in water under various thermodynamic conditions is currently in the stage of accumulation of experimental data. Therefore, it is definitely important to research the dependences of the composition, properties of heavy oil and the pathways of its transformation under hydrothermal conditions [13-29].

This article is devoted to the identification of the change regularities in the hydrocarbon composition and rheological properties of heavy oil under hydrothermal exposure in a vapor-air medium at temperatures from 360 to 420 °C and pressures from 4 to 24 MPa providing the transition of water to the sub- and supercritical states and steam water state.

2. Methodology

The Ashalchinskoye field is considered as one of the biggest natural reservoirs of heavy oils located in Russia. Oil form this field was chosen as an object of research. The oil is biodegraded and characterized by the absence of n-alkanes, and the predominance of isoprenoid hydrocarbons of the iC14-iC20 range. The oil is described by the following physicochemical properties: density - 0.9857 g/cm³, asphaltene content - 7.7 wt %, resins - 37.8 wt %, water content - 3.48 vol.%, sulfur content - 2.8 wt %. Distilled water was used for hydrothermal influence on the primary oil; the thermodynamic conditions of experiments ensured its transition to the superheated steam state, as well as to the sub- and supercritical state. Water in the processes of aquathermolysis can act not only as a heat carrier and solvent for organic compounds, but also as a reagent due to the decrease in the logarithm of the ion product indicating an increase in acidic and basic properties.

The used oil to water ratio was 1: 1. The experiments were conducted in a 250 cm³ isothermal reactor at 360-420 °C and 4-24 MPa (table 2). The heating time to the required temperature was ~ 40 minutes, then for a specified temperature the experiment conducted ~ 20 minutes. In the 1st experiment, water was in a state of superheated steam at 360 °C and 4 MPa. In the 2nd experiment, at the same temperature, the pressure was 21 MPa, which pointed out subcritical conditions of a water state. In the 3rd experiment, the conditions (T=420°C, P=24 MPa) corresponded to the presence of water in the supercritical state (D=0.7162 g/cm³). In the 4th experiment, there was an observation of oil cracking without water at 400°C and 24 MPa (D=0.9857 g/cm³).

Table 1. The results of fractional oil composition.

Exp. No.	Density at 20 °C, g/cm³	Yield of fraction, wt %

		v.o.c.-200°C	200°C-350°C	above 350°C
Oil	0.9857	6.5	18.8	74.7
1	0.9723	9.5	17.6	72.9
2	0.9125	11.3	18.8	69.9
3	0.7162	15.9	19.4	64.7
4	0.9150	8.1	18.1	73.8

3. Results and Discussions

The processes of heavy oil hydrothermal formation in the medium of heated water (T=360°C, P = 4 MPa) are conditioned with fractional content increase and (v.o.c.-200 °C at 46 rel.%) and increase of parafinonaphthetic compounds at 31 rel.% in comparison to initial oil. Mostly, the process of resins and asphaltenes destruction happens: its content decreases at 31 rel.% and 8 rel.% appropriately.

Water impact on oil in subcritical state (T = 360 °C, P = 21 MPa) is a consequence of maximal asphaltenes content decrease in the liquid (at 53 rel.% in comparison to initial oil), what is conditioned with fact that, apart from agent functions preventing compaction reactions and decreasing partial pressure of oil hydrocarbons, subcritical water owns functions of reagent participating in conversion reactions. In subcritical state, water ionization constant is higher in 3 times relatively ionization constant in standard conditions and it is able to participate in hydration reactions of heteroatomic oil compounds as hydrogen proton donor. Aquathermolysis reaction is characterized with such kind of condition.

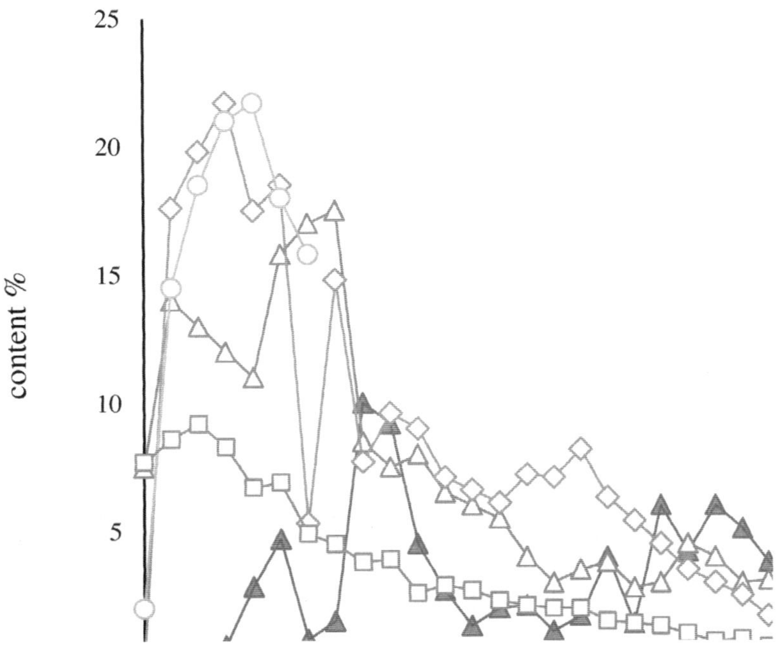

Figure 1. Molecule-mass distribution of oil n-alkanes and liquids of experiments.

An initial oil is determined with bimodal hydrocarbon distribution character – low content of low-molecular and high content of high-molecular alkanes. According to data of gas-liquid chromatography, upgraded oils contain more quantity of low-molecular alkanes C_{10}-C_{17} (figure 1). After hydrothermal water impact in sub- and supercritical states in conditions of experiments 2 and 3, n-alkanes with composition C_{11}-C_{15} content increases up to 22 % relatively 5 % of initial oil. In the 4th experiment implemented in identical thermodynamic conditions of 3rd experiment (supercritical water presence), oil destruction without water participation leads to accumulation of n-alkanes with composition C_{10}-C_{15} till 5% with maximal peak pertained to C_{12} of final liquid. Influencing oil with heated steam (T=360 °C, P = 4 MPa) in the 1st experiment, molecule-mass n-alkanes distribution peak is inherent to C_{15}-C_{17} in the final product. An initial oil is characterized with equal n-alkanes C_{10}-C_{16} and C_{16}-C_{20} sum content and more quantity of C_{10}-C_{16} n-alkanes in comparison to C_8-C_{10} n-alkanes. In the 1st experiment ,influencing oil with heated steam (T=360 °C, P=4MPa), ratio $(C_8+C_9+C_{10})/(C_{10}+C_{11}+C_{12}+C_{14}+C_{16})$ in new-formed oil increases from 0.24 to 0.66, and ratio $(C_{10}+C_{11}+C_{12}+C_{14}+C_{16})/(C_{16}+C_{17}+C_{18}+C_{19}+C_{20})$ - from 1.09 to 1.54; in 3 experiment, in supercritical water conditions (T = 420 °C, P = 24 МПа), these ratios change to 1.01 and 1.44 appropriately, in subcritical water conditions (T = 360 °C, P = 21 MPa) - to 1.03 and 1.46 appropriately. Thus, we conclude that character of molecule-mass distribution in new-formed oil mostly depends on temperature and pressure and at least on thermodynamic water condition in analyzed intervals.

According to infrared spectroscopy data and elemental analysis, in consequence of thermal heavy oil destruction in all experiments, there is a conduction of desulphurization reaction characterized with decreased indexes of sulphur availability (from 2.8 % to 1.8-0.4 %), and spectral coefficients C_5, describing sulfur content in functional groups (from 0.16 to 0.09) (Table 2). There is an assumption that hydrogen sulfide and sulfur oxide formation is possible depleting sulfur.

Table 2. Data of elemental and infrared-spectral analysis of oil and liquids.

experiment, №	Elemental analysis , wt %				Spectral index*			
	C	*H*	*S*	*N*	C_1	C_3	C_4	C_5
Oil	80.6	12.8	2.8	0.4	1.16	0.62	4.23	0.16
1	82.5	10.9	0,4	0.4	0.69	0.55	7.04	0.27
2	81.3	11.2	1.8	0.5	0.43	0.58	9.00	0.13
3	77.8	11.7	1,0	0.5	0.94	0.65	2.43	0.09
4	78.6	12.1	1.1	0.5	0.20	2.50	5.00	0.11

* $C_1=D_{1600}/D_{720}$; $C_3=D_{1380}/D_{1465}$; $C_4=(D_{720}+D_{1380})/D_{1600}$; $C_5=D_{1030}/D_{1465}$

It is important to indicate that the presence of nitrogen after the experimental processes remains permanent, which is caused by finding the nitrogen of the original oil not in aliphatic and aromatic amines, but in thermally stable tetrapyrrole compounds and vanadyl-porphyrin complexes.

Based on IR spectroscopy data, the formation of alkyl hydrocarbons is detected, confirmed by an increase in paraffin values C_4 from 4.23 to 9.0 in the final products of hydrothermal exposure, peak falls on product of 2 experiment. The content of isoprenoid structures in the transformed oils varies in different directions, which is reflected in the values of the spectral indicators of branching C_3: for experiments 1-3 with water participation it wobbles in diapason 0.55–0.65, whereas ,in experiment without water, the content of isoprenoid structures increases noticeably : index C_3 increases from 0.62 to 2.5. It is necessary to note a decrease in the C_1 coefficient characteristic of converted oil, indicating the presence of hydrocarbon polycondensation reactions containing an aromatic ring with the organization of supramolecular structures based on polycondensed aromatic compounds. The greatest difference in coefficient C_1 is observed in the final product of the 4 experiment, conducted without water (T = 420 °C, P = 24 MPa) – 0.2 versus 1.16 of initial oil, reformed oil is also high in asphaltenes – 9.4 versus 7.7 of initial oil.

In the 3 experiment (T = 420°C, P = 24 MPa), the reformed oil shows a decrease in viscosity both structural (from 4.50 to 2.99 Pa·s) and Newtonian (from 2.20 to 2.11 Pa·s) relative to the initial oil (Figure 2).

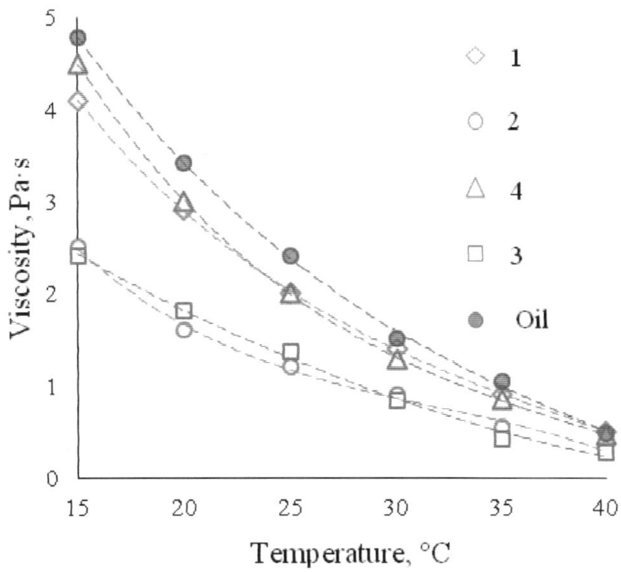

Figure 2. Viscosity-temperature dependences of oil and liquid products of its hydrothermal exposure after experiments №1-4.

This change in the rheological properties of newly formed oil is a result of both an increase in the number of low-boiling fractions, and a consequence of the rupture of C–S bonds in its resinous asphaltene components. The viscosity of the hydrothermal effect of 2 experiments (T = 360 °C, P = 21 MPa) with subcritical water is also different by a decrease in viscosity by more than 2 times compared with the initial oil.

4. Conclusion

In the range of hydrothermal influence experiments in the temperature 360-420°C and pressure 424 MPa, the decrease of aromatic and resins content was indicated and detection of decreased density of final liquids was implemented.

Acknowledgments

This work was supported by grant 18-48-160053/18 from the Russian Foundation for Basic Research.

References

[1] Iwamura H et al 2014 *Inst. Sci. Tech. Nihon. Univ.* **132** 1
[2] Huang S et al 2018 *Int. J. Heat and Mass Transfer* **122** 1074
[3] Kök M V et al 2018 *Energy Sources Part A: Recovery Utilization and Environmental Effects* **1** 645
[4] Kapadia P R et al 2014 *Energy* **64** 1035
[5] Xu K et al 2014 *Advanced Materials Research* **1008** 312
[6] Guo K et al 2016 *Fuel* **185** 886
[7] Kapadia P R et al 2015 *Fuel Processing Technology* **131** 270
[8] Ayache S V et al 2016 *SPE journal* **28** 1
[9] Jia N et al, 2016 *Energy Fuels* **30** 5291
[10] Al-Aulaqi T et al 2018 SPE-190445-MS SPE EOR Conference at Oil and Gas West Asia 2
[11] Shijun H et al 2017 *Fuel* **207** 568
[12] Zhao P et al 2017 *Petrol. Sci.Technol.* **34** 1452
[13] Ayache S V et al, 2017 *SPE Canada Heavy Oil Technical Conference* 2
[14] Mukhamatdinov I I, Vakhin A V, Sitnov S A et al 2018 *Chem. Technol. Fuels and Oils* **54** 574
[15] Huang S et al 2018 *Energy Fuels* **1** 5
[16] Sun F et al, 2017 *Energy* **125** 795
[17] Cuijpers M C M et al 2018 *J. Petrol. Explor. Prod. Technol.* **8** 291
[18] Arcelus-Arrillaga P et al 2017 *Energy Fuels* **31** 4571
[19] Cheng Z M et al 2009 *Energy Fuels* **23** 3178
[20] Petrov S M, Ibragimova D A, Safiulina A G et al 2017 *J. Petrol. Sci. Eng.* **159** 497
[21] Yang M G, Nakamura I, Fujimoto K 1998 *Catal. Today* **43** 273
[22] Timko M T et al 2014 *J. Supercritic. Fluids* **5** 1

[23] Li N et al 2015 X-M *Energies* **46** 8962
[24] Hassanzadeh H et al 2017 *AIChE Journal* **63** 42
[25] Adam G Carr et al 2015 *Energy Fuels* **29** 5290
[26] Watanabe M et al 2010 *J. Supercrit. Fluids* **53** 48
[27] Morimoto M et al 2010 *J. Supercrit. Fluids* **55** 223
[28] Sitnov S A, Vakhin A V, Mukhamatdinov I I et al 2019 *Petrol. Sci. Technol.* **37** 687
[29] Kayukova G P, Mikhailova A N, Kosachev I P et al 2018 *Energy and Fuels* **32** 6488

Features of the change in the composition heavy oils of various types and solid asphaltite in the hydrothermal-catalytic conditions

G Kayukova[1,2], I Abdrafikova[1], A Vakhin[1], V Antipenko[3]

[1]Kazan Federal University, 18 Kremlevskaya Str., Kazan, Russia
[2]Arbuzov Institute of Organic and Physical Chemistry, FRC Kazan Scientific Center, Russian Academy of Sciences, Kazan, Russia
[3]Institute of Petroleum Chemistry, Siberian Branch, Russian Academy of Sciences, Tomsk, Russia

E-mail: vahin-a_v@mail.ru

Abstract. The paper reveals the distinctive features of the change in the composition heavy oils of various types and of natural bitumen - asphaltite from the sandstones of the Permian sediments of the Spiridonovsky field in the territory of Tatarstan, in model experiments at a temperature of 250°C, characteristic for steam-thermal impact on the formation, and the pressure in the system of not more than 2 MPa. The choice of low pressure is due to the fact that bituminous rocks in the Permian sediments occur at relatively shallow depths in the near-surface deposits of the sedimentary strata, which necessitates the development of special technologies for their extraction. The experiments used natural iron oxide - hematite. As a part of the products of asphaltite conversion under the influence of hydrothermal-catalytic factors, the content of hydrocarbons almost 4-fold, the content of alcohol-benzene resins and asphaltenes decreases. The content of benzene resins increases somewhat. The composition of asphaltite proved to be less stable than of heavy oils from age-old sediments, apparently due to special conditions for the formation of its composition in natural conditions under the influence of secondary low-temperature biochemical processes that led to the accumulation of thermally unstable resinous-asphaltene components. Compared with the original asphaltite, in the products of the experiment, the content of alkanes of composition C_{10}-C_{30} substantially increases with an obvious predominance of homologues with an even number of carbon atoms in the molecule. The product is characterized by a lesser degree of aromaticity and content of sulfoxide groups. The results of the studies showed the possibility of upgrading the composition of solid bitumen - asphaltite in in situ conversion by diluting its composition with lighter newly formed hydrocarbons, due to the destruction of its high molecular weight components.

1. Introduction

In Permian sediments of Tatarstan region there are many bituminous rock formations, containing high viscous and solid bitumen – asphalts and asphaltites [1-3]. According to many experts [4-5] bitumen reservoirs in that region represent disrupted oil formations and its formation process is related with generation of oil during ancient Paleozoic era. The fractured zones of bedding rocks and sedimentary cover assisted migration of hydrocarbon fluids from the deep of Paleozoic horizons to surface laydown of sedimentary formations. In the process of geological history under secondary recover techniques, the oil reservoirs are destroyed resulting formation of bitumen with various composition and phase conditions. This prevents their genetical relation with primary source generation and provides a reason to assume about their initial bedding in other sediments. Thereby, determination of natural factors and processes, which bring to formation of hydrocarbon accumulations on surface of sedimentary rocks in Tatarstan region is very important and relevant fundamental issue. The perspective of oil saturated sedimentary formations of given territory depends on solving upper mentioned problems. Apart depletion of light hydrocarbon resources alternative resources like heavy oil and natural bitumen have been already developing by many oil producing countries including Russia in industrial scale [1]. However, hydrocarbon resources concentrated in bituminous rocks are resting untouched underground. Concerning this potential, developing new upgrading technologies of heavy oil and natural bitumen is of big interest. The primary target of investigated technology is conversion of high molecular components to light components in situ by steam treating and introducing various conversion catalysts [6-7].

2. Materials and Methods

In the work [8] we have determined the absence of direct relation between technological properties of heavy oil and natural bitumen from sheshminskoye layers (Permian deposits, prepared by PJSC "Tatneft" for industrial exploitation) and formation depth, effective oil saturation and oil productive area of reservoirs, including flow properties. This stands for occurrence of Permian deposits in secondary bedding and their composition depends on intensive effect of secondary recover processes. To study the genesis of hydrocarbon cumulations in Permian deposits and seeking for possible exploitation ways, the heavy oil composition properties of Ekaterina, Ashalcha and Olimpiada oil fields are investigated [9], [10]. These reservoirs are characterized by geological-geochemical and physic-chemical properties. Also hydrocarbon compositions and their components are different before and after hydrothermal-catalytic influences. In this work for comparative analysis rock sample of bituminous sediments from Spiridonovsky reservoir was included. Extract from the given rocks contain 69.8 % asphaltenes and corresponds to solid asphaltite (Table 1). The influences of hydrothermal-catalytic processes to composition of investigated fluids are experimentally investigated at 250°C, peculiar to steam injection techniques. The pressure of steam mixture in system is not exceeded 2 MPa with 30 wt % of water content in reactive system. Natural iron-oxide catalyst – hematite is applied in conversion of heavy oil as a heteroatom catalyst [11].

Oils before and after experiments were analyzed by the method of column-liquid-adsorption chromatography on ASK silica gel with separation of the hydrocarbon part and two groups of resins - benzene and alcohol-benzene. Before the adsorption separation according to the standard procedure, asphaltenes precipitated in 40 times the amount of petroleum ether (bp. 40–70°C).

The analysis of saturate fractions were carried out by highly effective gas chromatography method in Clarus 500 (PerkinElmer, Waltham, MA), which is equipped by auto dosing gas phase HS 40 (PerkinElmer). The capillary column with a length of 25 m, diamater of 0.2 mm and immobile phase PE-XLB (0.33 mm) was used. The temperature regime of chromatography procedure was in the range of 100°C – 300°C. From 50°C to 150°C, the rate of temperature increase was 10°C/min and starting from 150°C, the rate of temperature declines up to 3°C/min. The temperature in vaporizer was 300°C, while in detector it was 250°C and the gas-carrier was helium.

3. Results and Discussions

Based on relative distribution data (Fig. 1) of gas chromatography (GC), the normal and isoprenoid alkanes in heavy oil structure belong to different chemical types [12]. In common chromatogram of saturated hydrocarbons of Ekaterina oil field (Fig. 1a) the residue concentration of n-alkanes is clearly observed comparing to naphthenic “bunch”. This is corresponds to biodegraded type of oil in B^1 stage, when the major quantity of n-alkanes are already oxidized, mainly prevailing triterpenes hopane lines of C_{27}-C_{29}. In B^2 type oil from Ashalcha field (Fig. 1b) there are n-alkanes neither. However isoprenoid hydrocarbons C_{14}-C_{20} prevail, among which prystane (C_{19}) and phytane (C_{20}) have a maximum concentration.

a

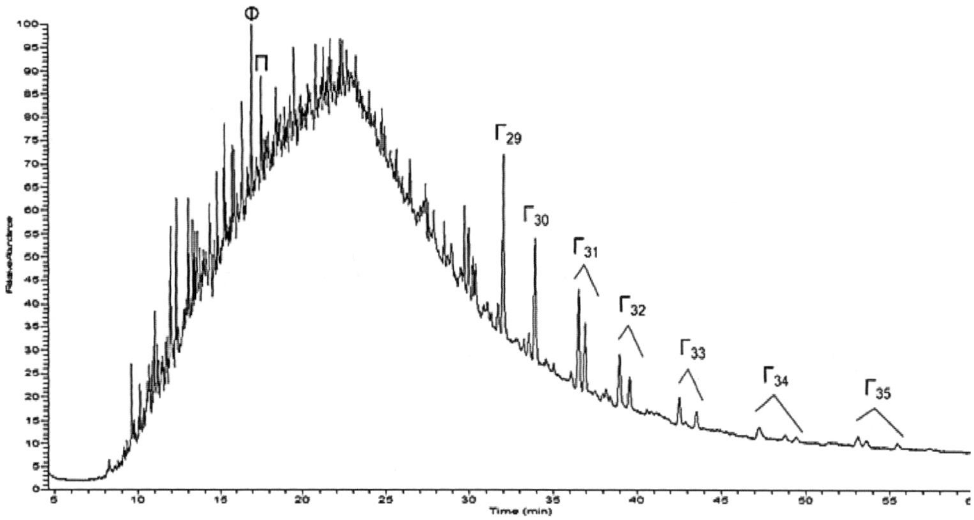

b

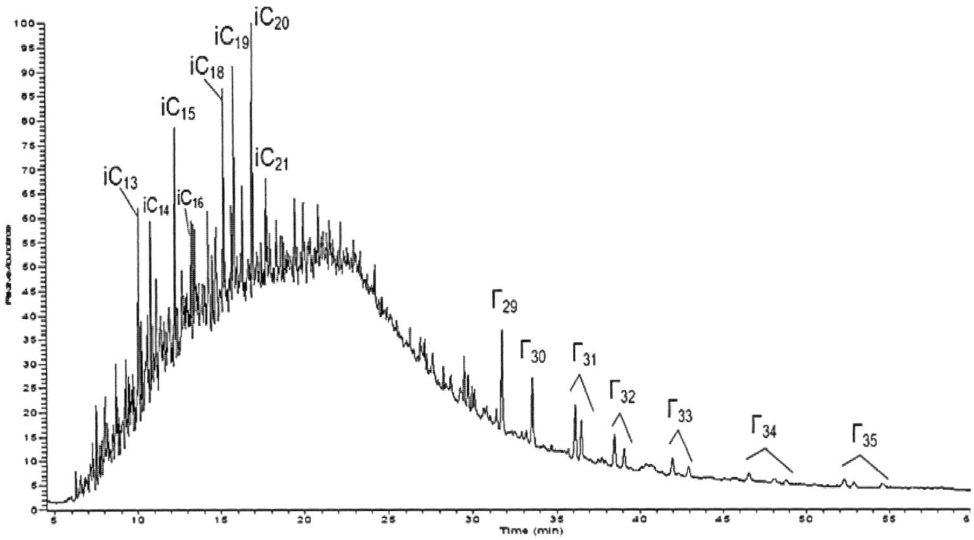

Figure 1. GC chromatograms saturated fractions of heavy oils: a – Ekaterina, b – Ashalcha.

Table 1. Heavy oil composition and Spiridonovsky asphaltite before and after hydrothermal-catalytic treatment.

Reservoir, Well number	Sampling Interval, m	**Composition, wt %				
		HC	BR	ABR	Σ resins	Asph.
		Heavy Oil				
Ekaterinovo, 6072	315-325	48.7	18.9	16.2	35.1	16.4
Ekaterinovo, 6072*		56.0	17.3	9.9	27.2	16.8
Ashalcha, 195	82	63.2	18.7	11.6	30.3	6.5
Ashalcha, 195*		65.0	18.7	10.1	28.8	6.2
Olimpiadov, 247	128-228	64.0	15.9	15.4	31.3	4.7
Olimpiadov, 247		69.0	16.9	8.8	25.7	5.3
		Asphaltite				
Spiridonovsky	rock outcrop	5.4	2.2	22,6	24.8	69.8
Spiridonovsky *		20.13	4.11	15.8	19.91	59.96

*Oil and asphaltite after experiments; **HC – Hydrocarbons; BR – Benzene resins; ABR – Alcohol-benzene resins; Asph. – asphaltenes

Oil from Olimpiadov field refers to paraffinic type A[1]. N-alkanes of C_{12}-C_{35} and bigger molecules prevails among its components with a maximum concentration of C_{17} (Fig. 2). The given type corresponds to paraffinic oil from deep Devonian coal formations. Independent of oil type in hydrothermal-catalytic processes the same processes intensify the composition: the light components of hydrocarbons increase because of decreasing the resin content 1.5-2 times, especially thermally less stable – alcohol-benzene resins. The asphaltene content slightly increases.

The changes do not have a specific character (Tab. 1). But some distinctive features are revealed in changing composition. The main change in composition of oil is observed in Ekaterina heavy oil, which is differed by high density (1.0174 g/cm^3) and high resin content and asphaltene. The least conversion is seen in Ashalcha oil with a density of 0.97 g/cm^3.

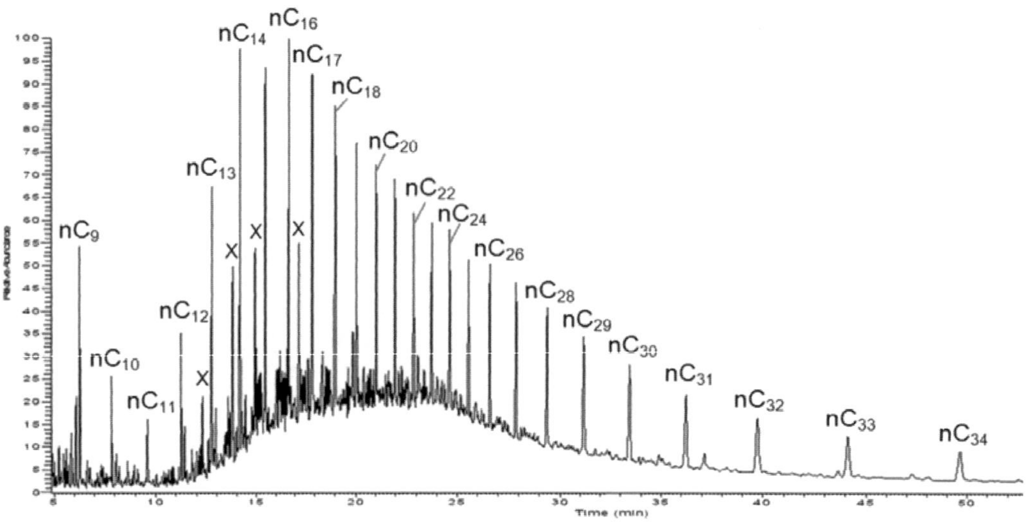

Figure 2. GC chromatograms saturated fractions of heavy oils from the Olimpiada deposit.

In Olympiad oil with a density of 0.9598 g/cm^3, as well as Ashalcha oil, alcohol-benzene resin content decreases. The light hydrocarbons are increased slightly. In the products of experiments new formed n-alkanes are observed which resemble C_{10}-C_{16}, among which homologue of even Carbon atoms prevail.

Apart oils, in the same experimental conditions the composition of Spiridonovsky bitumen-asphaltite is less stable to thermo-catalytic processes (Tab.1). It is explained by the change in natural conditions of secondary low-temperature biochemical processes. That results to accumulation of thermally unstable resin-asphaltene components in its composition [4]. As a result of decreasing alcohol-benzene resins, the light hydrocarbon content increases 4 times in asphaltite composition under hydrothermal-catalytic influence. The asphaltene content decreases as well. The benzene resin content slightly increases

Spiridonovsky asphaltite corresponds to B^1 chemical type and they contain almost no alkanes (Fig. 3). Among isoprenoid alkanes which are in slight concentrations, one can identify Prystane (C_{19}) and Phytane (C_{20}). The coefficient (K_i=Pr+Ph/n-C_{17}+n-C_{18}) representing character and degree of naturally conversed oil for initial crude oil is 0.64.

Generally in Spiridonovsky asphaltite composition the pentacyclic hydrocarbon – hopane components C_{27}-C_{29} prevail (Fig. 3a). Probably in initial composition of bitumen a bacterial material was prevailing. This is supported by low sterane content, rather than pentacyclic triterpanes [12]. The composition of Spiridonovsky bitumen was generated under surface condition effects: in the intensive contact with formation and surface waters, where a good

IOP Conf. Series: Earth and Environmental Science **282** (2019) 012011 doi:10.1088/1755-1315/282/1/012011

environment exists for oxidation and the sustainability of bacteria. In the product of samples (Fig. 3b) the significant increase in n-alkane C_{10}-C_{39} is observed. The prevailing component belongs to homologue of carbon atoms with even numbers in molecules. The maximum concentration has n-alkanes with even carbon atom in C_{14} and C_{16} compositions

a

b

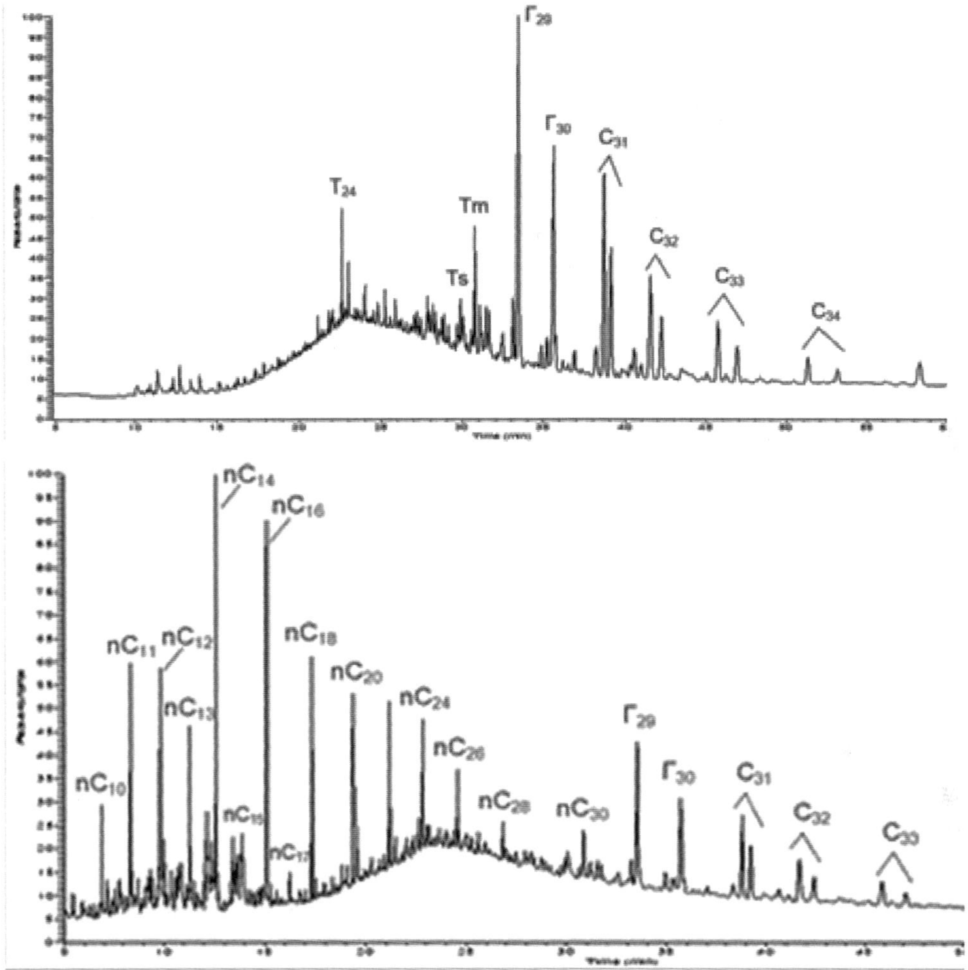

Figure 3. GC chromatograms hydrocarbon fractions of Spiridonovsky asphaltite: a - before experiment, b - after thermal treatment.

The K_i parameter after experiments declines till 0.2, which is related with n-alkane formations with prevailing even numbered Carbon atom. This is supported by significant decrease in Ph/n-C_{18} value. However, the value Ph/n-C_{17} doesn't change.

The mono-, di-, and tri methyl naphthalene are observed in products of samples by mass-spectrometer. Tetra methyl naphthalene can be observed. However, in initial native asphaltites they were not identified. There are phenanthrenes and new compounds, both in the low-molecular and high-molecular regions of the mass fragmentogram. According to IR spectroscopy the product of asphaltite conversions generally characterized by less aromaticity and low degree of sulfur content comparing with initial asphaltites. The increase in oxygen containing groups, corresponding to alcohol and ethers indicate oxidation reactions. It is also reflecting the group composition of conversion products of resin and asphaltenes.

4. Conclusion

The distinctive features of heavy hydrocarbon with different chemical types and natural solid asphaltite conversions under hydrothermal-catalytic processes at 250°C with using catalysts based on hematite mineral, containing iron oxides are revealed. Increasing light hydrocarbon and significant decrease of alcohol-benzene resin content in compositions of investigated sample products confirm destruction of high molecular components of oil. The different stability of investigated fluid components to conversion processes in hydrothermal-catalytic conditions is demonstrated. The destructive and hydrogenation processes in hydrothermal-catalytic systems play a clue role in changing the structure of asphaltite compositions. It is supported by significant increase of light hydrocarbon amount and tremendous decrease in resin and asphaltene contents. It is revealed that new formed n-alkanes in sample products have homologue of Carbon atom with even number. This can be used as an indicator feature for hydrothermal-catalytic conversion of heavy oil and natural bitumen, particularly solid asphaltites. Thus, results of experimental investigations revealed the composition conversion of not only heavy crude oil with significant high viscosity, but also solid bitumen – asphaltite types. In contrast in easy hydrothermal-catalytic condition the light hydrocarbon fractions are gained. Hereby, the given processes are perspective to apply for upgrading compositions of heavy hydrocarbon resources.

Acknowledgments

The work is performed according to the Russian Government Program of Competitive Growth of Kazan Federal University.

References

[1] Khisamov R S, Bazarevskaya V G, Tarasova T I et al 2016 *Neftyanoe Khozyaystvo - Oil Industry* **7** 10
[2] Khisamov R S, Bazarevskaya V G, Tarasova T I et al 2017 *Neftyanoe Khozyaystvo - Oil Industry* **6** 10
[3] Ostroukhov S B, Plotnikova I N, Nosova F F et al 2015 *Chem. Technol. Fuels Oils* **50** 561
[4] Kadiev Kh M, Oknina N V, Gyul'Maliev A M et al 2015 *Petrol. Chem.* **55** 563
[5] Kayukova G P, Uspensky B V, Abdrafikova I M et al 2016 *Petrol. Chem.* **56** 572
[6] Maity S K, Ancheyta J, and Marroquín G 2010 *Energy Fuels* 2010 **24** 2809
[7] Tumanyan B P, Petrukhina N N, Kayukova G P et al 2015 *Rus. Chem. Rev.* **84** 1145
[8] Sahu R, Song B J, Im J S, Jeon Y P, and Lee C W 2015 *J. Ind. Eng. Chem.* **27** 12
[9] Katritzky A R, Lapucha A R, and Siskin M 1990 *Energy Fuels* **4** 510
[10] Kayukova G P, Foss L E, Feoktistov D A et al 2017 *Petrol. Chem.* **57** 657
[11] Kayukova G P. Mikhailova A M, Feoktistov D A et al 2017 *Energy Fuels* **31** 7789

[12] Petrov A A 1984 *Hydrocarbons of oil.* - Moscow: Nauka, 263

IOP Conf. Series: Earth and Environmental Science **282** (2019) 012012 doi:10.1088/1755-1315/282/1/012012

Physical model experiment of heavy oil aquathermolysis with cobalt-iron based catalyst

S Sitnov, E Baigildin, M Khelkhal, A Chemodanov, and A Vakhin

Kazan Federal University, 18 Kremlevskaya Str., Kazan, Russia

E-mail: sers11@mail.ru

Abstract. Increasing steam-thermal methods efficiency for heavy using various catalysts is generating considerable interest in scientific research all over the world. Thus, it is worth to optimize catalyst effectiveness during their in-situ formation from oil soluble precursors using different transition metals. In this paper, a physical simulation of heavy oil catalytic aquathermolysis from the Ashalchinskoye field was carried out. The process was carried out in the presence of a hydrogen donor and a mixture of oil-soluble iron and cobalt metal tallate (1:1 weight ratio) at temperatures 250 °C and different exposure time (6 and 24 h). It was found that the most effective conditions for the thermal conversion of the oil under study are a temperature of 250 °C at 24 hours. In this case, a significant decrease in the proportion of high molecular weight components, mainly resins (by 39 %) and viscosity (about 45 %) occurs due to the course of destructive processes. At the same time, GCMS analysis of the aromatic oil fraction showed that an increase in time promotes the redistribution of mono- and polyaromatic structures.

1. Introduction

It is common knowledge that Enhanced Oil Recovery (EOR) is a widely used technique which increases oil production and lowers its high price. It is designed to recover the remaining oil in well after natural extraction process [1]. The thermal energy released during the application of thermal methods decreases the physical properties such as viscosity, density and interfacial tension which promotes oil flow in the reservoir [2]. In addition, it changes the chemical properties of reservoir fluids by cracking and dehydrogenation occurring during the process [3].

Despite the efficiency of aforementioned techniques, they are still limited by significant challenges, such as low sweep efficiency, high costs and potential formation damage. Thus, catalysts application is believed to be a potential solution to most of the challenges associated with these traditional EOR techniques.

Today, scientists all over the world are carrying out current research aimed at increasing the efficiency of steam-thermal methods of extracting high-viscosity oils due to various catalysts. When the oil-soluble precursors are decomposed, an active catalyst form is formed directly in the formation [4-6]. It is important to optimize the efficiency and reduce the cost of such catalysts by using a combination of various transition metals [7-9].

In all cases, the test results showed a high oil extraction degree (15 %) in the presence of catalyst, compared with the non-catalytic process. In addition, a significant improvement in the produced oil quality have been observed where oil viscosity decreased by 26 % at significant decrease in sulfur content [8]. An increase in lighter components associated with low content of asphaltenes after catalytic aquothermolysis has been established [9].

In this work, we carried out a physical model of heavy oil (Ashalchinskoye field) aquathermolysis process under reservoir conditions in the presence of a mechanical mixture of catalyst precursors (iron and cobalt tallates).

2. Materials and Methods

Laboratory modeling of the aquathermolysis process is carried out by using a high-pressure reactor (300 ml volume) manufactured by Parr Instruments, USA. Physical modeling of the process of catalytic and non-catalytic aquathermolysis was carried out on a sample of high-viscosity oil from the Ashalchinsk field of the Republic of Tatarstan at 250 °C for 6 and 24 hours. As a catalyst precursor, a mechanical mixture of iron and cobalt tallates in a mass ratio of 1: 1 was used, obtained on the basis of tall oil and sulphates of the corresponding metals, which was introduced into the hydrogen donor medium at 2.0 wt % to oil. A ratio of the aqueous phase and oil has fixed at 30:70.

Separation by the SARA method is carried out by taking into account the methodical recommendations of ASTM D4124-09 and GOST 32269-2013 in several stages. This method is based on the separation of oil into four groups of compounds: saturated hydrocarbons, aromatic compounds, resins and asphaltens (SARA) according to their solubility and polarity.

Dynamic viscosity (at 10 °C) was determined on a Fungilab Alpha L rotational viscometer (Spain) in combination with a Huber K-6 cryostat thermostat (Germany).

The saturated fraction of oil samples was studied on a gas chromatography– mass spectrometry system including a gas chromatograph “Chromatec-Crystal 5000” with an ISQ mass-selective detector, Xcalibur software for processing results. Compounds identification was carried out using the electronic library of NIST mass spectra and from literature data.

3. Results and Discussions

Figure 1 shows the viscosity-temperature characteristics of catalytic and non-catalytic aquathermolysis products at 250 °C for different exposure times.

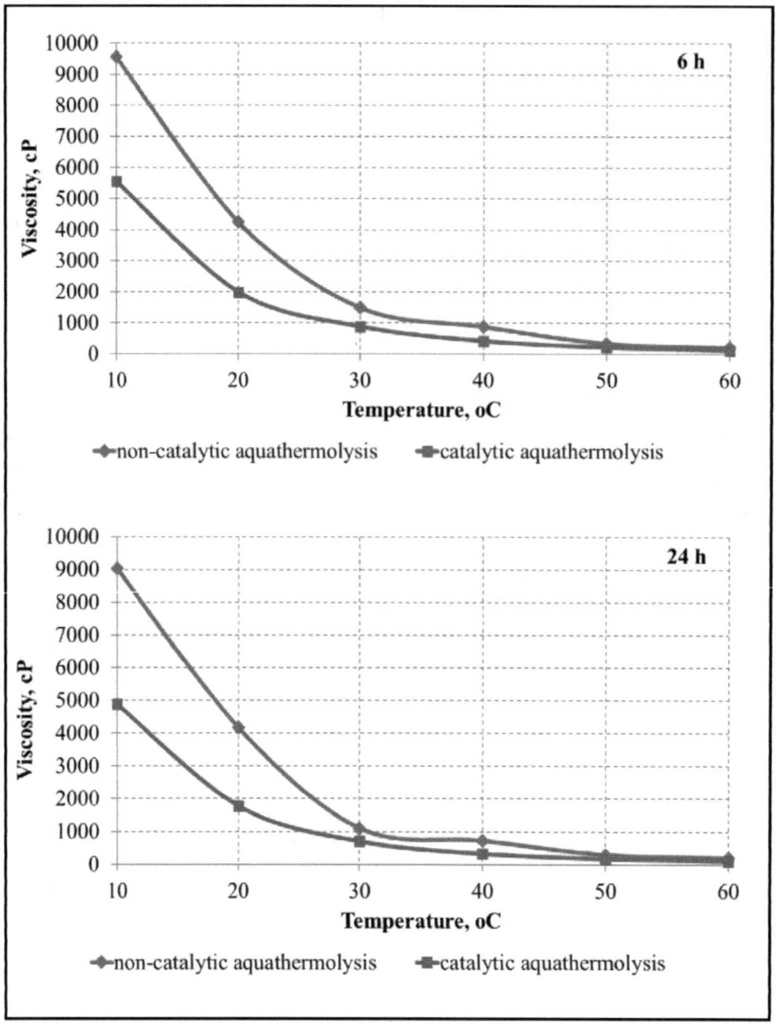

Figure 1. Viscosity-temperature characteristics of catalytic and non-catalytic aquathermolysis products at 250 °C at different times.

As can be seen from the figure, the introduction of the catalyst favorably affected the rheological characteristics of the converted oil. The viscosity (6 hours) of oil in the presence of catalyst compared to the non-catalytic process decreased by 42 %, and the effect accumulates over time - the daily effect leads to an additional decrease in viscosity by an additional 12 %. It is noteworthy that increasing exposure time during the non-catalytic process did not lead to an increase in the mobility of the studied oil.

The results of the determination of the composition group of the test samples by the SARA method are given in Table 1.

Table 1. Composition group (SARA) of the samples oil after aquathermolysis at 250 °C.

Objects	Group composition (SARA), wt %			
	Saturates	Aromatics	Resins	Asphaltenes
Oil after non - catalytic aquathermolysis (6 h)	32.72	31.47	30.06	5.75
Oil after catalytic aquathermolysis (6 h)	32.24	37.73	24.76	5.27
Oil after non - catalytic aquathermolysis (24 h)	31.29	28.98	33.93	5.80
Oil after catalytic aquathermolysis (24 h)	35.08	40.33	19.42	5.17

SARA-analysis data indicate that viscosity decrease under catalytic thermal is the result of destructive processes occurred to resinous compounds which content decreased by 18 % and 43 % at 6 and 24 hours, respectively due to the catalyst intensification of cracking reaction and hydrogenolysis comparing to the non-catalytic process. Also, there is an increase light hydrocarbon content of the converted oil, namely, aromatics – 20 % (6 h) and 39 % (24 h). This is apparently due to the fact that resins can form associated molecular complexes [10]. Consequently, basing on oil chemical composition change data, and exactly resin content decrease, it can be assumed that the catalyst ensures the destruction of associated complexes of resin molecules, thereby affecting the decrease in viscosity of oil in general [11].

The effect of the catalyst is also confirmed by GC-MS results of aromatic fraction of the studied oil samples (Figure 2).

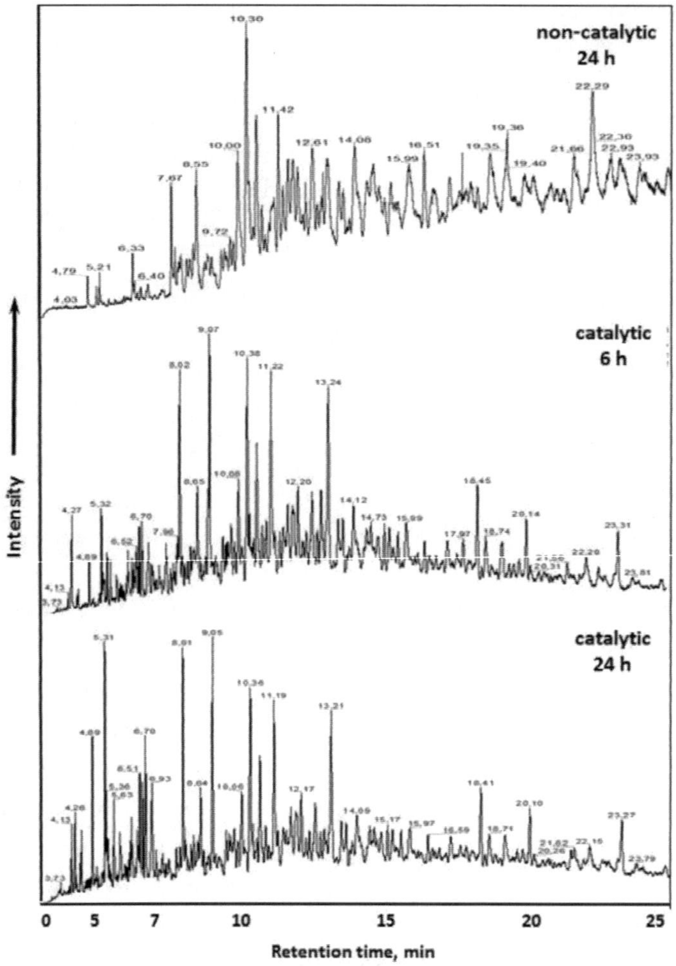

Figure 2. GC-MS of aromatic compounds change in time at 250 °C.

The presented mass spectra show that after 6 hours of thermal impact, the aromatic fraction gets enriched with the products of resinous compounds destructive hydrogenation. Time increasing contributes to the saturation of the lighter hydrocarbon compounds.

4. Conclusions

In summary, our work has investigated the effect of iron and cobalt tallates catalysts on extra-heavy oil aquathermolysis for enhanced oil recovery methods. According to dynamic viscosity and group chemical composition results, it was found that in the absence of catalyst, a noticeable decrease in viscosity does not occur while the introduction of catalyst reduces the content of high-molecular resinous compounds by 18 %, and as a result, decreases oil viscosity by 42 %. Over time, this effect accumulates. In addition, correlating GC-MS and SARA analysis results led to reveal the enrichment of aromatic fraction by the products of resinous compounds destructive hydrogenation.

Acknowledgments
The work is performed according to the Russian Government Program of Competitive Growth of Kazan Federal University.

References
[1] Zhao X, Liu Y, Xu C et al. 2013 *Energy Fuels* **27** 2874
[2] Fan H and Liu Y 2002 *Oil & Gas Journal* **100** 60
[3] Wen S, Zhao Y, Liu Y et al. 2007 *SPE International Symposium on Oilfield Chemistry, Houston*
[4] Khelkhal M A, Eskin A A, Sharifullin A V et al 2019 *Petrol. Sci. Technol.* In press
[5] Feoktistov D A, Kayukova G P, Vakhin A V et al 2018 *Chem. Technol. Fuels Oils* **53** 905
[6] Wang Y Q, Chen Y L, He J et al 2010 *Energy Fuels* **24** 1502
[7] Sitnov S A, Mukhamatdinov I I, Vakhin A V et al 2018 *J. Petrol. Sci. Eng.* **169** 44
[8] Shuwa S M, Al-Hajri R S, Mohsenzadeh A et al 2016 *Society of Petroleum Engineers - SPE EOR Conference at Oil and Gas West Asia, OGWA*
[9] Yusuf A, Al-Hajri R S, Al-Waheibi Y M et al 2016 *Journal of the Taiwan Institute of Chemical Engineers* **67** 45
[10] Verkhovykh A A, Ermeev A M, and Elpidinsky A A 2015 *Vestnik of the Technological University* **18** 64
[11] Mukhamatdinov I I, Vakhin A V, Sitnov S A et al 2018 *Chem. Technol. Fuels Oils* **54** 574

IOP Conf. Series: Earth and Environmental Science **282** (2019) 012013 doi:10.1088/1755-1315/282/1/012013

A comparative study of heavy oil oxidation thermal behavior in the presence of Cu-Mn tallates

M Khelkhal, A Eskin, and A Vakhin

Kazan Federal University, 18 Kremlevskaya Str., Kazan, Russia
E-mail: vahin-a_v@mail.ru

Abstract. Heavy oil in-situ combustion process has many benefits among other thermal enhancing oil recovery methods. However, its application is still a complex task due to the irregular propagation of combustion front in the reservoir and its early breaking down. Recent finding regarding in-situ combustion development have led to use catalysts based on transition metals in order to stabilize and enhance combustion front flow. Transition metals was chosen rather than other types of catalysts because of their practical, feasible and economic impact in a wide range of industrial processes.

Current solutions to catalytic in-situ combustion are inconsistent with their application in real fields due the problems associated to the injection of catalysts in the reservoir from one side and to the high cost that can be generated from the use of certain type of catalysts. In this context, we studied the effect of copper and manganese tallates in different metal ratios on the oxidation of heavy oil by applying various physical and physical-chemical methods. As a result, we have obtained accurate results demonstrating the potential effect of Cu-Mn tallates in stabilizing combustion front during high temperature oxidation reactions.

1. Introduction

Unconventional oil especially heavy and extra-heavy oils resources are generating considerable interest in terms of being an alternative way to replace conventional oils for at least the end of the current century. However, one of the major drawbacks in their exploitation is their high viscosity, which complicate their extraction. Thus, various methods, known as enhanced oil recovery methods have been proposed to solve the problem of high viscosity [1]. Thermal methods are among the most widely used techniques in the field and which are commonly investigated in the lab scale [2].

Among enhanced oil recovery methods, in-situ combustion process is believed to be an effective, economic and environmentally friendly way to increase heavy oil recovery factor [3]. It is based on injecting air in the reservoir to initiate heavy oil oxidation reactions and hence to generate heat in-situ which propagates further and leads to increase reservoir temperature and thus decreasing viscosity. However, regardless the simplicity of the process in the theory as it seems, it is considered as one of the most complexes processes [4] and its application has witnessed few successful examples, which could be counted on the fingers of one hand like the Bellevue (Louisiana) and Rumania fields [5,6]. Researchers have always associated the process failing to front combustion earlier breakdown. Therefore, combustion front propagation was set to become an important component in a successful application of in-situ combustion process [7]. Various approached has been suggested to enhance combustion front propagation using catalysts based on transition metals due the important role played by them in industrial world [8]. In the literature of catalytic oil oxidation process, different catalysts in non-organic or organic salts form [9] have been used while other studies [10,11] applied micro and nanoparticle catalysts based on transition metals.

Several studies for example [9,12,13] have been performed on oil oxidation in the presence of non-organic catalysts based on various types of transition metals. It has been found that catalysts based on manganese and copper could be promising metals for in-situ combustion due to their low cost, positive impact on the environment and their availability [14,15]. The same idea has been obtained when applying oil soluble catalysts based also on transition metals on heavy oil oxidation. In addition to this, even when applying microparticles or nanoparticles based on transition metals, enhancing high temperature oxidation were obvious in the presence of manganese and copper [15,16].

Many experts now believe that rather than using catalysts in ionic or particle forms it might be useful to apply oil soluble catalysts since they can highly disperse in oil phase. The only problem limiting their use is cost of some organic salts. For this reason, some researchers have suggested the use of metal tallates as an economic alternative for use [17]. It is commonly known that tall oil is an environment friendly precursor with low cost and broad availability.

Our first studies on heavy oil oxidation in the presence of metal tallates [18] found that they are able to enhance metal dispersion in the oil phase which lead to a better combustion front stabilization. In addition, we have observed that during high temperature oxidation, copper and manganese effect the reactions in different temperature ranges. At the beginning of high temperature region, the effect resulted from copper was more than manganese's one. But at a certain temperature (about 745 K) the situation gets reversed and the effect of manganese becomes better that copper's one.

In order to understand the dual effect that could be generated from the application of copper tallates and manganese tallates we applied a complexe of physical and physic-chemical methods, which provided interesting information about oxidation kinetic behavior and its relation to copper and manganese oxide nanoparticles obtained during the process.

2. Materials and Methods

The crude oil (13.8 API°) was obtained from Ashalcha oilfield (Volga-Ural basin, Republic of Tatarstan, Russia). Organic solvent (dichloromethane - 99.5 % pure), was supplied from

IOP Conf. Series: Earth and Environmental Science **282** (2019) 012013 doi:10.1088/1755-1315/282/1/012013

Component-Reactiv and used without additional purification. Inorganic salts and pure quartz sand fraction of 43-64 μm were purchased from Sigma-Aldrich.

Copper and manganese tallates were prepared according to a well-known method with some modifications. The experimental set up is detailed in [17]. Samples for DSC analysis were prepared by mixing heavy oil (10.0 wt %) with pure quartz sand fraction of 43-64 μm, ([90.0-ω] wt %) and symbolized as S(0). In the case of catalytic experiments, initial oil sample contained 2 wt % of Cu-Mn tallates in different Cu tallates:Mn tallates ratios and symbolized as follow: S(1) = (0:1), S(2) = (1:3), S(3) = (1:1), S(4) = (3:1), S(5) = (1:0).

STA 449 F1 Jupiter thermal analyzer supplied by Netzsch company have been used at a temperature range of 30-600 °C to provide differential scanning calorimetric data about the process of heavy oil oxidation. A linear heating rate of 5, 10, 15, 20 K×min^{-1} was maintained during experiments, under an airflow of 50 ml×min^{-1}. The software package used to analyze the data was Proteus Analysis v5.2.1, NETZSCH Peak Separation (version 2010.09), and NETZSCH Thermokinetics 3.1 (version 06.08.2014).

3. Results and Discussions

Differential scanning calorimetry is a less time consuming, less costly approach to describe the kinetic behavior of heavy oil oxidation process. Its application in our current research allowed to shed light on the effect generated from Cu-Mn tallates on the general kinetic behavior of heavy oils. Figure 1 compares the heavy oil oxidation in the presence of catalysts at different Cu-Mn tallates ratios at 5 °C/min. It is apparent from figure 1 that the oxidation process includes two separated regions referred as low temperature oxidation (LTO) and high temperature oxidation (HTO) respectively. It is generally accepted that LTO region results in oxygenated compounds with minimal amount of carbon oxide and carbon dioxide. On the other hand, HTO regions includes coke oxidation and the formation of carbon oxides and water [19]. Peak temperature shifting is presented in figure 2.

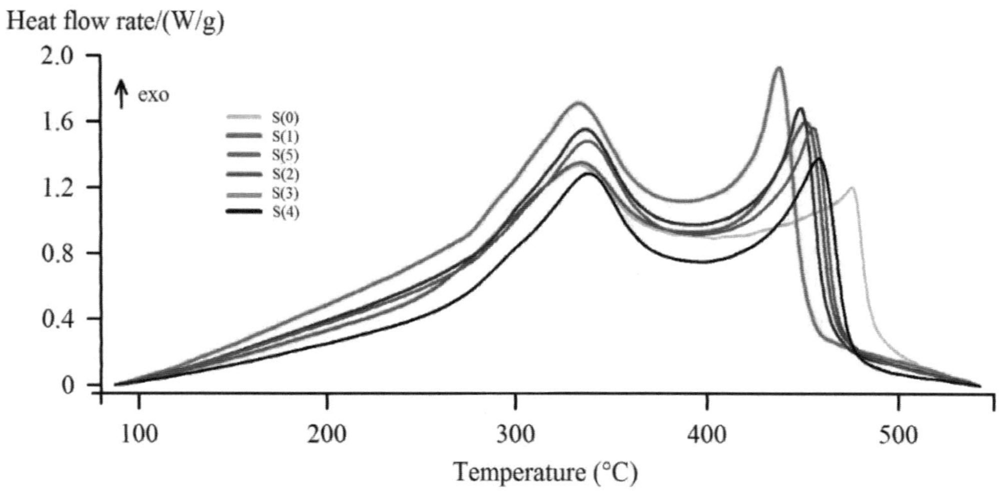

Figure 1. DSC curves for non-catalytic S (0) and catalytic heavy oil oxidation S (1-5).

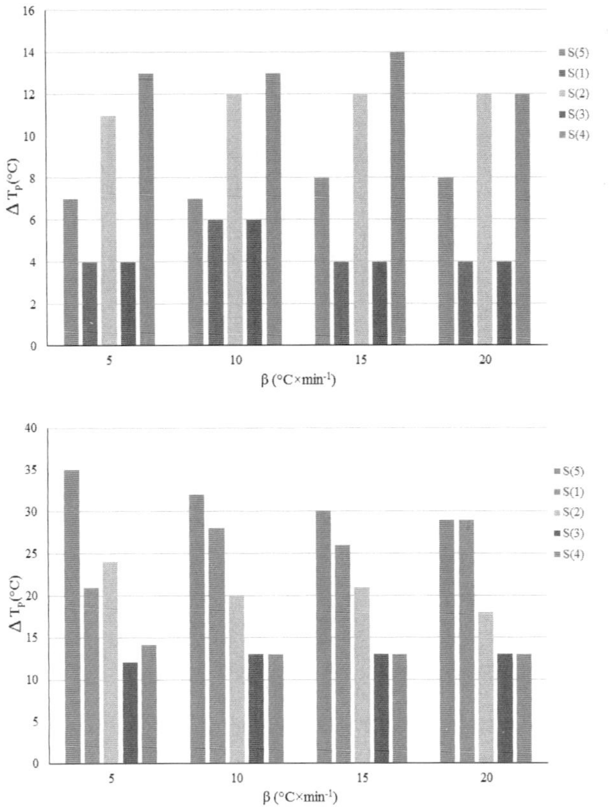

Figure 2. Differences between peak temperatures (ΔT_p) of non-catalytic s(0) and catalytic heavy oil oxidation S(1-5) in porous media for low (top) and high (bottom) temperature oxidations at different heating rates.

The correlation between figure1 and figure 2 is noteworthy because it demonstrates the efficiency resulted from applying Cu-Mn tallates in the process of heavy oil oxidation at different Cu tallates-Mn tallates ratios. As can be seen from this correlation, HTO peak has shifted significantly comparing to LTO peak shifting in the catalytic oxidation respectively especially in the case of S(1) and S(5) which were higher than the other catalysts S(2), S(3) and S(4). Further calculations have been performed on activation energy and pre-exponential factors of catalytic and non-catalytic oxidation in both LTO and HTO regions to confirm our observations.

It is common knowledge that the multiphasic nature of heavy oil oxidation process and its complexity make its kinetic study quite difficult comparing to traditional kinetics in homogeneous mediums. Therefore, several studies, for example [19] have been suggested that oil combustion rate is dependent on two essential parameters: conversion degree α which can be obtained from DSC data and oxygen partial pressure (Equation 1).

$$\frac{d\alpha}{dt} = k(T)P_{O_2}^{a}(1-\alpha)^{b} \quad (1)$$

Where *k(T)* is the rate constant which can be expressed by Arrhenius law (Equation 2):

$$k(T) = Ae^{-E/RT} \quad (2)$$

P_{O_2} is the oxygen partial pressure

$(1-\alpha)^b$ is the reaction model function

In order to simplify equation (1), we can consider the oxygen partial pressure constant during experiments because of the small sample size (less than 1 mg in each run), the high flow rate (50 ml/min) and the large furnace volume (approx. 250 ml). Beside this, many studies [20,21] have demonstrated the first order *(b=1)* in the model function $f(\alpha) = (1-\alpha)^b$. Thus, equation 1 becomes:

$$\frac{d\alpha}{dt} = k_{eff}(1-\alpha) \quad (3), \text{ where } k_{eff} = kP_{O_2}$$

In our research we made call to Kissinger's method [22]. This method was chosen rather than other methods because it is one of the most rapid ways to estimate the kinetic behavior of oil oxidation process. In addition, it is independent on baseline choice and peak profile influence, simply because it connects kinetic parameters to peaks temperatures. Its logarithmic form is presented by equation 4 as follows:

$$ln(\frac{\beta}{T_p^2}) = -\frac{E}{R} \times \frac{1}{T_p} + ln(Af'(\alpha)) \quad (4)$$

It is consistent with the chosen kinetic model (Equation 3) since it assumes first-order dependency of reaction rate on (1-α).

Figure 3 indicates the related Kissinger's plots and table 1 presents kinetic parameters for non-catalytic S(0) and catalytic heavy oil oxidation S(1-5). The interpretation of obtained results suggests the evident effect performed by the presence of Cu-Mn talltes on oxidation kinetic behavior where we almost note a significant decrease in energy of activation of both processes in the presence of catalyst comparing to energies of activation in its absence. However, in the presence of S(1) activation energy oppositely increases. On the other hand, the pre-exponential factor in both regions changed differently depending on the ratio of copper tallates to manganese tallates in the catalyst comparing to its absence where it increased in the presence of S(1) and S(4) while in the presence of S(2), S(3) and S(5) it decreases in both regions.

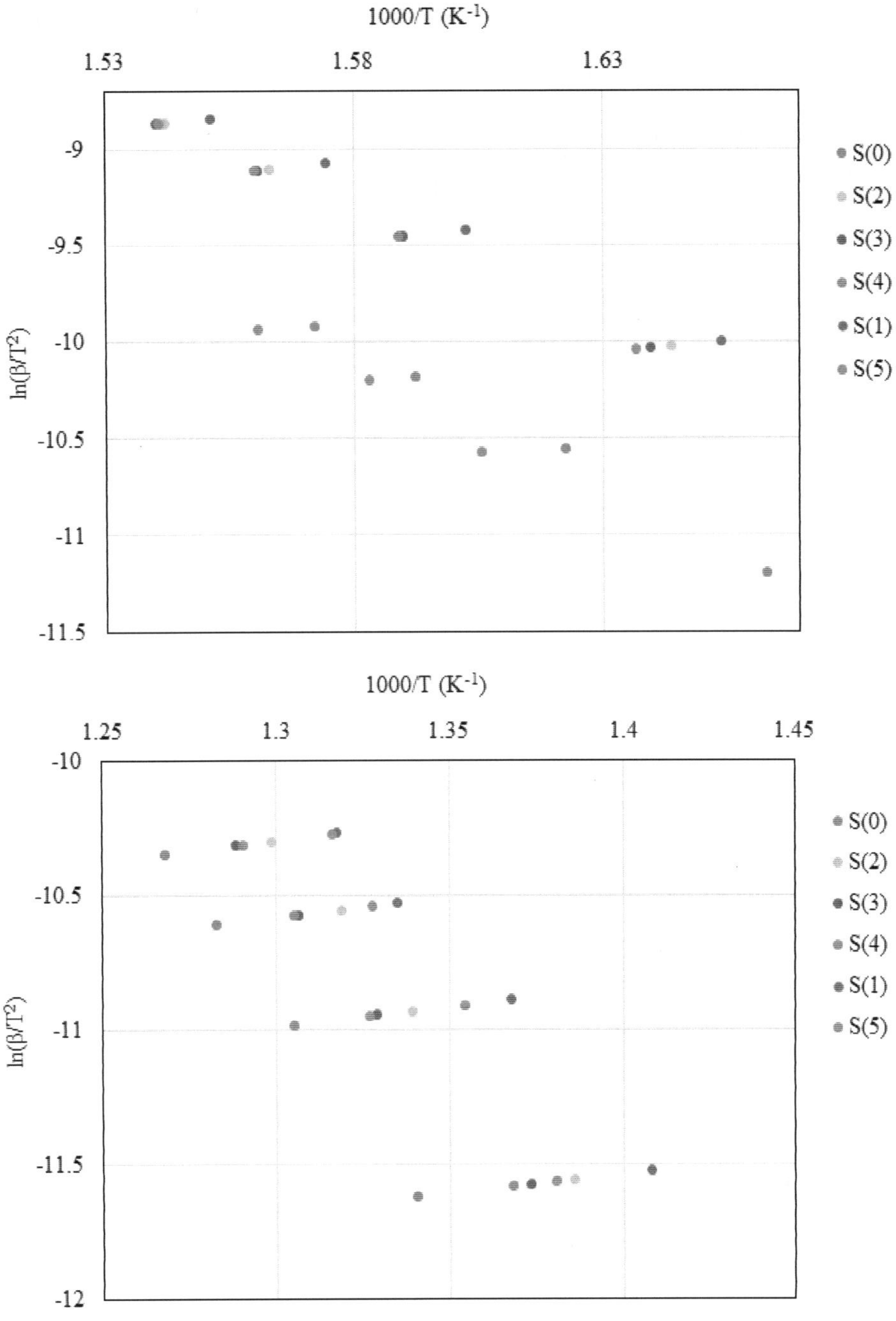

Figure 3. Kissinger's plots for catalytic and non-catalytic high temperature oxidation (top is for LTO, bottom is for HTO).

IOP Conf. Series: Earth and Environmental Science **282** (2019) 012013 doi:10.1088/1755-1315/282/1/012013

Table1. Kinetic parameters of the oxidation processes.

Kinetic parameters	Sample	LTO	HTO
E_a, kJ/mol	S(0)	104.6±1.6	145.8±8.5
	S(1)	103.4±1.3	163.3±8.1
	S(2)	95.17±2.4	120.6±7.4
	S(3)	97.48±1.4	123.57±5.3
	S(4)	101.4±2.6	135.1±6.6
	S(5)	94.3±2.7	112.4±4.8
$log_{10}A$, A in min^{-1}	S(0)	9.4	11.8
	S(1)	9.8	15.1
	S(2)	8.7	8.5
	S(3)	9.1	8.8
	S(4)	9.9	10.6
	S(5)	8.7	7.5

Despite the fact that energy of activation decrease is a strong indicator of catalyst effect, we decided to calculate reaction effective rate constant in all cases (from S(0) to S(5)) to remove any doubt related to pre-exponential decrease in some catalytic oxidation processes.

Figure 4 presents the variation of effective rate constants with temperature for catalytic S(1-5) and non-catalytic S(0) oxidation (top is for LTO, bottom is for HTO) in the presence of copper and manganese tallates in different ratios.

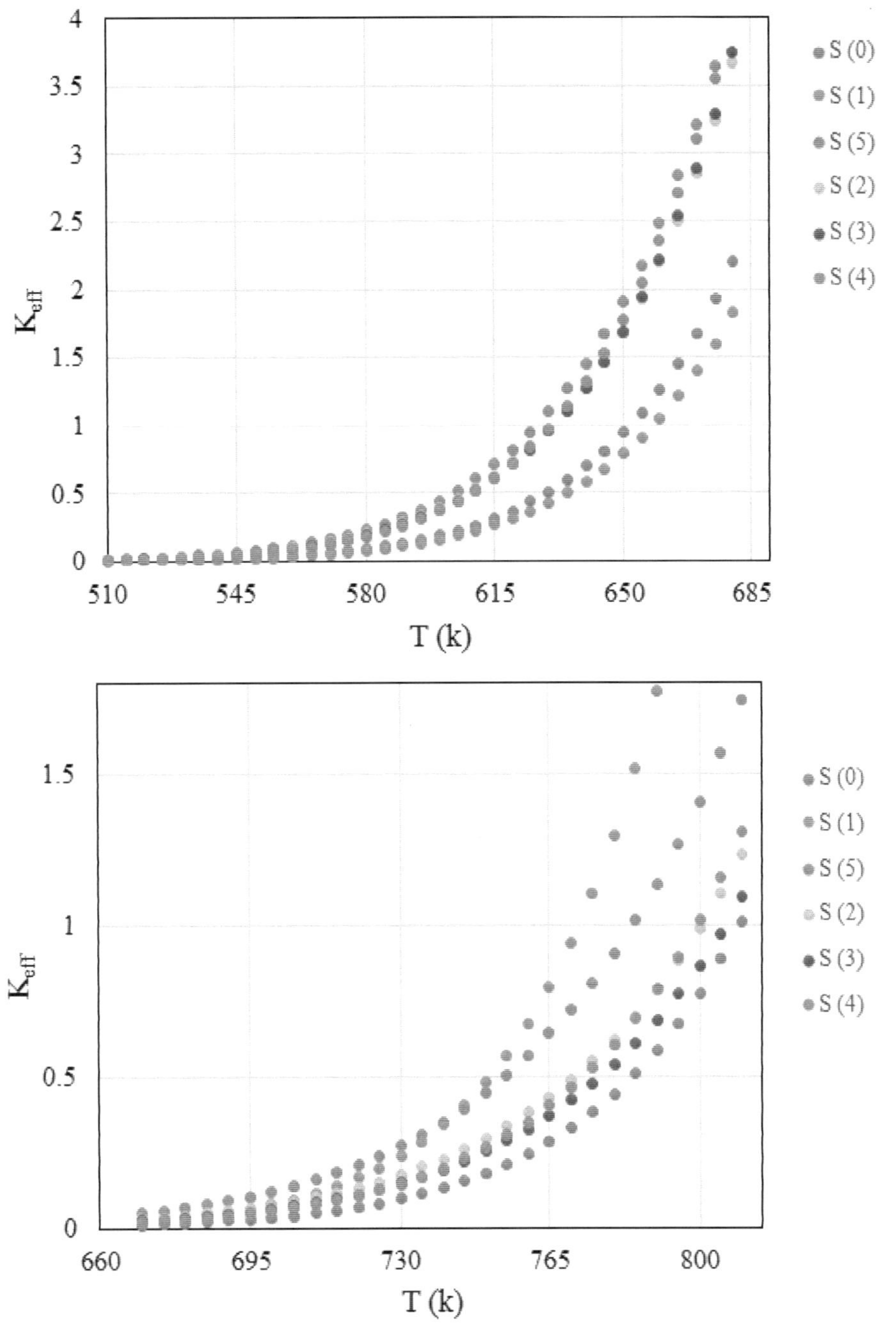

Figure 4. Variation of effective rate constants with temperature for catalytic and non-catalytic oxidation (top is for LTO, bottom is for HTO).

As expected, especially in HTO region, the effective rate constant was higher in the catalytic oxidation (S(1)≥ S(1)≥ S(5)≥ S(4)≥ S(2)≥ S(3)) comparing to non-catalytic oxidation S(0) which indicates the stabilization of combustion front in the presence of Cu-Mn tallates rather than in their absence.

4. Conclusions

To sum up our work has led us to propose a new effective, less costly and eco-friendly catalyst to enhance heavy oil oxidation process. The evidence from this study points towards the idea of using catalysts based on tallates because they provide a high solubility in oil with less cost, which can be a powerful solution for catalyst injection problem into reservoirs during field-testing. We have obtained comprehensive results proving the potential effect of Cu-Mn tallates by applying thermal analysis methods especially in the case of using manganese tallates or copper tallates separately. The current research has also revealed the negative effect on hight temperature oxidation reactions rate generated from using copper and manganese tallates together. However, the present study has only compared the effect of copper-manganese tallates on the process of heavy oil combustion using thermal analysis which takes into account the thermal aspect of the process. Consequently, further investigation in medium and larger scales should be performed to study other aspects related to its application in the field.

Acknowledgments

The work is performed according to the Russian Government Program of Competitive Growth of Kazan Federal University.

References

[1] Shah A, Fishwick R, Wood J et al 2010 *Energy Environ. Sci.* **3** 700–14
[2] Drici O and Voasoughi S 1987 *SPE J.* **2** 591–5
[3] Rabiu Ado M, Greaves M and Rigby S P 2018 *J. Pet. Sci. Eng.* **166** 94–103
[4] Shen C 2002 *Canadian International Petroleum Conference* (Petroleum Society of Canada) pp 1–14
[5] Cadelle C P, Burger J G, Bardon C P et al 1981 *J. Pet. Technol.* **33** 2–57
[6] Long R E and Nuar M F 1982 *SPE Enhanced Oil Recovery Symposium* (Society of Petroleum Engineers) p 6
[7] Baud G, Salvador S, Debenest G and Thovert J F 2015 *Energy and Fuels* **29** 6780–92
[8] Abuhesa M B and Hughes R 2009 *Energy Fuels* **23** 186–92
[9] Nares R, Schacht-Hernandez P, Ramirez-Garnica M A et al 2007 *International Oil Conference and Exhibition* (Society of Petroleum Engineers) p 5
[10] Hendraningrat L, Souraki Y and Torsater O 2014 *SPE/EAGE European Unconventional Resources Conference and Exhibition*
[11] Hernandez-Perez J R, Cabrera-Reyes M C, Schacht-Hernandez P et al 2008 *The SPE International Thermal Operations and Heavy Oil Symposium- Canada* (Society of Petroleum Engineers) p 11
[12] Shallcross D C, de los Rios C F, Castanier L M et al 1991 *SPE Reserv. Eng.* **6** 287–94

[13] Rana M S, Sámano V, Ancheyta J et al 2007 *Fuel* **86** 1216–31
[14] Suwaid M A, Emelianov D A, Al-Muntaser A A et al 2018 *Appl. Catal. A Gen.* **564** 79–89
[15] Galukhin A V., Khelkhal M A, Eskin A V et al 2017 *Energy Fuels* **31** 11253–7
[16] Galukhin A, Khelkhal M A, Gerasimov A et al 2016 *Energy Fuels* **30** 7731–7
[17] Feoktistov D A, Kayukova G P, Vakhin A V et al *Chem. Technol. Fuels Oils* **53** 905–12
[18] Khelkhal M A, Eskin A A, Sharifullin A V et al 2019 *Pet. Sci. Technol.* 1–7
[19] Sarathi P S 1999 *In-situ combustion handbook--principles and practices* (National Petroleum Technology Office, Tulsa, OK (US))
[20] Fassihi M R, Brigham W E and Ramey Jr H J 1984 *Soc. Pet. Eng. J.* **24** 399–407
[21] Bousaid I S and Ramey Jr H J 1968 *Soc. Pet. Eng. J.* **8** 137–48
[22] Kissinger H E 1956 *J. Res. Natl. Bur. Stand. (1934).* **57** 217–21

IOP Conf. Series: Earth and Environmental Science **282** (2019) 012014 doi:10.1088/1755-1315/282/1/012014

Fractionation of asphaltenes for determination of their composition and structure in case of heavy oil from Ashal'cha field

I Salih, I Mukhamatdinov, R Mukhamatdinova, and E Shmeleva

Kazan Federal University, 18 Kremlevskaya Str., Kazan, Russia

E-mail: iimuhamatdinov@gmail.com

Abstract. The asphaltene fractions from initial crude oil of Ashal'cha field were extracted for the further investigation. Moreover, the non-catalytic and catalytic products of aquathermolysis in the presence of cobalt-containing catalyst precursor and hydrogen donor dependent on treatment time were studied. Five fractions of asphaltenes and carben-carboids were obtained by batch extraction method. The catalytic hydrothermal treatment revealed a redistribution of group composition and the reduction of resins and asphaltenes because of destructive hydrogenation processes mainly on carbon – heteroatom bonds. In this paper, we have obtained comprehensive results regarding the redistribution between the asphaltene fractions (extracted by solvents with various compositions) due to catalytic aquathermolysis.

1. Introduction

In the modern oil refining industry, the tendency of crudes is toward the heavy hydrocarbons. In its term, it requires development and introduction of additional oil processing units that are based on structural destruction of heavy molecules of oil and thus, increasing the content of light fractions of crude oil. The most well-known methods of increasing the yield of light fractions such as thermal, radio-thermal and etc. is based on the destruction of heavy oil components. Therefore, it is important to know the structure of resins and asphaltenes, as well as content and the nature of molecule fragments that contain heteroatoms. The role of last is crucial in destruction of high-molecular compounds. However, the problem of heavy components of crude oil can be solved in reservoir formations by introduction of water- and oil-soluble catalyst precursors prior to steam injection techniques. The catalyst precursor activates only in reservoir conditions and leads to conversion of oil without significant constructive changes in downstream installations [1-5]. Injection of catalytically active metals into the productive formations intensifies the ability of mineral matrices to convert the crude oil in-situ. It is not only increasing the sweep of reservoir due to decreasing molecular masses of resins and asphaltenes, but also irreversibly reducing the viscosity of produced oil and the content of hard-to-refinery components. Using the catalysts at oil recovery stage increases the energy efficiency of thermal recovery methods and significantly decreases the environmental risks during the development of heavy oil fields. Currently, most studies on the structure of asphaltenes are carried out with the whole mass of asphaltenes and resins. However, fractionation of asphaltenes provides to determine the structure of molecules more accurately [6–11].

The fractionation method of complex mixtures of high-molecular compounds were well known and widely used. The main goal of which was separation of mixtures that compose of many various components (in case of carbene-carboids – several thousands) to the narrower fractions, each of which has a close composition to homogeneous one (set of molecules with similar properties). Consequently, the separate fraction may differ enough sharply by physical and chemical characteristics [12].

2. Materials and Methods

The model system was consisted of oil and water with a mass ratio of 70:30. This water-in-oil emulsion was exposed to the catalytic and non-catalytic thermal treatment processes (250 °C). The oil-soluble Co carboxylates obtained by means of exchange reaction between sodium salt of tall oil and non-organic salts of cobalt were used as the catalyst precursor. Then the catalyst precursor (1.5 wt % by oil) was injected to the oil sample.

The initial pressure was fixed at 0.3 MPa. The duration of the experiment was 6, 12, 24 hours. At the end of thermal process, the oil was separated from water by precipitation for 16 hours. After that, it was centrifuged on a laboratory centrifuge Eppendorf 5804r at 3000 rpm for 1 hour. Then, the content of the water in all samples were determined by the Dean and Stark method. The asphaltenes were precipitated from the oil samples (5-10 gr) by warm isooctane (200 ml) through filter paper. Then they were placed in Soxhlet apparatus with further stepwise solvent treatment by the mixture of isopropyl alcohol and benzene (the share of latter was 20, 40, 60, 80 %), as well as pure benzene. The given solvents are characterized by almost similar boiling temperature (isopropyl alcohol -82°C and benzene – 80°C) and gradually increasing dissolving abilities [13]. In each separation, maltene fraction (M), five fractions of asphaltenes and residue on filer were obtained. Maltene fractions were separated into 3 compositional groups by SARA-method (saturates, aromatics and resins) according to ASTM D4124-09. The glass chromatography column was used to gradually elute the aliphatic and aromatic hydrocarbons from adsorbent – initially dehydrated neutral aluminum oxide. The elemental analysis (CHNS) was carried out in CHNS-analyzer 2400 Series II (USA, Perkin Elmer) at temperature of 1000 °C.

3. Results and Discussion

3.1. SARA-analysis and fractionation of asphaltenes

According to the Figure 1, the content of saturate hydrocarbons increased, while resins and asphaltenes decreased after adding catalyst and hydrogen donor. In order to obtain the detail and reliable information about the composition and the structure of high-molecular oil components, we have to separate a set of thousand different molecules - resins and asphaltenes to the possible narrower fractions that contains molecules with similar properties. The batch extraction of each fraction was continued until the solution in Soxhlet extractor became bleach. After that, the obtained fractions were distilled off from the solvent and evaporated in rotational evaporator until the constant mass was achieved.

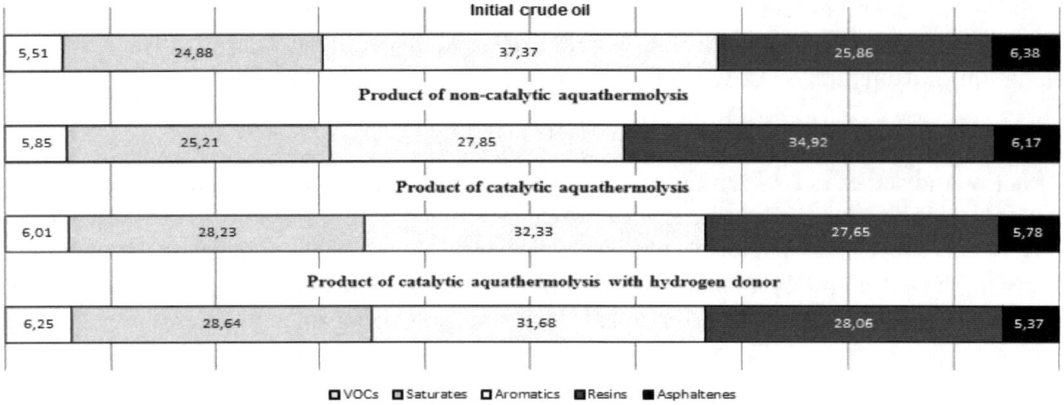

Figure 1. The SARA-analysis of initial crude oil and the products of non-catalytic and catalytic aquathermolysis with hydrogen donor (treatment time -6 hours).

According to the Figure 2, the main mass of asphaltene is concentrated in A2 fraction (60% isopropyl alcohol + 40 % benzene) that probably justifies the polar characteristics of asphaltene components of oil. The stimulation of catalytic hydrothermal treatment in the presence of hydrogen donor leads to transition of the most high molecular compounds from solvation layer – A1 toward the most heaviest fraction – A5 and carben-carboid (residue in filter paper). Moreover, the least heavy fractions of asphaltenes (A1-A2) shift into aromatic and saturate hydrocarbons. The conversion can be presented by the following scheme:

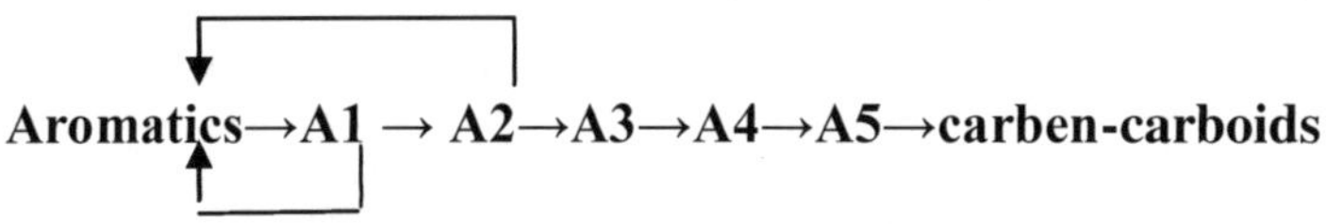

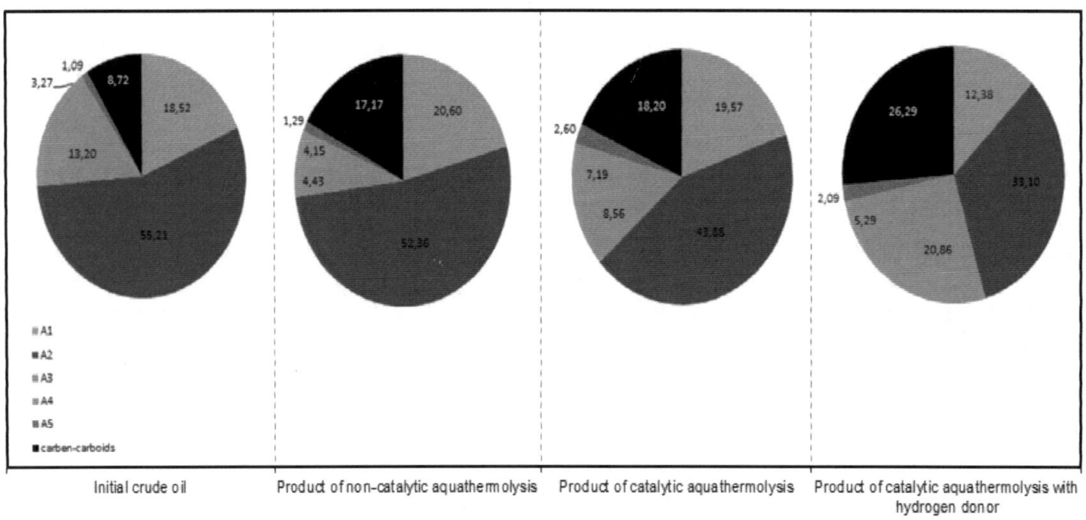

Figure 2. The fractional composition of asphaltenes from initial crude oil, products of non-catalytic and catalytic aquathermolysis, and in the presence of hydrogen donor.

The next step was investigation of elemental composition of asphaltenes. The elemental composition of oil is necessary, particularly for choosing the refinery method and compiling material balance of some refining processes. The data of elemental composition and structural group composition of narrow fractions of saturates, aromatics and heavy residues, from which extraction of individual substances is not possible, provides significantly enlarge the understanding of structures and build their "average" molecule model [14]. Table 1 presents the elemental composition of asphaltene fractions.

Table 1. Elemental composition of asphaltene fractions

Asphaltene fractions	Content, wt %				
	C	H	N	S	O
Initial crude oil					
Oil	83.88	11.35	0.37	1.76	2.64
A1	80.76	8.24	1.63	3.75	5.62
A2	81.43	8.1	1.54	3.66	5.27
A3	79.78	7.33	1.61	4.51	6.77
A4	79.32	7.44	1.66	4.98	6.6
A5	78.89	6.36	1.58	5.79	7.38
carben-carboids	78.69	6.5	1.76	5.87	7.18
Product of non-catalytic aquathermolysis (6 hours, 250 °C, 30 bar)					
A1	75.93	7.64	1.62	6.22	8.59
A2	80.07	7.87	1.48	4.34	6.25
A3	80.26	7,97	1.55	4.29	5.93
A4	75.08	5.8	0.6	7.96	10.56
A5	76.47	6.28	0.59	7.33	9.33
carben-carboids	77.54	7.42	0.56	6.52	7.96
Product of catalytic aquathermolysis with hydrogen donor (6 hours, 250 °C, 30 bar)					
A1	79.02	7.88	1.64	4.58	6,88
A2	80.43	7.94	1.54	4.04	6,05
A3	80.21	7.87	1.67	4.10	6,15
A4	80.03	7.63	1.70	4.26	6,38
A5	79.72	7.84	1.68	4.3	6,46
carben-carboids	75.04	7.71	1.67	7.01	8.57

From the results above, the "average" molecule of Ashal'cha oil is characterized by small content of heteroatoms, particularly sulfur and oxygen. According to the results of elemental analysis, H/C ratio significantly increased. In initial crude oil sample, the content of sulfur is higher in heavier fractions. The non-catalytic thermal treatment significantly increased the content of heteroatoms (S, O). However, the catalytic thermal treatment decreased the content of sulfur and oxygen to the initial crude oil values. The decrease in sulfur content under catalyst influence was observed mostly in high-molecular A5 fraction [15].

4. Conclusion

Upgrading of heavy oil from Ashal'cha field by introduction of oil-soluble catalyst precursor within hydrogen donor during hydrothermal treatment gave promising results. The content of high-molecular heteroatom substances in resins and asphaltenes was decreased in contrast with non-catalytic thermal treatment. Significant redistribution of group composition took place due to cracking and hydrogenolysis reactions, primary in C-S, C-O bonds of asphaltenes, that was justified by the data obtained from elemental composition of A1-A5 fractions.

Acknowledgments

The work is performed according to the Russian Government Program of Competitive Growth of Kazan Federal University.

References

[1] Sitnov S A, Vakhin A V, Mukhamatdinov I I et al 2019 *Petrol. Sci. Technol.* **37** 687
[2] Vakhin A V, Mukhamatdinov I I, Aliev, F A et al 2018 *Petrol. Sci. Technol.* **36** 1850
[3] Mukhamatdinov I I, Vakhin A V, Sitnov S A et al 2018 *Chem. Technol. Fuels Oils* **54** 574
[4] Sitnov S A, Mukhamatdinov I I, Vakhin A V et al 2018 *J. Petrol. Sci. Eng.* **169** 44
[5] Kayukova G P. Mikhailova A M, Feoktistov D A et al 2017 *Energy Fuels* **31** 7789
[6] Mullins O C 2010 *Energy&Fuels* **24** 2179
[7] Fossen M, Sjoblom J, Kallevik H, and Jakobsson J 2007 *J. Dispers. Sci. Technol.* **28** 193
[8] Spiecker P M, Gawrys K L, and Kilpatrick P K 2003 *J. Colloid Interface* **267** 178
[9] Strausz O P, Mojelsky T W, and Lown E M 1992 *Fuel* **71** 1355
[10] Andreatta G, Goncalves C C, Buffin G et al 2005 *Energy Fuels* **19** 1282
[11] Artok L, Su Y, Hirose Y, Hosokawa M, Murata S, and Nomura M 1999 *Energy Fuels* **13** 287
[12] Perez-Hernandez R, Mendoza-Anaya D, Mondragon–Galicia G et al 2003 *Fuel* **82** 977
[13] Khairudinov I R, Unger F G, and Syunyaev Z I 1987 *Chem. Technol. Fuels Oils* **6** 297
[14] Bazhenova O K, Bazhenova T K 2008 *Lithology and Mineral Resources* **43** 488
[15] Salih I Sh S, Mukhamatdinov I I, Garifullina E I, et al 2018 *Chem. Technol. Fuels Oils* **54** 44-50

IOP Conf. Series: Earth and Environmental Science **282** (2019) 012015 doi:10.1088/1755-1315/282/1/012015

Aquathermolysis of heavy oil in the presence of supercritical water

S. Petrov[1,2], R. Soldatova[1], A.Lakhova[1]

[1]Kazan National Research Technological University, 68 Karl Marx St., Kazan, Russia
[2]Kazan Federal University, 18 Kremlevskaya Str., Kazan, Russia

E-mail: psergeim@gmail.com

Abstract. The purpose of the work is to identify patterns of transformation of high-molecular components of super-heavy oil by hydrothermal-catalytic processes, namely in the presence of a supercritical state of water. The regularities of changes of the component, structural-group, fractional and elemental compositions of heavy oil during the conversion under the above conditions were established, rheological characteristics of the initial crude oil and converted oil were studied as well. As a result of carrying out aquathermolysis in the supercritical water environment and in the presence of carbonaceous substances , the high-molecular weight components of the initial crude oil were degraded with the formation of light distillate fractions, which were scarcely present in the initial crude oil.

1. Introduction

Recent academic studies have shown that supercritical solvents gives high level of quality control and productivity in the reaction chemistry and the processing of materials that is difficult to provide using conventional techniques and solvents. Supercritical fluids currently are becoming more and more relevant in including in the process of oil extraction and refining, especially in connection with development of unconventional hydrocarbon resources: heavy oil and natural bitumen [1-3]. Aquathermolysis in the presence of proton provides blockage of free radicals of high-molecular weight hydrocarbons and saturation of unsaturated hydrocarbons, produced by cracking reactions, and inhibiting of condensation reactions of aromatic macromolecules. The hydrogen protons also promote hydrogenation reactions in the crude oil. Water in a supercritical state is presumably a vehicle in the zones of subduction and flow of oil hydrocarbons to the reservoir [4-10].

The work is devoted to modeling of geothermal transformations of hydrocarbons in oil-bearing formations of deep horizons of the earth's crust under abnormally high pressure, in the presence of water and carbonaceous substances. The pressure and temperature in the experiments are typical for water in the supercritical state. The regularities of the conversion of heavy oil in supercritical water and in the presence of carbonaceous substances were shown.

2.Methodology

The object of this study is the crude oil of the Ashalchinskoye field, which is characterized by a density of 0.9715 g/cm^3 and the following group composition: oils 54.4 wt %, resins benzene 24.2 wt %, resins alcohol-benzene 13.3 wt %, asphaltenes 7.5 wt %, containing mechanical impurities of 0.6 wt % and a total water content of 2.5 wt % for oil. It is important to note that this oil is biodegraded, due to what the content of alkanes normal structure is extremely small (normal alkanes are destroyed bacteriologically). Model experiments on aquathermolysis heavy oil in supercritical water with carbonaceous substances (4 wt %) are carried out in a closed autoclave batch action made of stainless steel and with a capacity of 200 ml. He is supplied manometer, thermometer, safety valves. The experiments were carried out under conditions of high pressure 21-23 MPa and temperature 647.1-698 K, with parameters ensuring the transition of the aqueous phase to the supercritical fluid (supercritical point T_k = 647.1 K, P_k = 22.1 MPa).

After the experiment, the reactor was cooled naturally to room temperature, the residual pressure in the autoclave was observed in the range of 0.7-1.8 MPa.

3. Results and Discussions

In the products of the experiment there are significant changes in the component and fractional composition, which is accompanied by the formation of gasoline there were practically no fractions in the initial crude oil (table, figure). With increasing boiling point, the yield of individual fractions increases (table), while the distillation of bitumen above 573 K at atmospheric pressure leads to significant destruction of high molecular weight hydrocarbons. Degradation processes occur under the selected experimental conditions high molecular weight components (resins and asphaltenes) and hydrogenation processes with the formation of light hydrocarbon fractions (figure). Results of fractional distillation at atmospheric pressure showed that the effect of the carbonaceous substances in the medium of supercritical water on the crude oil leads to the formation of light fractions, boiling up to 573 K, which were present in an insignificant amount in the initial crude oil. Thus, the yield for fractions b.p.-473 K and 473-623 K for product of experiment No. 3 is 19.1 wt % and 24.5 wt %, respectively, against 9.8 wt % and 16.1 wt %. For samples a multidirectional change in the initial boiling point is characteristic: for the product of experiment 2 rises to 372 K, against 364 K, the initial boiling point of the initial agent.

IOP Conf. Series: Earth and Environmental Science **282** (2019) 012015 doi:10.1088/1755-1315/282/1/012015

Table 1. Fractional composition of the initial crude oil and products of aquathermolysis

The product of experiment No.	Experiment parameters		Density (293 K), g/sm^3	Boiling point of sample, K	Fraction yield, wt %		
	Temperature, K	Pressure, MPa			b.p.-473 K	b.p.-623 K	>623 K
Initial crude oil	-	-	0.9715	364	9.8	25.9	74.1
1	647	21	0.8723	367	11.7	27.4	72.6
2	647	22	0.9241	372	14.4	38.5	61.5
3	673	22	0.7777	360	19.1	43.6	56.4
4	698	22	0.7162	360	19.3	45.9	54.1

All products of supercritical aquatermolysis are characterized by an increase in the content of paraffin-naphthenic hydrocarbons (HC): up to 76.2 wt % and 68.2 wt % for the products of the model mixture 3 and the model mixture 4, respectively, against 40.8 wt % of the initial crude oil. At the same time there is a decrease in the content of aromatic and resinous substances (figure 1).

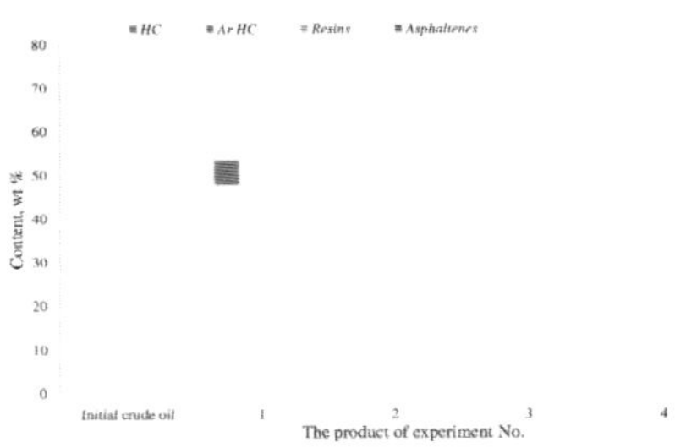

Figure 1. SARA-analysis initial crude oil and product of aquathermolysis in the supercritical water

The content of asphaltenes in the products of aquathermolysis varies in different directions: for the products of the experiment 1 increases to 9.9 and 9.2 wt % respectively. The destruction of the high-molecular part of the crude oil proceeds more efficiently with the increase process temperature, increasing the yield of light fractions in the product of the experiment. As a result of the experiment, the resin content significantly decreases, this can be explained by the fact that the resins not only collapsed to form parafinonaphthenic and aromatic structures, but also interacted with asphaltenes. For the products of the experiments 2 and 4, the opposite is observed: the content of asphaltenes decreases from 7.7 to 5.4 wt % and to 5.5 wt % respectively. In the product of experiment No. 1 at a pressure 21 MPa and a temperature 647 K, increase in the content of asphaltenes is observed as to simple aquathermolysis (without supercritical water).

4. Conclusion

The regularities of changes in fractional and component compositions of heavy oil as a result of supercritical aquathermolysis are revealed. It is shown that in the processes of aquathermolysis is the destruction of high-molecular components with the formation of low boiling fractions and paraffin-naphthene hydrocarbons. It can be assumed that heavy crude oil represents immature organic matter undergoes catagenetic maturation as a result of aquatermolysis in supercritical water.

Acknowledgments

This work was supported by grant 18-48-160053/18 from the Russian Foundation for Basic Research.

References

[1] Hyne J B et al, 1982 The Second International Conference on Heavy Crude and Tar Sands, Caracas, Venezuela 1
[2] Wang Y et al, 2010 *Energy Fuels*. **24** 1502.
[3] Nashaat N Nassar et al *Energy Fuels* 2011 **25** 1566
[4] Kun Chao et al, 2012 *Fuel Processing Technol.* **104** 174
[5] Kayukova G P, Mikhailova A N, Kosachev I P et al 2018 *Energy and Fuels* **32** 6488
[6] Liu Yongjian et al, 2002 *Energy Fuels* **16** 842
[7] Liu Y et al, 2002 *Energy Fuels* **16** 842
[8] Sitnov S A, Vakhin A V, Mukhamatdinov I I et al 2019 *Petrol. Sci. Technol.* **37** 687
[9] Kayukova G P, Mikhailova A N, Kosachev I P et al 2018 *Energy and Fuels* **32** 6488
[10] Petrov S M, Ibragimova D A, Safiulina A G et al 2017 *J. Petrol. Sci. Eng.* **159** 497

IOP Conf. Series: Earth and Environmental Science **282** (2019) 012016 doi:10.1088/1755-1315/282/1/012016

Multifrequency (9 and 95 GHz) EPR study of stable radicals in asphaltenes fractions of oils and bitumen

M. Gafurov[1], G. Mamin[1,*], Yu. Ganeeva[1,2], T. Yusupova[1,2], M. Volodin[1,3], I. Gracheva[1] and S.Orlinskii[1]

[1] Kazan Federal University, 18 Kremlevskaya Str., Kazan, Russia
[2] Arbuzov Institute of Organic and Physical Chemistry, FRC Kazan Scientific Center of RAS, 420088 Kazan, Russia
[3] Sakhalin Energy Investment Company, Ltd., Dzerzhinskogo, 35, 693020 Yuzhno-Sakhalinsk, Russia

* E-mail: george.mamin@kpfu.ru

Abstract. Structural characterization of asphaltenes in complex systems such as native hydrocarbons is in the focus of scientific and industrial interests since many years. Various analytical tools and approaches are used for that. We present the results of our study of asphaltene fractions A1 and A2 with the predominantly "island" type and "archipelago" type asphaltene molecules by conventional and pulsed electron paramagnetic resonance (EPR) at X-band (9 GHz) and W-band (95 GHz) with aim to expand the abilities of EPR technique for asphaltene characterization. Shift to the higher frequencies allows to separate spectrally the contributions from paramagnetic complexes of different origin, define the EPR parameters more accurately comparing to the conventional X-band EPR. Application of pulsed techniques allows (at least partially) to separate the radicals in time domain. Features of the obtained spectra are described. Electronic relaxation times are found to be different for stable "free" radical for A1 and A2 fractions. We suggest that the obtained results can be used for investigation of various petroleum systems.

1. Introduction

Asphaltene is known as the *cholesterol of petroleum* because of its ability to precipitate, deposit, and as a result, interrupt the continuous production of oil from underground reservoirs. Despite decades of research, asphaltene composition and molecular structure as well as their impact on the mechanisms of asphaltene stabilization are not completely understood. Chemical analyses of asphaltenes indicate that they contain fused ring structures, naphthenic rings, and some aliphatic chains with hydrogen and carbon; small quantities of heteroatoms, such as nitrogen, sulfur, and oxygen; and trace metals, such as vanadium and nickel [1]-[11].

Today it seems to be (due to the application of advanced analytical techniques like time-resolved fluorescence depolarization (TRFD), high-field mass spectrometry (MS), atomic electron microscopy (AEM)) that the consensus is that asphaltenes are monomeric and their molecular weights are in the range of between about 250 and 1200 g/mol is achieved [12], [13] though also not completely [14]. Disagreement remains on the molecular structure. There are two schools of thought as to the nature of aromatic core organization within asphaltene compounds. The classic island architecture, also referred as "like your hand" model, depicts asphaltenes consisting of one aromatic core with peripheral alkyl side chains. On the contrary, the archipelago model suggests that asphaltenes contain more than one aromatic core linked by alkyl bridges. The relative abundances of single- and multicore compounds in asphaltenes are still being debated despite decades of investigation [1]-[13].There are still challenges of how to determine the accurate molecular structure, constituent and structure-behaviour relationships of asphaltene.

Commercial availability of the pulsed high field electron paramagnetic resonance (EPR) spectrometers open new ways for the analysis of petroleum dispersed systems (PDS), investigation of the asphaltenes' structure, processes of their treatment, etc. using intrinsic paramagnetic centers as sensitive probes [15]-[17]. Paramagnetic properties and EPR spectra in PDS (such as oil rocks, crude oil, bitumen, kerogens, their fractions, etc) are mainly defined by the presence of complexes of d-metals (first of all V) and stable "free" radicals (FR). FR are mainly concentrated in asphaltenes and arise due to the delocalized π-electrons of the aromatic rings and stable organic radicals of the side chains. It can be suggested that relaxation and spectroscopic parameters of FR derived at various microwave (MW) frequencies can give additional information about the asphaltene structure but such type of investigation is rare [16], [18], [19] and does not cover even a part of all possible cases due to the complexity of PDS and the limited amount of PDS species studied so far. The aim of this paper is to expand on these objectives the abilities of the pulsed EPR at the W-band range ($\nu_{MW} \approx 95$ GHz) for asphaltene structure characterization.

2. Materials and Methods

W-band EPR spectra at T = 300 K were registered with Bruker ElexSys 680 EPR spectrometer by using standard electron-spin-echo (ESE) $\pi/2 - \tau - \pi$ with the $\pi/2$ pulse duration of 32 ns, time delay $\tau = 240$ ns pulse sequence. Transverse relaxation T_{2e} was studied by tracking the primary ESE amplitude with the same $\pi/2$–π pulse durations while varying τ. Longitudinal relaxation time constant T_{1e} was extracted from an inversion-recovery studies by applying the $\pi - T_{delay} - \pi/2 - \tau - \pi$ pulse sequence, where π pulse duration and τ were fixed (64 ns and 240 ns, correspondingly) while T_{delay} was varied. The value of the external magnetic field, B_0, during the relaxation times measurements was fixed at the particular point defined from the EPR spectra. Conventional X-band measurements ($\nu_{MW} \approx 9$ GHz) are carried out on Bruker ESP-300. Spectra simulation are done in Easyspin package for MatLab [20].

The list of the studied samples is given in Table 1. Asphaltenes ($A_{init.}$) were precipitated from the raw material by the addition of 40 ml·g^{-1} of the petroleum ether (b.p. 40 – 70°C). Precipitated asphaltenes were washed in a Soxhlet apparatus with petroleum ether until the filtrate became colorless. Then the asphaltenes with filter were washed out with the benzene, which was then evaporated.

Table 1. List of the studied samples.

Sample	Raw Material	Source	Fractions of asphaltenes
No. 1	Oxidized bitumen	Oil refinery of Republic of Tatarstan, Russia	A_{init}, A1, A2
No. 2	Crude oil taken from the Carboniferous deposits	Akan oilfield (well 2023) of Republic of Tatarstan, Russia	
No. 3	Crude oil taken from the Carboniferous deposits	Akan oilfield (well 2262) of Republic of Tatarstan, Russia	

The A1 and A2 fractions of asphaltenes were obtained as follows. One gram of A_{init} asphaltenes was first completely dissolved with 28 ml toluene. Petroleum ether (b.p. 40 – 70 °C) was then added in the amount of 52 ml. The resulting solution was kept in a dark place for 24 hours and then filtered. Precipitated material was washed in a Soxhlet apparatus with toluene until the solvent became colorless, dried and weighed until there was no change in mass. Obtained fraction was called A1.

Then the petroleum ether was added to the supernatant in the amount of 228 ml and precipitated for 24 hours. Using the operations described above, the fresh precipitated material was separated, washed with toluene in a Soxhlet apparatus and dried. A2 fraction of asphaltenes was recovered by evaporating the remaining solvent.

Literature data suggest that A1 fraction possess predominantly "island" type asphaltene molecules while A2 fraction consists mainly from the "archipelago" asphaltene molecules [21].

3. Results and Discussion.

EPR spectra detected at X-band (Figure 1) and W-band (Figure 2) for the all investigated samples are very similar to each other and are defined by the presence of stable "free" radicals (FR, single line, electron spin S = 1/2, g ≈ 2.003) and paramagnetic vanadyl complexes (vanadylporhypin, VP). Due to the nearly planar skeleton structure of single VP molecule (VO^{2+}, S = ½), the VP EPR spectra could be satisfactory described by the g-tensor of axial symmetry with $g_{||} \approx 1.96$, $g_{\perp} \approx 1.98$ and anisotropic hyperfine interaction with a magnetic moment I = 7/2 (therefore, 8 projections of I with the quantum numbers m_I = ±7/2; ±5/2; ±3/2 and ±1/2 are possible) for ^{51}V nuclei with hyperfine structure constants $A_{||} \approx 480$ MHz, $A_{\perp} \approx 157$ MHz. Detailed description and investigation of VP EPR and ENDOR spectra are given in papers [15], [22]. In the present EPR for FR are investigated in more details.

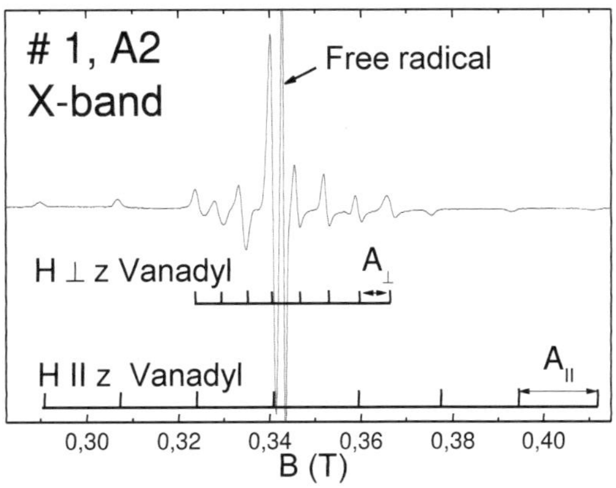

Figure 1. Conventional X-band EPR spectrum at T = 300 K of A2 fraction of sample #1 defined by the presence of vanadyl complexes and FR.

X-band concentration measurements allow for the estimation of the FR concentration in the range of (3–9) × 10^{18} spins per gram, while for VP, this value is in the range of (2–5) × 10^{18} spins per gram. Assuming that the molecular weight of asphaltenes is of 1000 amu, it implies one paramagnetic centre per roughly 100 asphaltene molecules.

In the case of asphaltenes deposition on various supports, we have found that the relative concentration ratio FR/VP is growing compared to the initial asphaltene [5], [6]. The results for the relative concentrations for the investigated samples are presented in Table 2. It follows that the (relative) amount of VP in A2 fraction can be significantly higher than in A1 (samples 1 and 2) or at least not less (sample 3). It gives a sign that organization of the asphaltenes molecular aggregates in thin films (i.e. deposited on surface) can differ from this in the bulk.

A usual way to fit the EPR lineshape of FR at X-band is a convolution (or sum) of Gauss and Lorentz with an isotropic g-factor and $\Gamma_G \approx \Gamma_L$ (the linewidth (Γ) for Gaussian (G) and Lorentzian (L) absorption lineshapes are connected with the full width at half maximum (FWHM) via $\Gamma_G = (2\ln 2)^{-1/2}$·FWHM and $\Gamma_L = (3)^{-1/2}$·FWHM). In ref. [19] it was shown that in contrast to X-band, in the conventional W-band EPR the lines from FR are not symmetric and their spectral EPR linewidths became larger. Our experiments on the investigated samples in pulsed mode confirm that conclusion

(Figure 2). But from other side, as we found recently [23], for some specific oils the linewidths in X- and W-bands can be the same.

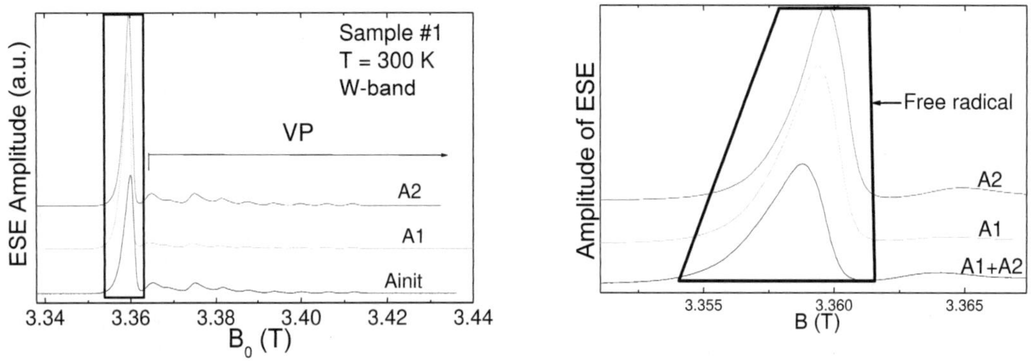

Figure 2. ESE detected EPR spectra for sample #1 in the W-band. Left panel – whole range; right panel – FR region.

Table 2. Relative concentrations [FR]/[VP] from X-band EPR measurements.

Sample	A_{init}	A1	A2
# 1	1.9(1)	3.6(2)	1.35(5)
# 2	1.15(9)	1.0(1)	0.50(5)
# 3	0.7(1)	0.7(1)	0.67(7)

There is still a problem of assigning the W-band spectrum either to one type of FR with the anisotropic EPR parameters or to a sum of contributions from several FR types. Probably, both scenario are fulfilled in complex petroleum systems. Time-domain separation can help to unravel the origin of EPR spectrum.

No significant difference in T_{1e} and T_{2e} measurements were obtained along the FR EPR spectrum. The data derived at B_0 corresponding to the maximum of the ESE signal are listed in Table 3.

Table 3. Relaxation parameters for FR derived from the W-band experiments.

Sample	A_{init}	T_{1e}, μs	T_{2e}, ns
# 1	A_{init}	102(4)	247(8)
	A1	16(1)	284(8)
	A2	27(1)	365(10)
# 2	A_{init}	28.5(7)	150(7)
	A1	17(1)	131(6)
	A2	31(1)	191(8)
# 3	A_{init}	71(3)	136(7)
	A1	54(2)	126(5)
	A2	87(4)	142(6)

As it was already noted in [15], for FR (in contrast to the vanadyl complexes) T_{1e} varies significantly from sample to sample and from fraction to fraction. For all studied samples fraction A2 is characterized by the longer T_{1e} than fraction A1. Taking into account that the molecules forming the asphaltenes in both fractions are of comparable size, the difference in the relaxation rates reflects the fact that the FR surrounding is different for fractions A1 and A2. Therefore, the corresponding ifference in the longitudinal relaxation times can be exploited for the selective investigation of the suggested asphaltene structures. Additional experimental and theoretical efforts are in need to understand the main mechanisms defining the longitudinal relaxation in asphaltenes.

The same trend is obtained for T_{2e} curves (Table 3). In ref. [15] an influence of VP on FR relaxation was investigated. Additionally, it can be supposed that the relative abundance of two asphaltenes type with the slightly different T_{2e} values can also be revealed. Indeed, as Figure 3 demonstrates, with the increase of the time delay (τ), the low-field shoulder of the EPR spectrum disappears (in A1 and A2 fractions), FR line becomes to be more symmetrical. Therefore, we suggest that "archipelago" type of asphaltenes is characterized by the longer relaxation times T_{1e} (see Table 3) and T_{2e} (of 700-800 ns for sample #1; schematically, P_1 in Figure 3) while "island" asphaltens contain FR with the slightly higher g-factor(s) and shorter relaxation times (T_{2e} of 250 ns for sample #1; schematically, P_1 in Figure 3).

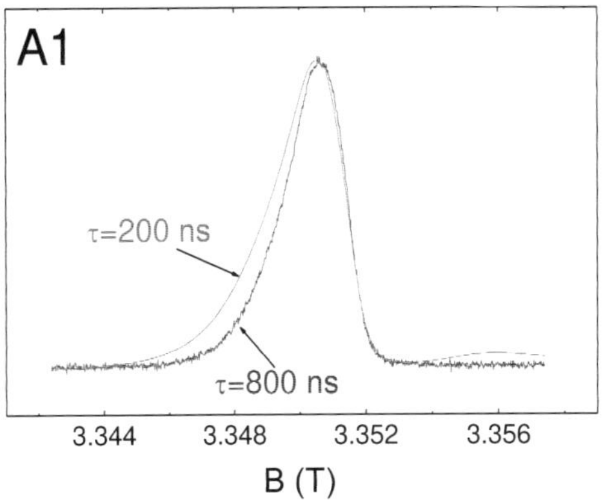

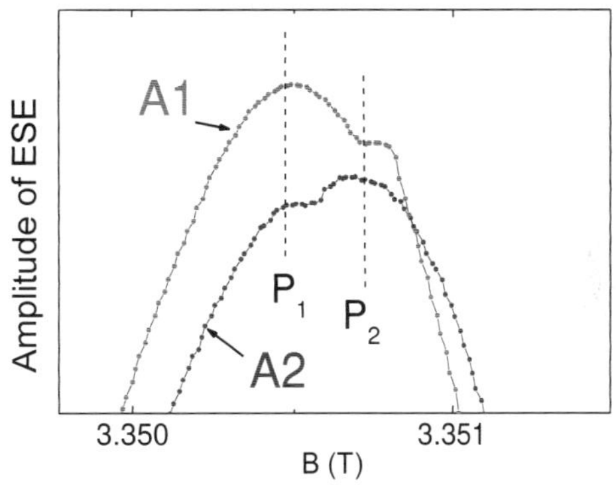

Figure 3. ESE detected EPR spectra for sample #1 in the W-band for τ = 200 and 800 ns. Left panel – FR range; right panel – in the vicinity of FR ESE maximum.

4. Conclusion

In this work we show that EPR spectra of asphaltenes and relaxation characteristics of stable carbonaceous "free" radicals are sensitive to the presence of different structural asphaltenes motifs. It demonstrates further potential capabilities of EPR techniques for characterization of asphaltenes and petroleum systems with aim to elucidate asphaltenes structure.

Acknowledgments
The authors are thankful to D. Sitdikov and A. Rodionov for EPR measurements. The work was done in the line with the Program of competitive Growth of Kazan Federal University among World's Leading Academic Centers. S.O. is also supported by Russian Science Foundation (Project No 17-73-10350).

References

[1] Zuo P, Qu S and Shen W 2019 *J. Energy Chem.* **34** 186
[2] Dong X *et al* 2019 *Fuel*, **246** 126
[3] Cheshkova T V, Sergun V P, Kovalenko E Y *et al* 2018 *Bulletin of the Tomsk Polytechnic University, Geo Assets Engineering* **329** 61
[4] Vargas F M and Tavakkoli M 2018 *Asphaltene deposition: Fundamentals, prediction, prevention, and remediation* (Boca Raton, FL: CRC Press) p 359
[5] Gafurov M *et al* 2019 *Magn. Reson. Solids* **21** 19101

[6] Gafurov M R *et al* 2018 *Russ. J. Gen. Chem.* **88** 2374
[7] Gafurov M R, Volodin M A, Rodionov A A *et al* 2018 *J. Petrol. Sci. Eng.* **166** 363
[8] Gracheva, I. N *et al* 2016 *Magn. Reson. Solids* **18** 16102
[9] Kayukova G P, Gubaidullin A T, Petrov S M *et al* 2016 *Energy Fuels* **30** 773
[10] Ganeeva Y M, Yusupova T N and Romanov G V 2011 *Russ. Chem. Rev.* **80** 993
[11] Yen T F and Chilingarian G V 2000 *Asphaltenes and asphalts.* Vol. 2 *(Developments in petroleum science)* 40B (Amsterdam: Elsiever Science) p 622
[12] Chac^n-PatiÙb M L, Rowland S M and Rodgers R P 2017 *Energy Fuels* **31**, 13509
[13] Chac^n-PatiÙb M L, Rowland S M and Rodgers R P 2018 *Energy Fuels* **32** 9106
[14] Hernández M S, Coll D S and Silva P J 2019 *Energy Fuels* **33**, 990
[15] Mamin G V, Gafurov M R, Yusupov R V *et al* 2016 *Energy Fuels* **30** 6942
[16] Ramachandran V, van Tol J, McKenna A M *et al.* 2015 *Anal. Chem.* **87** 2306
[17] Ben Tayeb K, Delpoux O, Barbier J *et al* 2015 *Energy Fuels* **29** 4608
[18] Volodin M A, Mamin G V, Izotov V V, Orlinskii S B 2013 *J Phys: Conf Series* **478** 012003
[19] Di Mauro E, Guedes C L B, Nascimento O R 2005 *Appl. Magn.Reson.* **29** 569
[20] Stoll S and Schweiger A 2006 *J. Magn. Reson.* **178** 42
[21] Acevedo S, Guzman K and Ocanto O 2010 *Energy Fuels* **24**, 1809
[22] Biktagirov T B *et al* 2014 *Energy Fuels* **28** 6683
[23] Gizatullin B *et al* 2018 *Energy Fuels* **32** 11261

IOP Conf. Series: Earth and Environmental Science **282** (2019) 012017 doi:10.1088/1755-1315/282/1/012017

95 GHz electron paramagnetic resonance of $LiYF_4$: Nd^{3+} crystal in parallel orientation

G Mamin, F. Murzakhanov, A. Razina, I. Kurkin, E. Baibekov and S. Orlinskii

Kazan Federal University, 18 Kremlevskaya Str., Kazan 420008, Russia

e-mail: George. Mamin@kpfu.ru

Abstract. We report the observation of the conventional and pulsed electron paramagnetic resonance (EPR), electron-nuclear double resonance (ENDOR) spectra from ^{19}F and ^{7}Li nuclei on impurity Nd^{3+} ions in $LiYF_4$ crystal doped with Nd^{3+} ions in parallel orientation at microwave frequency of $\nu \approx 95$ GHz (W-band). The resolved structure from the nearby and remote nuclei in spectra is observed. The outcome shows that $LiYF_4$:Nd^{3+} system can be exploited as a convenient matrix for performing spin manipulations and adjusting quantum computation protocols while ENDOR technique is usable for the investigation of electron-nuclear interaction with the nuclei of the system.

1. Introduction

Rare earth (RE) fluorides have been become anew a research focus in the material field due to their unique applications in optical communications, three-dimensional displays, solid-state laser, catalysis, solar cells, biochemical probes and medical diagnostics [1], [2]. It is proposed that RE doped fluoride glasses as parts of laser systems could be fruitfully used for determining identity and/or quantity of a component of a fluid at a remote location such as downhole in a wellbore or inside a pipeline, particularly at elevated temperatures up to 175° C, such as an oil based mud or crude oil, and may detect components such as methane, ethane, carbon dioxide, hydrogen sulfide, etc. [3]. RE fluorides are used as tribological additives in lubricating oils [4]. The potential for using RE doped crystals as scintillators (detectors) for radioactive radiation is not disclosed [5]. To make good use of the excellent properties of metal fluorides, their morphology and microstructure should be carefully controlled. One of the powerful method of the RE doped crystals inspection is an electron paramagnetic/spin resonance (EPR/ESR).

Apart from being the cornerstone of modern laser technology, crystals and glasses doped with RE metals are known to have excellent properties for classical and quantum information storage and for performing quantum computation protocols [6], [7], [8]. In addition, $LiYF_4$ crystals, having a relatively simple crystalline structure and growing almost without defects, are unique model systems for studying the crystal field, electron-phonon, interionic, hyperfine interactions, and also isotopic effects [9]. The lithium ions (Li^+) are in the centre of tetrahedra composed of four fluoride ions (F^-). The yttrium ions (Y^{3+}) are in the center of polyhedra composed of eight F^-. Y^{3+} can be substituted by rare earth metals presenting an oxidation state of +3.

In this work we present the high-frequency (high-field) study of $LiYF_4$: 0.01 % Nd^{3+}. crystal. It is known that such small amount of Nd^{3+} does not influence much the lattice parameters of $LiYF_4$ scheelite structure: $a \approx 0.517$ nm; $c \approx 1.074$ nm [10], [11]. (At ambient pressure, the crystal cell of $LiYF_4$ is tetragonal with space group $I4_1/a$ or C_{4h}^6. This phase is commonly named the scheelite structure, in

IOP Conf. Series: Earth and Environmental Science **282** (2019) 012017 doi:10.1088/1755-1315/282/1/012017

reference to the $CaWO_4$ crystal). The obtained results are compared with those obtained in low magnetic fields (conventional for EPR X-band microwave frequency $\nu \approx 9$ GHz) and on $CaWO_4$ crystals. In our recent papers [12], [13] we have studied $LiYF_4$: Ce^{3+} systems and performed pulsed electron-nuclear double resonance (ENDOR) and double electron-electron resonance (DEER) manipulations of electronic and nuclear spin states of trivalent gadolinium impurity ions in $CaWO_4$ host crystal at X-band frequencies (in the magnetic fields B_0 up to 1000 mT) [14].

2. Experiment and Discussion

EPR investigations were done by using helium flow cryostats at W-band Bruker Elexsys E580 spectrometer (T = 10-300 K, microwave frequency $\nu \approx 95$ GHz). Crystals of $LiYF_4$ doped with rare-earth ions were grown at Kazan Federal University by the Bridgman-Stockbarger technique in argon atmosphere. A piece of crystal with the linear sizes of 0.4×0.2×0.3 mm was cut and placed into the EPR cavity. The impurities of neodymium were added in the forms of ~0.01 at % of neodymium oxide.

It is worth mentioning that the synthesized LiYF4 sample exhibited the traces of various rare-earth ions including. Gd^{3+}. It is known that EPR of Gd^{3+} can be observed in a wide temperature range from liquid helium to room temperature, since the ground state $^8S_{7/2}$ of the Gd^{3+} ion is an orbital singlet weakly interacting with the crystal lattice. Presence of Gd^{3+} ion (which concentration was estimated as less than 0.005 wt % in X-band measurements) gives an opportunity to orient crystal in the external magnetic field B_0 by its rotating at T = 300 K and T = 50 K and observing the shifting the whole spectrum (and especially its high-field component) to the strongest magnetic field characterized by g-factor $g_{||}(Gd^{3+}) \approx 1.99$ [7].

Figure 1 demonstrates the EPR spectrum detected in conventional mode (cw) at T = 50 K and microwave power of 10 μW for $B_0 || c$. (Due to the saturation effect, the intensities of the main components are out of the analog-digital converter range while the lineshapes of the observed components are is closer to absorption lines rather than first derivatives). The spectrum at this temperature EPR lines is due to $S = 7/2$ for Gd^{3+} ions. The cw and pulsed EPR spectra of Gd^{3+} ions in various host crystals, particularly in $CaWO_4$ single crystal, had been extensively studied at lower microwave frequencies [7]. It was shown that the most intense lines are from the $M \leftrightarrow M \pm 1$ transitions ($M = -7/2, -5/2.... 7/2$). Lower intense lines are from $M \leftrightarrow M \pm 3$ and $M \leftrightarrow M \pm 5$ transitions. The smallest signals are from the "forbidden" transitions $M \leftrightarrow M \pm 2$ and $M \leftrightarrow M \pm 4$. The exact assignment of components of the experimental EPR spectrum to the calculated one for different crystal orientation is on the way.

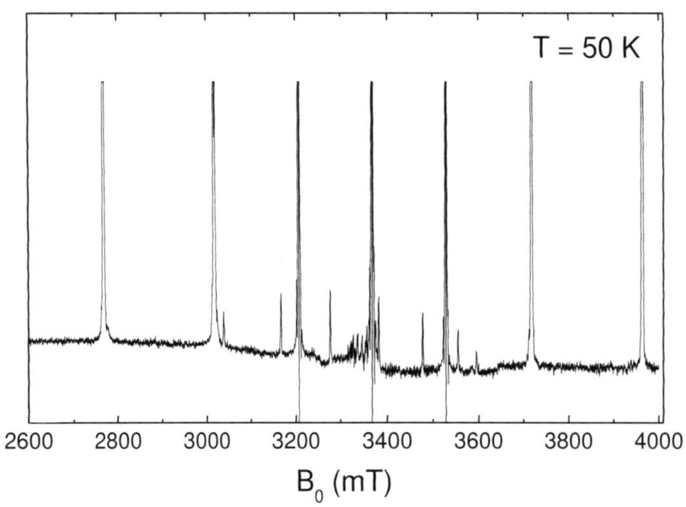

Figure 1. W-band EPR spectrum at T = 50 K.

We used two–pulse electron spin echo measurements for the primary Hahn–echo sequence ($\pi/2$–τ–π–τ–echo), where τ is the interpulse delay time of 240 ns, with initial $\pi/2$ and π pulse lengths of 12 and 24 ns, respectively, to detect the field-swept (FS) electron spin echo (ESE) EPR spectra at lower temperatures. Figure 2 shows an appearance of additional line in FS ESE with $g_{\|} = 1.987$ which we assign to Nd^{3+} ($g_{\|} = 1.987$; $g_{\perp} = 2.554$, see [9], [15]). The corresponding relaxation times for Nd^{3+} at T = 10 K are measured to be $T_{2e} = 2.0(2)$ μs and $T_{1e} = 26(1)$ μs.

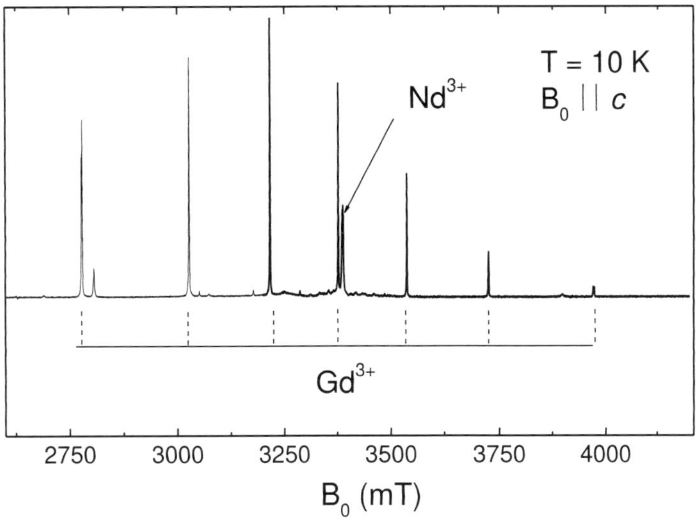

Figure 2. FS ESE EPR at T = 10 K.

Applying the pulses of various length one can directly prove that the appeared Nd^{3+} line ($S = 1/2$) does not belong to one of the possible ("forbidden") transitions for $Gd^{3+}$with $S = 7/2$: their ESE amplitudes and phases change differently with the length of $\pi/2$ pulse. The shape of the Nd^{3+} line is mainly defined by the unresolved superhyperfine structure which is known to be the most resolute exact for $B_0 \,||\, c$ at X-band frequency [9]. Conventionally, at W-band the lines become broader due to the various reasons and it seems to be quite problematic to extract the details of superhyperfine structure from the high-frequency EPR but in many cases it can be done from ENDOR experiments [12], [13].

For ENDOR experiments we used special double (for nuclei and electron) cavities and Mims pulse sequence - $\pi/2$-τ-$\pi/2$-T-$\pi/2$ with an additional radiofrequency (RF) pulse $\pi_{RF} = 18$ μs inserted between the second and third microwave $\pi/2$ pulses. RF frequency in our setup could be swept in the range of (1-200) MHz. Magnetic field value was kept constant, corresponding to the position of Nd^{3+}. Recent review of ENDOR applications for quantum computing is presented elsewhere [16]. Some ENDOR basics are briefly described below.

In the case of a "free" nucleus, RF pulse applied at the Larmor frequency

$$\nu_{Larmor} = |\gamma B_0| \equiv h^{-1}|g_I \beta_I B_0|, \qquad (1)$$

where γ is a gyromagnetic ratio of the nuclear spin I, h is a Planck constant, g_I is a nuclear g-factor and β_I is a nuclear Bohr magneton, can change the state of the nuclear spin (the population of the nuclear sublevels).

The ENDOR splitting, a_{ENDOR}, can help not only to identify a type of nuclei coupled with the electron spins but also provide spatial relationships between them. For the pure electron-nuclei dipole-dipole interaction in the point model, the electron-nuclear distance, r, from the ENDOR splitting can be estimated from

$$a_{ENDOR} \approx g \cdot g_I \cdot (1 - 3\cos^2\theta)/r^3, \qquad (2)$$

where g is a g-factor of electron spin *S*, *Θ* is an angle between directions of the parallel component of g ($g_{||}$) and B_0. As it follows from (2), a_{ENDOR} depends on the distance between the electron and nuclear spins and their mutual orientation. ENDOR splitting should be maximal for ions with the maximal g-factors and orientations Additionally, for *I* equals or larger than one, an electric nuclear quadrupole coupling exists that can split or shift the ENDOR lines: the nuclear quadrupole interaction is sensitive to the electric field gradient at the site of the nucleus.

All nuclei in $LiYF_4$ compound have sensible magnetic moment capable to create a perceptible additional magnetic field at the location of the rare earth impurity (see [9], [12], [13]). Well-resolved ENDOR spectra for ^{19}F nuclei in parallel orientation presented in Figure 3 which is due to the electron-nuclear interactions of Nd^{3+} and with ^{19}F. From ENDOR pattern for ^{19}F one can estimate the electron-nuclear distance to the most remote nuclei from Eq. (2) as $r \approx 0.66$ nm. Taking into account that in the YF_8 octahedra of $LiYF_4$ structure, the four first F neighbours are situated at 0.224-0.226 nm (a_{ENDOR} ≈ 4.2 MHz), the distance to the four second neighbours is of about 0.229-0.232 nm [12], one can see that the electron spin density of Nd^{3+} ions distributes well out of even the second coordination sphere. Detailed interpretation of ENDOR spectra from the angle and frequency dependencies is a matter of the ongoing research.

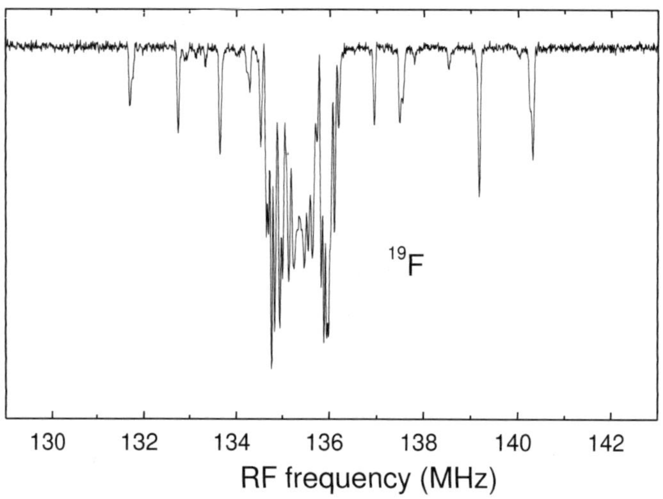

Figure 3. ENDOR spectra caused by interactions with ^{19}F nuclei in $LiYF_4$: Nd^{3+} monocrystal in $B_0 || c$ orientation. $B_0 = 3380$ mT (Nd^{3+} line), T=10 K, ν = 93.967 GHz.

3. Conclusions

The rich, well-resolved, intensive ENDOR spectra show that $LiYF_4$:Nd^{3+} system potentially can be used as an element (or model system) of the many qubit quantum computer. Indeed, it is known that electrons are natural candidates as physical qubits to be exploited for quantum computing and information processing (QC/QIP), and therefore magnetic resonance techniques, are among the most appropriate techniques to be exploited for quantum computing. Pulsed EPR enables manipulation of electron spin and nuclear spin qubits in an equivalent manner and ENDOR can be used as the most useful spin manipulation technology in implementing QCs/QIPSs [16]. The corresponding experiments are on the way.

Acknowledgments
The work is financially supported by the Russian Science Foundation, Project # 17-72-20053.

References

[1] Tressaud A 2010 *Functionalized Inorganic Fluorides: Synthesis, Characterization and Properties of Nanostructured Solids* (Chippenham, UK: Wiley)
[2] Nizamutdinov A S et al 2014 Opt. Spectrosc. **116** 732
[3] Csutak S 2008 *Method and Apparatus to Determine Characteristics of an Oil-Based Mud Downhole* US Patent 20080245960 A1
[4] Shahnazar S et al 2016 *Int.J. Hydrogen Energy*, **41** 3153
[5] Anderson D F and Sparrow R W 1991 *Divalent fluoride doped cerium fluoride scintillator* US Patent 5039858
[6] Bertaina S et al 2007 *Nature Nanotechnology* **2** 39
[7] Baibekov E I et al 2017 *Phys. Rev. B* **95** 064427
[8] Baibekov E I et al 2014 *Phys. Rev. B* **90**.174402
[9] Aminov L K et al 2013 *Phys. Solid State* **55** 1343
[10] Garcia E and Ryan R R 1993 *Acta Crystal. C* **49** 2053
[11] Ranieri I M et al 2002 *J.Alloys Compd.* **344** 203
[12] Gafurov M et al 2018 *IOP Conf.Ser.: Earth Environ. Sci.* **155** 012003
[13] Gafurov M R et al 2017 *Magnetic Resonance in Solids Electronic Journal* **19** 17206
[14] Gafurov M R et al 2018 *EPJ Web of Conferences* **195** 06003
[15] Aminov L K et al 2015 *Phys. Solid State* **57** 2400
[16] Takui T, Berliner L and Hanson G 2016 *Electron spin resonance (ESR) based quantum computing*. (New York: Springer)

IOP Conf. Series: Earth and Environmental Science **282** (2019) 012018 doi:10.1088/1755-1315/282/1/012018

Overhauser-driven dynamic nuclear polarization for petroleum systems: literature survey and comparing with experiments

A Alexandrov and M. Gafurov*

Kazan Federal University, 18 Kremlevskaya Str., Kazan, Russia

E-mail: marat.gafurov@kpfu.ru

Abstract. Overhauser dynamic nuclear polarization (ODNP) in solutions of various paramagnetic complexes has been studied for 60 years, but only in recent years has found applications of broad interest to biophysical and biomedical sciences, for the investigation of soft materials and biomolecules. Relatively few aplications are focused on the ODNP in petroleum dispersed systems (PDS) like oils, bitumen, their fractions and solutions. This work present a short review of the ODNP studies of PDS with aim to introduce the basic concepts and key values for the effective petroleum ODNP in low and high magnetic fields. Experimental results obtained in our Laboratory by using home-made spectrometer are included. The study can be used for designing ODNP spectrometers, proton precession magnetometers for geological and geophysical exploration, investigation of supramolecular organisation of PDS and their components.

1. Introduction

Nuclear magnetic resonance (NMR) is a powerful method for the routine analysis and detailed studies of diverse hydrocarbon systems (petroleum dispersed systems, PDS). For decades, NMR logging tools have been developed that allow in situ determination of a number of petrophysical parameters such as porosity, permeability, water content, gas, and oil saturation, oil viscosity etc. Analyses of the NMR relaxation times are used for the *in situ* and non-invasive investigations, to measure fluid properties in porous, oil containing media [1], [2]. Electron paramagnetic resonance (EPR) is also known since 1950s as a convenient tool for PDS detection and investigation due to the existence in PDS various paramagnetic centers [3], [4]. Application of modern pulsed and high-frequency EPR techniques last decade broaden significantly the EPR abilities for PDS analysis [5]-[7].

However, despite this body of literature, there is a surprisingly small amount of publications reporting on the use of double resonance techniques like dynamic nuclear polarization (DNP) in which both EPR (microwave, MW) and NMR (radiofrequency, RF) electromagnetic waves are applied to crude oils and their fractions [1], [8]-[15] and references therein. Possible reasons for this are, apart from the complexity of the natural samples, the lack of available instrumental tools and limited models available for interpreting data of double resonance experiments. Obviously, investigations of the coupling between the electronic spin reservoir S and the nuclear spin reservoir I in PDS and their components can provide insight into the structure and dynamics of these complex materials, but at the same time, the coupling itself can be used to enhance the NMR signal by spin polarization transfer from S to I by means of DNP techniques, thus increasing sensitivity by one or two orders of magnitude [10], [11]. Indeed, as it was firstly shown by Poindexter [12], 15 fold NMR enhancement could be achieved in the low-viscous oil in the magnetic field of B_0 = 1.55 mT while potentially up to 45 in solutions of asphalt in chlorobenzene at B_0 = 7.4 mT [15].

Since crude oil is conventionally considered as a viscous liquid, the Overhauser type mechanism (ODNP) for the hyperpolarization is supposed to be the most effective because relatively high mobility of oil molecules and compounds including paramagnentic radicals implies a sufficiently strong modulation of electron-nuclear interaction leading to effective polarization transfer between electrons and nuclei in oil [12], [16] . The key property of ODNP is the fact that the maximal effect of electron polarization transfer on the nuclear system occurs when the frequency of the microwave pumping is exactly in resonance with the EPR transition, i.e. $\nu_{MW} = \nu_S$. This work is a short review of the classical liquid state DNP papers [17]-[24] as applied to PDS [8], [9], [12]-[15] with aim to introduce the basic concepts and key values of ODNP between "free" radicals and protons in PDS. The presented results can be used for designing ODNP spectrometers, proton precession magnetometers for geological and geophysical exploration, investigation of supramolecular organisation of PDS and their components.

2. Basic concepts of ODNP in PDS, simple semi-calssical approach, and comparing with experiments

In the high permanent homogenous external magnetic field B (along which we choose z - axis of the laboratory coordinate system) the spins precess around z with the Zeeman (Larmor) angular frequency

$$\omega_S = -\gamma_S B,\ \omega_I = \gamma_I B, \qquad (1)$$

where γ_S and γ_I are the gyromagnetic ratios of electron and nuclear spins (γ_S is negative) which result in ν_S = 280 MHz for the unpaired electron and ν_I = 425 kHz for protons in the magnetic field B = 10 mT ($\omega = 2\pi\nu$). If the mean distance, r, between S and I is sufficiently large ($r \geq d \approx 10$ **Å,** where $d \approx 10$ **Å** is a typical distance of the closest approach for the asphaltenes solutions, [15]) to assume that dipole-dipole interaction dominates over scalar.

The majority of paramagnetic centers in PDS are concentrated in asphaltenes. They frequently exist in the form of stable carbonaceous "free" radicals (FR) – unpaired electrons delocalized over many conjugated or aromatic chemical bonds [3]. However, the exact nature and structure of the intrinsic paramagnetic centres in PDS remains unclear due to the complexity of the investigated material and the different forms of aggregation of its larger constituents. Usually in PDS, FR is detected by conventional X-band ($\nu_{MW} \approx 9.5$ GHz) EPR as a single line ($S = 1/2$) with the peak-to-peak linewidth of ΔHpp = 0.4-0.8 mT and a g-factor between 2.0008 and 2.0081 (the typical range of 2.0027-2.0036 – see Figure 1, [3], [9]). It is shown that in low magnetic fields of 1.44-10 mT the values of g-factor and ΔHpp are roughly the same [25]-[27].

To flip nuclear spin I, the energy of its interaction with the environment should be time-dependent, i.e. fluctuating local magnetic field, B_{local}, should exist Fourier component of which with ω_I would cause nuclear spin reorientation. One of the sources of such "modulation" field is an electron spin S generating on the nuclei B_{local}. In liquids these fluctuations could be caused by the further reasons:

- rotational motion of the paramagnetic molecule (characterized by the rotational correlation time τ_{rot});
- exchange (Heisenberg) interaction between the electron spins or between the molecules with $\tau_{ex}=\tau_H = 1/\omega_{ex}$;
- translational motion (diffusion) of the molecules in the solvent with τ_{tr}.

Then the „integrated" correlation time of the z-component of the electron spin, Sz, is (assuming that the values are independent and are not correlated)

$$1/\tau = 1/\tau_{tr} + 1/\tau_{rot} + 1/\tau_{ex}, \qquad (2)$$

i.e. the correlation time is mainly defined by the shortest term(s). Concentration of FR in asphaltenes of heavy oils and bitumen is in the range of 10^{16}-10^{20} spins/g or 1 FR on 10-10000 asphaltenes molecules

[3], [9] and, therefore, exchange interaction (that can be operative at high concentration of paramagnetic centres), can be excluded from further consideration.

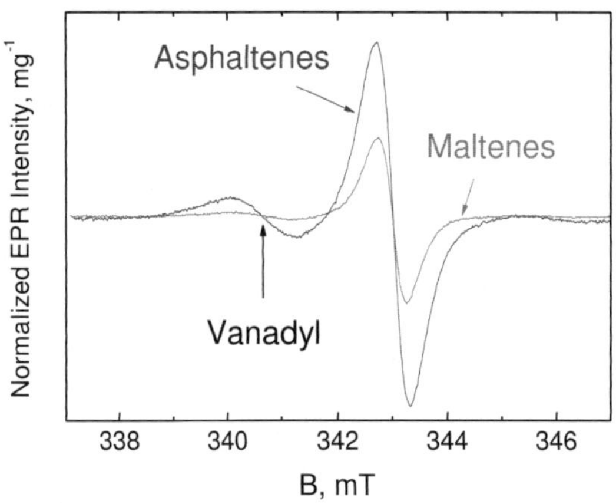

Figure 1. EPR spectra of FR (g = 2.0030, ΔHpp = 0.54 mT (15 MHz) in asphalenes (C = 9(1)·10^{18} spin/g) and maltenes (C = 4.0(5)·10^{18} spin/g) fractions of the oxidized bitumen at T = 300 K in X-band normalized to the sample masses. Additional line in the low-field part of the asphaltenes spectrum belongs to the paramagnetic vanadyl complexes [27].

The probability of nuclear flipping caused by the relaxation of $S = ½$ (per time unit) located at the distance r drops with r and magnetic field strength as [21]

$$\frac{3}{10}\frac{(g\beta)^2}{\tau B^2}\frac{1}{2r^6}. \quad (3)$$

It follows that high nuclear polarization via electron spin is achievable only in the case of sufficiently rapid relative motion of S and I in low magnetic fields. Conventionally, the condition for ODNP observation can be written as [18]

$$\omega_S\tau < 1. \quad (4)$$

Therefore, ODNP is most effective in (low-viscous) liquids, metals and semi-conductors [16]-[23]. From Eq. (4) for B = 10 mT (ν_S = 280 MHz, $\omega_S = 1.76\cdot10^9$ rad/s), $\tau < 560$ ps.

The most of the ODNP results in liquids, including those for PDS, were explained by using translational motions models only in which (for the isotropic diffusion)

$$\tau \approx \tau_{tr} \approx d^2/D_{solvent}, \quad (5)$$

where $D_{solvent}$ is the diffusion (self-diffusion) coefficient of diamagnetic particles of solvent molecules. Large sizes and high-molecular mass of asphaltenes (250-1500 a.m. u. [3], [25]), their propensity to self-aggregation lead to the fact that the radical diffusion is to slow to be accounted. Therefore, τ_{tr} is mainly defined by the movement of the solvent molecules rather than radical-containing sructures.

Poindexter [15] in 1972 reported extensive DNP studies of asphalts and asphaltenes in various solvents at near room temeperature in a magnetic field of 7.4 mT (the proton NMR frequency is 320 kHz and the EPR of 210 MHz). He found that the best ODNP enhancement could be observed in chlorobenzene and toluene and estimated τ_{tr} as varying from 36 ps for toluene to 700 ps in n-dodecylbenzene. The shortest time of 36 ps implies that perceptible ODNP effect can be potentially observed in PDS solvents in the magnetic fields up to 150 mT (ν_S = 4.3 GHz, ν_I = 6.5 MHz). It means that one can rarely expect ODNP detection in the conventional for EPR X-band range ($\nu_S = 9$ GHz, $B_0 = 320$ mT) even for the diluted solutions.

Figure 2 demonstrates the ODNP effect in toluene solvent of the oxidized bitumen in home-made DNP equipment operating in pulsed mode at $B_0 = 10$ mT. The technical description of NMR/DNP spectrometer is given in ref. [8]. The ODNP effect reveals as a disappearing of NMR signal with the EPR pumping as a sign of the prevailed dipole-dipolar electron-nuclear coupling [18], [24] and due to the solvation of stable asphaltenes-resins aggregates [3] while no ODNP effect is observed for the initial bitumen. It gives a potential opportunity to follow the processes of asphaltenes dissagreagation by magnetic resonance techniques [15].

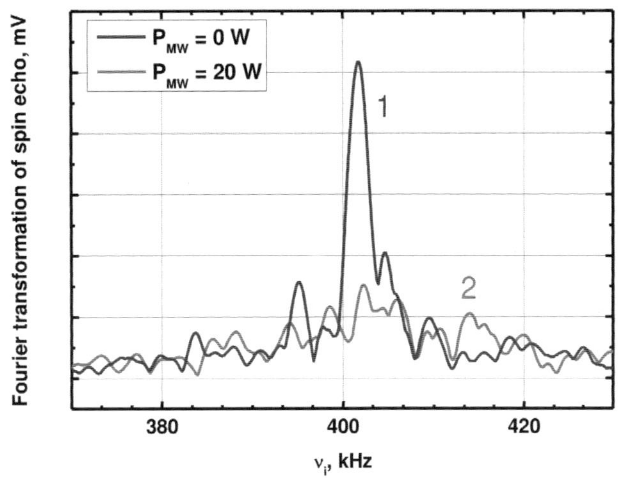

Figure 2. Change of the Fourier-transformed NMR spectrum (1) from the oxidized bitumen dissolved in toluene (1:10, v:v) in a magnetic field of 10 mT and $\nu_I = 420$ kHz with EPR pumping at $\nu_S = 280$ MHz (2).

Translational correlation time is a function of microspoic parameters such as temeperature and viscosity. For a spherical molecule of radius R in a pure solvent, one can write the dependence of the diffusion coefficient (Stokes-Einstein-Smoluchowski equation) on temperature and viscosity (η) as

$$D = k_B T / 6\pi\eta R, \quad (6)$$

where k_B ia a Bolznman constant. Then from the Eq. (5) it follows that

$$\tau_{tr} \propto \eta(T)/T. \quad (7)$$

Because viscosity is also a function of temperature and generaly decreases sharply with T [27], ODNP effect should be more pronounced at higher temperatures and, as it was already mentioned, for the low viscous systems. It gives a potential opportunity to monitor thermal and catalytic PDS changes in the well-boring operations by applying DNP tecniques. For the investigated bitumen the viscosity of the initial sample was measured to be 44 Pa·s at T = 313 K. Therefore, any ODNP effect could be hardly expected for the undiluted species.

3. Conclusion

Although NMR has been extensively used to characterize petroleum feedstocks, products, and catalysts while ODNP in low and high magnetic fields has found broad applications in biophysical, biomedical, surface characterization sciences, for the investigation of soft materials and biomolecules last decades, there is surprisingly small amount of PDS related ODNP research and utilization. We believe that PDS

DNP has a great potential to provide valuable information on supramolecular properties of PDS such as asphaltene aggregation and can be used to effectively enhance NMR signals for various PDS NMR employment.

Acknowledgments
The work was funded by the subsidy of the Russian Government to support the Program of competitive Growth of Kazan Federal University among World's Leading Academic Centers. EPR measurements are done on the equipment of Centre of the Shared Facilities for Physical and Chemical Research (KFU).

References
[1] Pradhan A, Ovalles C and Moir M 2018 *Characterization of Heavy Petroleum Fractions by NMR Techniques The Boduszynski Continuum: Contributions to the Understanding of the Molecular Composition of Petroleum* ed Ovalles C, Moir M (Washington: OUP) pp. 73-86
[2] Coates G R, Xiao L and Prammer M G 1999 *NMR Logging: Principles and Applications*; (Houston: Halliburton Energy Services)
[3] Yen T F and Chilingarian G V 2000 *Asphaltenes and asphalts. Vol 2 (Developments in petroleum science)* 40B (Amsterdam: Elsiever Science) p 622
[4] Yen T F and Chilingarian G V 1994 *Asphaltenes and Asphalts. Vol. 1 (Developments in Petroleum Science)* 40A (Amsterdam: Elsiever Science) p 459
[5] Mamin G V, Gafurov M R, Yusupov R V *et al* 2016 *Energy Fuels* **30** 6942
[6] Ramachandran V, van Tol J, McKenna A M *et al.* 2015 *Anal. Chem.* **87** 2306
[7] Ben Tayeb K, Delpoux O, Barbier J *et al* 2015 *Energy Fuels* **29** 4608
[8] Alexandrov A S, Archipov R V, Ivanov A A *et al* 2014 *Appl. Magn. Reson.* **45** 1275
[9] Gizatullin B, Gafurov M., Rodionov A *et al* 2018 *Energy Fuels* **32** 1126
[10] Sapunov V, Denisov A, Saveliev D *et al* 2016 *Magn. Reson. Solids* **18** 16209.
[11] Sapunov V A, Narkhov E D, Fedorov A L *et al* 2015 *International Multidisciplinary Scientific GeoConference Surveying Geology and Mining Ecology Management, SGEM*, 215
[12] Poindexter E 1958 *Nature* **182** 1087
[13] Firat Y E, Yildirim H and Peksoz A 2016 *J. Disper. Sci. Technol.***37** 1349
[14] Yalçiner A 1981. *J. Colloid Interface Sci.* **79** 114
[15] Poindexter E 1972 *J. Colloid Interface Sci.* **38** 412
[16] Overhauser A W 1953 *Phys. Rev.* **92** 411
[17] Abragam A 1961 *The Principles of Nuclear Magnetism.* (Oxford: Clarendon Press) p 599
[18] Hausser K H and Stehlik D 1968 *Adv. Magn. Reson.* **3** 79
[19] Atsarkin V A 1978 *Sov. Phys. Usp.* **21** 725
[20] Khutsishvili G R 1966 *Sov. Phys. Usp.* **8** 743
[21] Khutsishvili G R 1961 *Sov. Phys. Usp.* **3** 285
[22] Dweck R A, Richards R E and Taylor D 1969 *Ann. Rev. NMR. Spectrosc.* **2** 293
[23] Potenza J. 1972 *Adv. Mol. Relax. Proc.* **4** 229
[24] Lingwood M D and Han S 2011. *Ann. Rep. NMR Spectrosc.* **73** 83
[25] Dolomatov M Yu, Rodionov A A, Gafurov M R *et al* 2016 *Magn. Reson. Solids* **18** 16101
[26] Mehrabi-Kalajahi S S, Varfolomeev M A, Yuan C *et al* 2018 *J. Petrol. Sci. Eng.* **169** 673
[27] Gafurov M R, Volodin M A, Rodionov A A *et al* 2018 *J. Petrol. Sci. Eng.* **166** 363

IOP Conf. Series: Earth and Environmental Science **282** (2019) 012019 doi:10.1088/1755-1315/282/1/012019

Lattice distortions in hydroxyapatites with size as follows from the electronic relaxation time measurements

D. Shurtakova[1], G. Mamin[1,*], M. Gafurov[1], L. Kuznetsova[2], V. Putlayev[3,4], E. Klimashina[3,4], S. Orlinskii[1]

[1] Kazan Federal University, 420008 Kazan, Russia
[2] Kazan State University for Architecture and Engineering, 420043 Kazan, Russia
[3] M.V.Lomonosov Moscow State University, Chemistry Department, 119991 Moscow, Russia
[4] M.V.Lomonosov Moscow State University, Faculty of Materials Science, 119991 Moscow, Russia

E-mail: george.mamin@kpfu.ru

Abstract. Pure and substituted hydroxyapatites ($Ca_{10}(PO_4)_6(OH)_2$, HAp) have been recently gaining special importance in catalysis for selective reactions where the synergetic functioning of acid–base active site pair facilitates the molecular transformations. Transition to the nanostructured HAp materials and ion incorporation can cause lattice distortions and influence physicochemical properties of HAp. We have investigated powders of HAp synthesized by wet techniques from the nitrogen-containing solutions with the average particle sizes of 30 (nanosized) and 1000 nm (microsized) with pulsed electron paramagnetic resonance (EPR). Electronic relaxation times (T_{1e}) for the stable impurity paramagnetic NO_3^{2-} centers were measured in the temperature range of 10-300 K. It is revealed that T_{1e} shortens with the size reduction of factor two in the whole temperature range. The results are discussed in the framework of the modified Debye model taking into account non-Debye HAp phonon density. The obtained results show the feasibility of pulsed EPR techniques for characterization HAp based materials.

1. Introduction

Hydroxyapatite (HAp, $Ca_{10}(PO_4)_6(OH)_2$) belongs to the apatite family. It is one of the most usual forms of calcium phosphates (CaP) materials, more commonly known and studied as biomaterial being a major constituent of teeth and bones [1], [2]. The special attribute of HAp is its ability to form solid solutions and to accept a large number of anionic and cationic substituents exhibiting both acid (PO_4^{2-} ion) and base (Ca^{2+} ion) active sites together. In addition to its various medical and biological applications, HAP is applied such as fluorescent markers, material for fuel cell, or an adsorption and stabilization matrix for radioactive waste and harmful metals [1]-[6] .

Since recently, due to their mentioned ion-exchange capacity, acid base adjustability, adsorption capacity, non-toxicity, thermal stability, ability to synthesize species with a controlled size and porosity, substituted by various cations and anions, HAP based materials attract scientific attention as heterogeneous catalysts for biodiesel production and natural gas-methane-C_2 hydrocarbon conversion [7]-[16]. Aside from economic considerations, another driving reason for the development of heterogeneous catalysis is its ecological impact forcing to develop manufacturing processes that release the minimum amount of possible waste. In this context, HAp (that can be derived also from the natural sources such as eggshell, animal bone, mollusk shells, etc. and is a renewable one) is considered as a cost-effective catalyst.

Adsorption and incorporation of impurity ions can manifest itself in HAp lattice distortion. It could be also expected that size reducing (necessary for increasing surface-to-volume ratio) can lead to the changes in the vibrational spectrum of a crystal lattice. Changes in crystal lattice, in their turn, often cause changes in crystallinity, thermal stability, solubility and other physicochemical properties of the material [1], [3]. Intrinsic natural or artificially created / incorporated paramagnetic centres can serve as sensitive probes to follow these changes. Investigation of temperature dependence of the longitudinal (spin-lattice) electronic relaxation time, T_{1e}, is the most informative for these purposes [4], [17]. However, to the best of the authors' knowledge, there are no convincing experimental proofs of the size effect on the spin lattice relaxation (SLR) rate of paramagnetic centres. In this work we present study of temperature dependence of T_{1e} for NO_3^{2-} stable radicals in the microsized and nanosized powders of HAp.

2. **Materials and Methods**

Two samples of hydroxyapatite – microsized (with the crystallites of about 1 μm) and nanosized (30 nm) were studied. The NO_3^- species in the trace amounts were incorporated into the structure of hydroxyapatite powder with the chemical formula $Ca_{10}(PO_4)_6(OH)_2$ to obtain the crystallites according to the equation

$$10Ca(NO_3)_2 + 6(NH_4)_2HPO_4 + 8NH_4OH = Ca_{10}(PO_4)_6(OH)_2 + 20NH_4NO_3 + 6H_2O \qquad (1)$$

by dropwise addition of 0.3 M stock solution of $(NH_4)_2HPO_4$ (99.9%, Labteh, Russia) to 0.5 M stock solution of $Ca(NO_3)_2{\cdot}4H_2O$ (99.9%, Labteh, Russia).

The obtained samples were characterized by infrared spectroscopy, X-ray diffraction technique, electron microscopy, elemental analysis. Details of the synthesis, post-synthesis treatments and analytical characterization of HAp powders are given in paper [3].

X-ray irradiation of the synthesized nanopowders was performed using a URS-55 tube ($U = 55$ kV, $I = 16$ mA; W-anticathode) at room temperature (RT) with an estimated dose of 5 kGy to create stable paramagnetic centers. Pulsed and continuous wave (cw) electron paramagnetic resonance (EPR) measurements were done using X-band (9 GHz) Bruker Elexsys 580 spectrometer equipped with the liquid helium temperature controller. Electron spin echo (ESE) EPR spectra were recorded by means of field-swept two-pulse echo sequence (FS ESE) $\pi/2 - tau - \pi$ by sweeping the external magnetic field B_0 with the pulse length of π pulse of 32 ns and time delay $tau = 300$ ns. To extract the spin-lattice (longitudinal) relaxation times T_{1e} the Inversion-Recovery pulse sequence $\pi - T_{delay} - \pi/2 - tau - \pi$ with the T_{delay} varied was applied.

3. **Results and Discussion**

The synthesized HAp powders at RT are EPR silent. ESE EPR spectrum of the microsized HAp after irradiation is shown in Figure 1. No significant changes in EPR spectrum is observed with the size decrease. The obtained stable paramagnetic centres are due to the NO_3^{2-} of axial symmetry with $S = 1/2$, $g_\parallel = 2.0015(5)$, $g_\perp = 2.0055(5)$, $A_\parallel = 6.65(40)$ mT, $A_\perp = 3.35(40)$ mT caused by the near planar symmetry of radical and $I = 1$ for ^{14}N nuclei [18]. No other paramagnetic impurities were detected within the limits of sensitivity of our setup.

We have checked the difference in T_{1e} along the spectrum in the magnetic fields corresponding to the three local maxima in ESE signal (Figure 1), Our T_{1e} measurements did not show anisotropy within the acquisition accuracy. Therefore, the results for the central peak ($g_\perp$) are presented. Experimental data for the inverse for T_{1e} value (relaxation rate) along with the corresponding fittings discussed below are drawn in Figure 2.

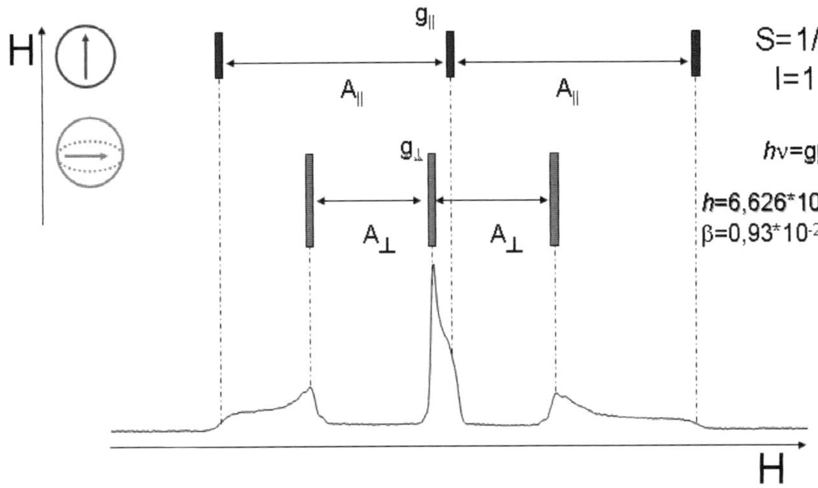

Figure 1. EPR spectrum for the microsized HAp sample and its schematic explanation for the NO_3^{2-} centre of the axial symmetry.

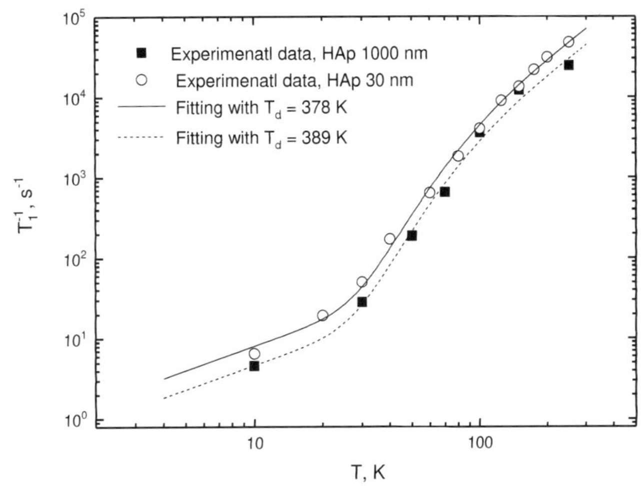

Figure 2. Experimental T_{1e} data for the microsized (solid squares) and the nanosized (open circles) HAp powders and their approximations with $T_D = 378$ K (solid line) and $T_D = 389$ K (dashed line) by using Eq. (5).

The observed temperature dependences of the relaxation rates $1/T_{1e}$ can be fitted as a sum of direct and two-phonon processes [19], [20]:

$$T_{1e}^{-1} = T_{dir}^{-1} + T_{Raman}^{-1} = C_{dir}T + C_{Raman}T^9 f(\tfrac{\Theta_D}{T}), \qquad (2)$$

where the first term is responsible for the direct process of relaxation at low temperatures, the second one defines the Raman spin-lattice relaxation process with Debye approximation for lattice vibrations,

Θ_D is a Debye temperature, C_{dir} and C_{Raman} are coefficients which values were fitted , $f(\frac{\Theta_D}{T}) \equiv f(z) \equiv I_8(z)/I_8(\infty)$, $I_8(z) = \int_0^z x^8 e^X / \left(e^X - 1\right)^2 dx$.

HAp are non-Debye solids, i.e. the HAp phonon density does not obey Debye law [21]. Taking into account the last one calculated in the previous work of our group [19], we have fitted the experimental data by using Eq. (2). The best approximations are achieved for $\Theta_D = 389(5)$ K for the microsized sample and $\Theta_D = 378(5)$ K for the nanosized one. The first value is in a good correspondence with the data derived from the heat capacity measurements [19], [21]. For the microsized sample $C_{dir} = 0.46$; $C_{Raman} = 6.13$ while for the nanosized one $C_{dir} = 0.81$; $C_{Raman} = 9.70$. Analysis of the fitting coefficients is a matter of the separate research.

We have to note that the size reduction leads to the increase of the relaxation rate of up to 2.5 times in the whole temperature range that was hardly foreseen *a-priori*. Indeed, with the reducing crystal size to the nanometer scale in one, two or three dimensions, because of the absence of translational symmetry, periodicity can no longer be used in analysing the dynamics of lattice vibrations. As a result, first of all, the waves with $\lambda/2 > d$ are not realized (here d is the smallest characteristic size of the system and λ is a vibrational wavelength). In this case, certain frequencies fall from the spectrum of the lattice vibrations, which leads to a change in the Debye formula. Therefore, it can be expected that a decrease in particle size should shift the phonon spectrum into the high-frequency region making T_{1e} longer (or relaxation rate less) than for the bulk sample. Experimentally, from the comparison of the specific heats of massive single-crystal samples and nanopowders of palladium, Si, MgO, etc., a decrease of the Debye temperature was observed with the sample size [22].

It could be speculated that the surface and the surface paramagnetic centres rather than bulk paramagnetic species could play a significant role in the case of the nanomaterials. But as we have already mentioned, no significant changes in EPR spectra and T_{2e} values for stable NO_3^{2-} complexes were found with the size or HAp surface treatment in a series of our previous papers (see [18] and references therein). Therefore, we suppose that for the both types of the investigated species we have deal with the intrinsic (bulk) NO_3^{2-} radicals.

4. Conclusions and Perspectives

Rational design of heterogeneous catalysts requires application of various theoretical and analytical tools for their study and characterization. Electron paramagnetic resonance spectroscopy, which includes continuous wave and pulsed methods, is frequently used in the study of catalytically active systems. The role of EPR in these studies varies considerably from, on the one hand, a comprehensive description of the electronic states of the paramagnetic active site in the catalytic cycle to, on the other hand, a simple analytical confirmation that a radical centre is present [23]-[26]. In this paper we show some abilities of pulsed EPR for investigation and analysis of HAp based materials to encourage the specialists in different scientific and industrial branches to use different EPR approaches in their studies and routine analyses.

Acknowledgements

M.G. and G.M. are thankful to RFBR grant 18-29-11086 for the financial support. S.O. appreciates a support of the Program of the competitive growth of Kazan Federal University. E.S. are thankful to RFBR grants 18-08-01473, 19-38-70042 (also was funded by Moscow city Government) for the financial support and V.P. appreciates a support of the Program for Development of MSU.

References

[1] Dorozhkin S V 2018 *Nanodimensional and nanocrystalline calcium orthophosphates.* In *Developments and Applications of Calcium Phosphate Bone Cements* (Singapore:Springer) pp. 355-448.

[2] Shurtakova D *et al* 2019 *Magn. Reson. Solids* **21** 19105
[3] Gafurov M, Biktagirov T, Mamin G, Klimashina E, Putlayev V, Kuznetsova L and Orlinskii S 2015 *Phys. Chem. Chem. Phys.* **17** 20331
[4] Gafurov M R, Biktagirov T B, Mamin G V, Shurtakova D V, Klimashina E S, Putlyaev V I and Orlinskii S B 2016 *Phys. Solid State* **58** 469
[5] Fadeeva I V et al 2017 *BioNanoScience* 7 434
[6] Chelyshev Y, Gafurov M, Ignatyev I *et al* 2016 *BioMed Res. Int.* **2016** 3706280
[7] Fihri A *et al* 2017 Coord. Chem. Rev. **347** 48
[8] Sebti S *et al* 2008 Curr. Org. Chem. **12** 203
[9] Gruselle M 2015 *Journal of Organometallic Chemistry* **793** 93
[10] Oh S *et al* 2016 *Fuel* **167** 208
[11] Viswanadham N *et al* 2015 *RSC Adv.* **5** 67380
[12] Lee K-Y et al 1998 *Stud. Surf. Sci. Catal.* **119** 385
[13] Abbaszaadeh A *et al* 2012 *Energy Convers. Manage* **63**, 138
[14] Chen G *et al* 2015 *Energy Convers. Manage* **98**, 463
[15] Shan R *et al* 2018 *Energy Convers. Manage* **178** 277
[16] Bennett J A *et al* 2016 *EJ. Mater, Chem. A* **4** 3617
[17] Gabbasov B et al. 2019 *J. Magn. Magn. Mater.* **470** 109
[18] Gafurov M *et al* 2014 *Appl. Magn. Reson.* **45** 1189
[19] Biktagirov T *et al* 2016 *J. Low Temp. Phys.* **185** 627
[20] Aminov L K *et al* 2001 *Physica C* **349** 30
[21] Slepko A and Demkov A 2015 *J. Appl. Phys.* **117**, 074701
[22] Escalera-Moreno L *et al* 2018 *Chem. Sci.* **9** 3265
[23] Spencer J, Folli A, Richards E and Murphy D M 2018. *Applications of electron paramagnetic resonance spectroscopy for interrogating catalytic systems. Electron Paramagnetic Resonance* **26**, 130-170.
[24] Mukhambetov I N *et al* 2014 *J. Phys. Chem. C* **118** 14998
[25] Gafurov M R *et al* 2015 *J. Phys. Chem. C* **119** 27410
[26] Gafurov M *et al* 2019 *Magn. Reson. Solids* **21** 19101

Catalytic Conversion of Tatarstan Heavy Oil using Copper Based Catalysts at 250 °C

M. Suwaid[1,2], M. Varfolomeev[1,2], A. Almuntaser[1,2], R. Djimasbe, S. Khramov[1], A. Al-Rubaye[1], L. Khaziakhmetov[1]

[1]Kazan (Volga region) Federal University, Institute of Geology and Petroleum Technologies, Russian Federation

[2] Kazan (Volga region) Federal University, Butlerov Institute of Chemistry, Russian Federation

E-mail: Suwaidmuneer@gmail.com

Abstract. In this work, the in-situ upgrading of heavy oil from Tatarstan oilfield using steam and copper stearate as a catalyst at 250 °C and 35 bar for 24 hours was studied in stainless-steel reactor. Changes in the density/API gravity, viscosity, chemical composition (SARA-analysis), elemental composition and structure (FTIR-spectroscopy) of heavy oil before and after conversion were determined. Generally, we can summarize, that the content of resins and asphaltenes as well as the average molecular weight of heavy oil were reduced after conversion process. Whereas, amount of saturated and aromatic hydrocarbons was increased due to the destruction of its high molecular weight components. Also, irreversible decrease of viscosity was fixed.

1. Introduction

Currently, under the conditions of the deterioration of the hydrocarbon reserves structure and the increase in energy consumption, there is a crucial problem of developing non-conventional hydrocarbon reserves, which, in particular, include heavy oil and natural bitumen [1–3]. The development of heavy oil field is complicated by their high viscosity, density and significant content of resin and asphaltene compounds, which make oil extraction process more difficult and expensive [1].

Recently, various technologies have been proposed and implemented to increase the efficiency of the extraction of heavy oils, based on the injection of chemicals reagents, thermal effects, etc. [1]. Among them, steam injection plays an important role in the extraction of heavy oils and natural bitumens, but with the passage of time efficiency of this method has been decreased, and steam generation requires high energy costs, and the viscosity of the extracted oil on the surface increases as a result of degassing at high temperatures and generation of free radicals that initiate polymerization reactions [2]. These problems of steam and thermal effects can be solved by applying catalytic aquathermolysis technology.

In this work we use injection of steam and copper-based oil-soluble catalysts for in-situ upgrading of heavy oil (partial processing of heavy fractions in the reservoir) and reduce the viscosity of the extracted oil, as well as improve its qualitative composition.

2. Materials and Methods

Heavy oil from one of the Tatarstan oilfields was in this work. The main characteristics of studied oil sample are shown in Table1.

Table1. Properties of studied heavy crude oil

API gravity	14.1
viscosity at 25 °C	2073 mPa·s
Saturates, % mas.	28.79
Aromatics, % mas	44.32
Resin, % mas	20.98
Asphaltenes, % mas	5.91

A stainless-steel 350-ml batch Parr reactor (4575/76 HP/HT Reactor) equipped with a heating mantle and a temperature controller was used to conduct experiments. The reactor was loaded with oil and water at a weight ratio of 2:1. The experiment with steam was conducted in inert atmosphere by addition of small amount of nitrogen. The initial pressure was 2 bar. The temperature of reaction was 250 °C and pressure during the 24 hours experiments was practically the same and equal to 35 bar. The experiment with steam and catalyst (2 % by crude oil mass) was conducted at the same conditions of experiment without catalyst.

Viscosity, API gravity and composition of the crude oil (SARA-analysis) before and after conversion in the presence and absence of catalyst were studied. Viscosity and API gravity of oil were measured by SVM-3000 Stabinger viscometer. The elemental composition of oil before and after conversion was determined by Perkin Elmer 2400 Series II analyzer. Vertex 70 FTIR-spectrometer (Bruker, Germany) was used to study structure of oil samples before and after conversion.

3. Results and Discussions

3.1 Viscosity and API gravity of oil before and after upgrading at 250 0C

The viscosity and API gravity of crude oil before and after conversion by steam injection in the presence and the absence of the catalyst at 250 ^{0}C are shown in the table 2.

Table 2. Viscosity and API gravity of crude oil before and after conversion

Sample	Viscosity 25°C,mPa·s	^{0}API, gravity
Initial oil	2073	14.1
Crude oil with steam	1758	14.8
Crude oil with steam and catalyst	1685	15.0

The results show that the viscosity (at 25 °C) was reduced from 2073 to 1685 and the API gravity increased from14.1 to 15. As shown in the table 2, the catalyst could improve the conversion quality compared with steam alone.

3.2 SARA-analysis of crude oil before and after conversion

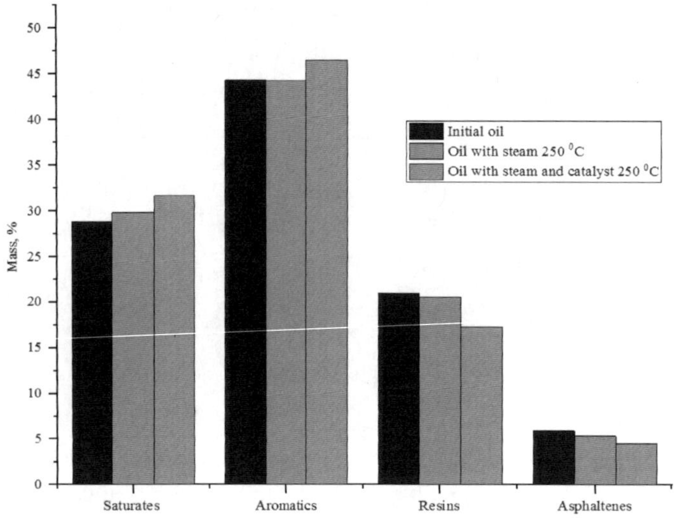

Fig.1. SARA-analysis of oil before and after conversion

Figure 1 shows the results of SARA-analysis of crude oil before and after conversion. An increase in saturated hydrocarbon fraction was observed after conversion in the presence of catalyst. Also, a decrease in resin and asphaltene amount was observed in the presence of catalyst as shown in fig.1

3.3 Elemental analysis of crude oil before and after conversion

Table 3 shows the elemental analysis results of oil before and after conversion. The content of sulfur and nitrogen were reduced due to the desulfurization and denitrogenation process as shown in table 3. After conversion in the presence of catalyst, a slight increase in hydrogen and a decrease in the carbon content were observed, and this shows how much the upgrading process is effective.

Table 3. Elemental analysis of oil before and after conversion

Sample	**Content, wt %**				
	C	H	N	S	N_H/N_C
Initial oil	83.68	11.44	0.36	4.55	1.641
Oil with steam	82.37	11.31	0.36	4.24	1.648
Oil with steam and catalyst	82.32	11.45	0.29	4.20	1.669

3.4 Fourier Transform Infrared Spectra (FTIR) spectroscopy analysis of oil

FTIR spectra of oil before and after conversion in the absence and the presence of catalyst is shown in Fig. 2.

Spectral indices were calculated from peak heights of selected infrared bands (Table 4) to determine the chemical composition of crude oil before and after conversion in the absence and in the presence of catalyst. Spectrometric indices (C_{1-4}) were calculated according the following equations [4-7]: C_1= D1600/D720 (aromaticity index); C_2= D1380/D1465 (branching index); C_3= (D720+D1380)/D1600 (aliphatic index); C_4= D1030/D1465 (index of sulfurization). According to result in Table 4, a decrease in the index of aromaticity (C_1) and sulfurization (C_4) was observed. About index of aliphaticity (C_3) an increase was observed. For index of branching (C_2) small changes were observed.

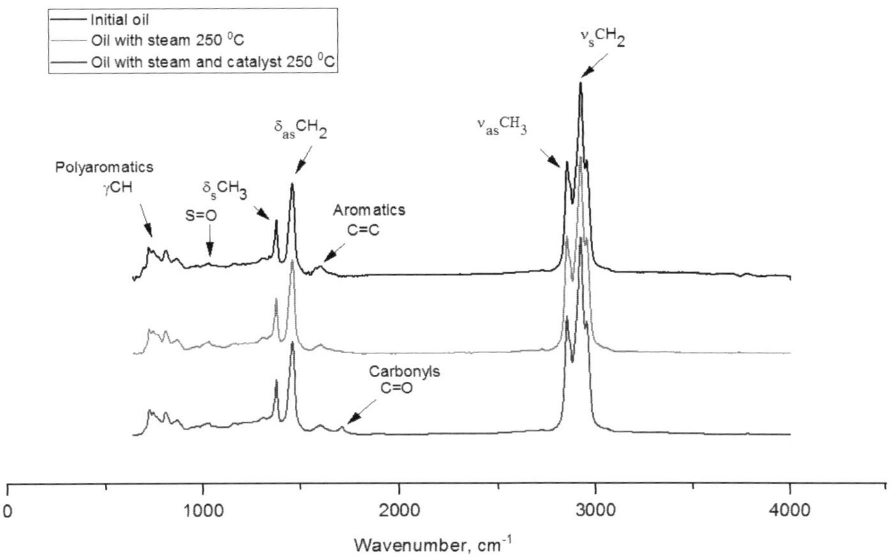

Fig.2. FTIR of oil before and after conversion

Table 4. FTIR analysis of crude oil before and after conversion

Sample	Spectral indices			
	C_1	C_2	C_3	C_4
initial oil	0.4242	0.6088	5.8334	0.2144
Oil with steam	0.4391	0.5684	7.0411	0.1586
Oil with steam and catalyst	0.3894	0.5634	7.325	0.1458

C_1 – aromaticity index, , C_2 – branching index, C_3– aliphaticity index and C_4 – index of sulfurization

4. Conclusion

In this work, it was found that the content of high molecular weight like resins and asphaltenes decrease, and the amount of saturates increase due to thermal treatment of heavy oil at 250 ^{0}C by steam and copper-based catalyst. These data show that catalyst could improve the conversion quality compared with the steam injection alone.

The elemental content of sulfur and nitrogen were reduced due to the desulfurization and denitrogenation process. These processes occur due to catalytic effect of copper stearate. These effect is based on the weakening of C−S and C−N bonds in heavy oil by intermolecular interactions with copper in catalyst. After conversion in the presence of catalyst, a slight increase in hydrogen and a decrease in the carbon content were observed. This shows how much the upgrading process is effective.

The heavy hydrocarbons component with long chain had broken into small alkanes due to aquathermolysis reactions of resin and asphaltene molecules.

Funding

This work has been partly performed according to the Russian Government Program of Competitive Growth of Kazan Federal University.

References

[1] Shah A., Fishwick R., Wood J. et al., , Energy Environ. Sci., 2010, V. 3, pp.700–714.
[2] Suwaid M. A. Yuan, C., Almuntaser, A. A., & Varfolomeev, M. A. //International Multidisciplinary Scientific GeoConference: SGEM: Surveying Geology & mining Ecology Management. – 2018. – T. 18. – C. 283-289.
[3] Maity S.K., Ancheyta J., Marroqun G., Energy & Fuels, 2010, V. 24, pp. 2809–2816.
[4] Permanyer, A., Douifi, L., Lahcini, A., Lamontagne, J., Kister, J., 2002. Fuel 81, 861–866.
[5] Rakhmatullin, I. Z., Efimov, S. V., Tyurin, V. A., Al-Muntaser, A. A., Klimovitskii, A. E., Varfolomeev, M. A., &Klochkov, V. V. (2018). 168, 256-262.
[6] Al-Muntaser, A.A., Suwaid, M.A., Rakhmatullin, I.Z., Varfolomeev, M.A., 2017. In: SGEM2017 Vienna GREEN Conference Proceedings, vol. 17. pp. 75–82.
[7] Al-Muntaser, A.A., Yuan, C., Suwaid, M.A., Varfolomeev, M.A. SGEM, 10.5593/sgem2018/1.4/S06.044, Volume 18, Issue 1.4, 2018, Pages 335-342

IOP Conf. Series: Earth and Environmental Science **282** (2019) 012021 doi:10.1088/1755-1315/282/1/012021

Effect of pressure, temperature and gas phase composition on the rheological properties of heavy crude oil

A A Akhmadiyarov[1], A A Petrov[1], I T Rakipov[1], M A Varfolomeev[1]
[1]Institute of Chemistry, Kazan Federal University, Kazan, Russia
E-mail: aydaru@ya.ru

Abstract. Influence of pressure, temperature and the nature of gas on the viscosity of heavy crude oil from the Tatarstan field was studied using rheology method (rheometer MCR 302). The effect of temperature from 25 to 125 °C and pressure from 0.1 to 12.5 MPa on crude oil viscosity was investigated. It was found that an increase in temperature of 125 °C leads to a decrease of viscosity by 8 times. At the same time, an increase of pressure up to 10 MPa leads to an increase in viscosity by 38% at low temperatures. Also the influence of the nature of gas (nitrogen, argon and air) on the viscosity of crude oil at different temperatures and pressures was analyzed. It is shown that crude oil viscosities depend on gas composition. On the other hand, the influence of the composition of binary gas mixtures in a wide range of temperatures and pressures on the viscosity of heavy oil has been studied. The activation energies of flow are calculated for two types of gas mixture. It is shown that the gas phase at different ratios of the components does not significantly affect the rheological parameters of crude oil.

1. Introduction.

The recovery of heavy crude oil across the world grows due to the severe depletion of traditional light oil reserves. Heavy crude oils and bitumens present about a quarter of total hydrocarbon resources [1, 2]. Russia along with Venezuela and Canada own largest discovered deposits of heavy crude oil in the world [3]. A third part of heavy crude oil and bitumen reserves of Russia is located in Tatarstan. The production of hydrocarbons from heavy oil fields is limited by high viscosity of crude oils and bitumens, which does not allow using conventional methods of recovery. Therefore, methods and technologies, which can help to reduce viscosity inside the reservoir, have great practical interest. In order to increase

IOP Conf. Series: Earth and Environmental Science **282** (2019) 012021 doi:10.1088/1755-1315/282/1/012021

the efficiency of heavy oil recovery different technologies can be used like gas injection [4–6], thermal treatment [7] and chemical flooding [8, 9].

A viscosity is important parameter which determines possibility of the extraction of heavy crude oils by these methods. On the other hand, it is known that the viscosity of crude oil substantially depends on the gas phase. Therefore, the aim of the work was to study the effect of the gas phase on the viscosity of heavy crude oils at different pressures and temperatures. The influence of the nature and composition of gas mixtures on the rheological characteristics is analyzed and the activation energy of flow is estimated. The influence of nature of gas in mixture on viscosity is discussed.

2. Experimental part.

2.1 Physical-chemical properties of crude oil.

Measurements of viscosity, density and API-gravity (American Petroleum Institute Gravity) of oil samples were carried out on Anton Paar SVM 3000 Stabinger viscometer at 20 °C. Crude oil was thoroughly stirred before measuring. Water content determined by Karl Fischer titration was shown not to exceed 0.05 wt%. Dynamic viscosity of crude oil from one of the Tatarstan oilfields was measured to be 2480 mPa·s. Density and API-gravity was measured to be 0.9617 and 15.2 correspondingly [10].

2.2 Viscosity of crude oil at reservoir temperature and pressure

Viscosities of heavy crude oil samples at simulated reservoir conditions (high pressure and high temperature) were measured using Physica MCR302 rheometer (Anton Paar, Austria) equipped with pressure cell up to 150 bar with two measuring geometries. Sample volume of CC25/PR concentric cylinders geometry was 15 ml. The rheometer was checked at different pressures and temperatures on reference fluid squalane ($C_{30}H_{62}$), which was purchased from Sigma-Aldrich (purity over 99 wt%). A good agreement between obtained experimental and literature values is observed [11–15]. The apparatus and experimental procedures have been described in detail in our previous publication [10].

3. Results and discussion

IOP Conf. Series: Earth and Environmental Science **282** (2019) 012021 doi:10.1088/1755-1315/282/1/012021

3.1. Influence of pressure and gas nature on the viscosity of heavy crude oils.

The effect of pressure and temperature on the viscosity of crude oil in a nitrogen atmosphere was studied. Measurements were carried out in a wide range of temperatures from 25 to 125 °C and pressures from 0.1 to 10 MPa, Table 1.

Table 1. Viscosity of crude oil at different pressure and temperature, (mPa·s, P=MPa, T=^{0}C).

Pressure	η_{25}	η_{50}	η_{75}	η_{100}	η_{125}
0.1	1600	225	69	32	23
2.5	1757	234	74	34	23
5.0	1896	249	81	37	24
7.5	2054	266	87	38	23
10.0	2214	283	91	40	22

Strong dependence of crude oil viscosity from the temperature was observed. Increasing of temperature from 25 to 50 ^{0}C leads to decreasing of viscosity in 8 times (Table 1). Increasing of pressure from 0.1 to 10.0 MPa leads to increasing of viscosity by 38% at 25 ^{0}C. It should be pointed out that crude oil viscosity does not depend on pressure at 125 ^{0}C (Table 1). Since viscosity is highly dependent on pressure, it is necessary to investigate the effect of the gaseous medium on the physical-chemical properties of heavy crude oils. Thus, we analyzed the influence of the gaseous medium (argon, nitrogen and air) on the viscosity of crude oils at 25 ^{0}C, Table 2. The viscosity in nitrogen and air atmosphere was found to be close, in contrast to argon medium.

Table 2. Viscosity of crude oil at 25 °C at various pressures in the atmosphere of nitrogen, argon and air.

Pressure, MPa	η_{N_2}	η_{Ar}	η_{Air}
2.5	1674	1704	1686
5.0	1809	1826	1836
7.5	1958	1973	1985

10.0	2121	2141	2139
12.5	2287	2312	2308

3.2. Effect of gas mixture composition on crude oil viscosity. Estimation of the activation energy of flow.

In order to determine the effect of the gas phase composition on the viscosity of crude oil samples, gas mixtures of carbon dioxide (1) with oxygen (with a molar/volume fraction of carbon dioxide x_1 in the range from 0.4 to 1) and nitrogen (1) with oxygen (with a molar/volume fraction of nitrogen x_1 in the range from 0.6 to 1) were employed. For all gas mixtures, pressures of up to 15 MPa were chosen, with the exception of the atmosphere of pure carbon dioxide, in which the pressure did not exceed 6 MPa, due to the transition of the gas into the liquid phase. Results are presented in Tables 3, 4 and in Figures 1, 2.

Table 3. The dependence of crude oil viscosity (mPa·s) from the composition of the gas mixture $CO_2(1) – O_2$.

x_1	0.4			0.6			0.8			1	
Pressure, MPa	3	9	15	3	9	15	3	9	15	3	6
Temperature, °C											
30	1079	1302	1535	1093	1314	1550	1106	1328	1571	1110	1197
45	368.9	447.4	529.3	372.5	451.8	534.5	379.0	459.5	543.5	378.3	410.1
60	144.0	174.5	206.5	145.4	176.2	208.5	147.8	179.2	211.9	147.7	159.9
75	75.1	88.7	105.0	73.8	89.6	106.0	75.1	91.1	107.8	74.9	81.3
90	36.3	44.0	52.0	36.6	44.4	52.5	37.2	45.2	53.4	37.2	40.3
105	21.0	25.5	30.1	21.2	25.7	30.4	21.6	26.1	30.9	21.6	23.3
120	12.8	15.0	17.3	12.5	15.2	17.4	12.5	15.5	18.4	11.9	13.1
135	6.9	8.4	9.9	6.9	8.4	10.0	7.1	8.6	10.2	7.1	7.7
150	4.1	4.8	5.8	4.2	4.8	5.8	4.2	4.9	5.9	4.2	4.6

Table 4. The dependence of crude oil viscosity (mPa·s) from the composition of the gas mixture $N_2(1) - O_2$.

x_1	0.6			0.8			1		
Pressure, MPa	3	9	15	3	9	15	3	9	15
Temperature, °C									
30	1103	1326	1564	1104	1326	1568	1101	1335	1570
45	375.9	455.9	539.3	378.3	458.7	542.5	375.8	455.9	539.4
60	146.7	177.8	210.4	147.5	178.9	211.6	146.8	177.8	210.6
75	74.5	90.4	107.0	75.0	90.9	107.6	74.4	90.4	107.0
90	36.9	44.8	53.0	37.2	45.1	53.3	36.9	44.8	53.0
105	21.4	26.0	30.6	21.5	26.1	30.9	21.4	26.0	30.6
120	12.6	15.3	17.6	12.5	15.5	18.4	12.3	15.2	18.1
135	7.0	8.5	10.1	7.1	8.6	10.1	7.0	8.5	10.1
150	4.2	4.9	5.9	4.2	4.9	5.9	4.2	4.9	6.0

From Tables 3, 4 and Figures 1, 2 it can be concluded that the composition and nature of the gas phase has little effect on the viscosity of crude oil samples without a catalyst. The main factors affecting viscosity are temperature and pressure.

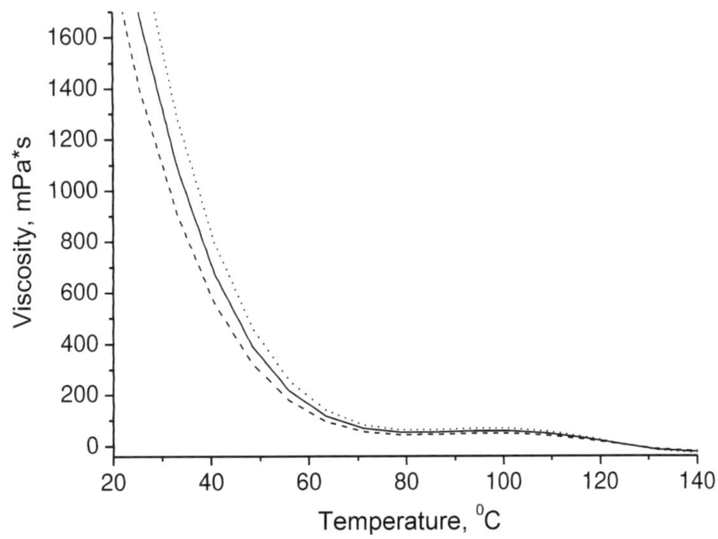

Figure 1. Effect of temperature and pressure on the viscosity of crude oil in the atmosphere CO_2–O_2 (60:40): dashed line – 3 MPa; solid line – 9 MPa; dotted line – 15 MPa.

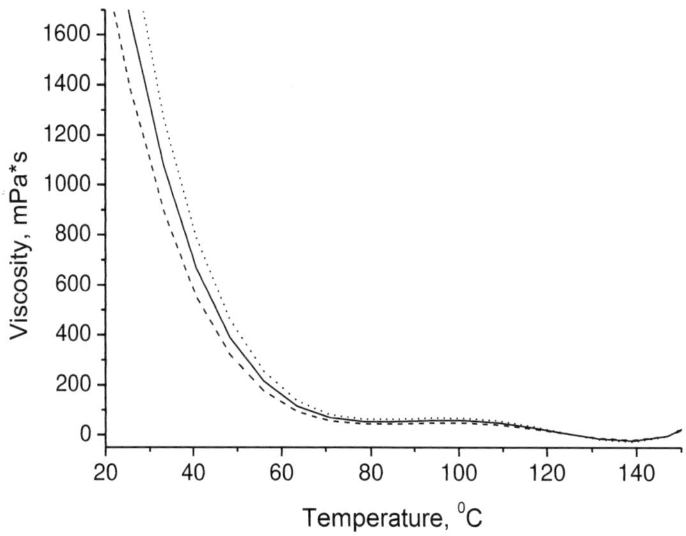

Figure 2. Effect of temperature and pressure on the viscosity of crude oil in the atmosphere N_2–O_2 (60:40): dashed line – 3 MPa; solid line – 9 MPa; dotted line – 15 MPa.

In order to estimate the intermolecular interaction energy of the studied crude oil, the values of the activation energy of flow E_{act}^{flow} were calculated. The activation energy of flow is usually determined by the Frenkel–Andrade equation

$$\eta = Ae^{\frac{E_{act}^{flow}}{RT}}$$

where η – viscosity; A – pre-exponential factor considered independent of temperature; R – gas constant; T – absolute temperature, K.

The results of calculation are shown in Table 5.

Table 5. Activation energy of flow of crude oil.

Sample		E_{act}^{flow}, kJ/mol
Crude oil		20.8
medium N_2–O_2 (60:40)	3 MPa	21.0

	9 MPa	21.0
	15 MPa	21.0
medium N_2–O_2 (80:20)	3 MPa	21.0
	9 MPa	21.0
	15 MPa	21.0
medium N_2	3 MPa	21.0
	9 MPa	21.1
	15 MPa	21.0

The activation energy of flow was found to be practically the same in different gas mixtures.

4. Conclusions

The effect of pressure, temperature and gas nature on the viscosity properties of heavy crude oil from the one heavy oil field of Tatarstan has been studied. The influence of temperature from 25 to 125 °C and pressure from 0.1 to 10 MPa on the viscosity of crude oil has been established. It was found that an increase in temperature by 125 °C leads to a decrease in viscosity by 8 times. While an increase in pressure up to 10 MPa leads to an increase in viscosity by 38% at low temperatures. The influence of the nature of gas (nitrogen, argon and air) on the viscosity of crude oil at different temperatures and pressures was analyzed. On the other hand, the influence of the composition of gas mixtures in a wide range of temperatures and pressures on the viscosity of heavy crude oil has been studied. The activation energies of flow state for two gas mixtures are calculated. It is shown that the gas phase at different ratios of components does not significantly affect the rheological parameters of crude oil.

Funding

The reported study was funded by RFBR and Tatarstan Republic according to the research projects № 18-45-160012. The reported study was funded by RFBR according to the research project № 18-35-00710. This work has been partly performed according to the Russian Government Program of Competitive Growth of Kazan Federal University.

References

[1] Head I M, Jones D M, Larter S R 2003 Biological activity in the deep subsurface and the origin of heavy oil *Nature* **426** 344–52

[2] Speight J G 2006 Chemistry and Technology of Petroleum. Marcel DekkerInc: New York.

[3] Myer R F, Attanasi E D, Freeman P A 2007 Heavy oil and natural bitumen resources in geological basins of the world *U.S. Geological Survey Open-File Report* 2007 **1084** 1–36

[4] Nguyen P, Fadaei H, Sinton D 2014 Nanoparticle stablized CO2 in water foam for mobility control in enhanced oil recovery via microfluidic *method Society of Petroleum Engineers - SPE Heavy Oil Conference Canada* 1928–36

[5] Talebian S H, Masoudi R, Tan I M, Zitha P L 2014 Foam assisted CO2-EOR: A review of concept, challenges, and future *prospects J. Petrol Sci. Eng.* **120** 202–15

[6] Hu R, Crawshaw J P, Trusler J P, Boek E S 2015 Rheology of diluted heavy crude oil saturated with carbon dioxide *Energy & Fuels* **29** 2785–9

[7] Muraza O, Galadima A 2015 Aquathermolysis of heavy oil: A review and perspective on catalyst development *Fuel* **157** 219–31

[8] Fletcher P D, Savory L D, Woods F, Clarke A, Howe A M 2015 Model study of enhanced oil recovery by flooding with aqueous surfactant solution and comparison with theory *Langmuir* **31** 3076–85

[9] Chen H X, Tang H M, Wu X Y, Liu Y G, Bai J H, Zhao F 2016 Synthesis, characterization, and property evaluation of a hydrophobically modified polyacrylamide as enhanced oil recovery chemical *J. Dispersion Sci. Technol.* **37** 486–95

[10] Akhmadiyarov A A, Rakipov I T, Khachatrian A A, Petrov A A, Sitnov S A, Gerasimov A V, Osin Yu N, Varfolomeev M A Thermocatalytic upgrading of heavy oil by iron oxides nanoparticles synthesized by oil-soluble precursors *J. Petr. Sci. Eng.* **169** 200–4

[11] Harris K R 2009 Temperature and pressure dependence of the viscosities of 2-ethylhexyl benzoate, bis(2-ethylhexyl) phthalate, 2,6,10,15,19,23-hexamethyltetracosane (squalane), and diisodecyl phthalate *J. Chem. Eng. Data* **54** 2729–38

[12] Pensado A S, Comuñas M J, Lugo L, Fernández J 2006 High-pressure characterization of dynamic viscosity and derived properties for squalane and two pentaerythritol ester lubricants: Pentaerythritol tetra-2-ethylhexanoate and pentaerythritol tetranonanoate *Ind. Eng. Chem. Res.* **45** 2394–404

[13] Bair S 2006 Reference liquids for quantitative elastohydrodynamics: Selection and rheological characterization *Tribol. Lett.* **22** 197–206

[14] Krahn U G, Luft G 1994 Viscosity of several liquid hydrocarbons in the temperature range 298–453 K at pressures up to 200 MPa *J. Chem. Eng. Data* **39** 670–2

[15] Bair S, McCabe C, Cummings P T 2002 Calculation of viscous EHL traction for squalane using molecular simulation and rheometry *Tribol. Lett.* **13** 251–4

Construction of geological model of a shallow super-viscous oil deposit based on core analyses and geophysical monitoring data

A N Delev[1], D K Nurgaliev[1], S V Sidorov[1], R R Akhmadullin[2], N A Aslyamov[2]

[1] Institute of Geology and Petroleum Technologies, Kazan Federal University, Kazan, Russia
[2] PJSC Tatneft, Almetyevsk, Russia

E-mail: AlNDelev@163.com

Abstract. Deficiency of fossil minerals, limited reserves of conventional hydrocarbon raw materials indicates the need to involve in the fuel and energy complex of other sources of hydrocarbons. Fields of super-viscous oil are one of the sources. Development of super-viscous oil deposits opens new challenges in the field of building geological models associated with the shallow occurrence of these deposits. This article is devoted to the study of sediments of the Permian sedimentary complex containing super-viscous oil deposits. Study object is Sheshmian sandstone pack of the Ufimian stage. The pack is composed of sands and sandstones with different degree of carbonate cementation with small interlayers of siltstones. Cementation is the main factor which affect reservoir quality. In the process of analysis and preparation of the initial data and, subsequently, the geological model construction, the main features of the super-viscous oil deposits were revealed.

1. Introduction, general information about deposit

In this paper, the Lower Karmalsky uplift of the Cheremshansky oil field, is considered. The uplift controls super-viscous oil deposit and territorially confined to the western Zakamye. In tectonic terms, it is located on the western slope of the South Tatar arch, which is a sloping monocline, step-wise submerging in the direction of the Melekess depression.

The Lower Karmalsky positive structural form is an integral part in the ridge of elongated sand bodies of the northwest strike. Sand bodies are paleobars; they have the form of local sedimentary domes of the brachiantical type. The chains of the uplifts are separated from each other by sections of reduced thickness of the sand pack [1-2].

The Sheshmian sandstone pack of the Ufimian stage is the main stratum containing hydrocarbons. The pack is composed of sands and sandstones with different degree of carbonate cementation with small interlayers of siltstones. Cementation is the main factor which affect reservoir quality [3]. The thickness of the pack varies from 3 to 10 meters on the wings and up to 45 meters in the arch of the structure.

Deposits of Sheshmian sandstone pack of the Ufimian stage are overburden by the lower pack of the Baytuganskian horizon, composed of clays, dark gray, with a bluish tinge and is a reliable impermeable layer (reference horizon – "lingula clays"). The sections with the maximum thickness of the Sheshmian sandstones correspond to the minimum thickness of the "lingula clays" pack.

The upper pack of the Baytuganskian horizon, lying over the "lingula clays", is composed of limestones, bluish-gray, dark gray, steel-gray, porous, cavernous, fractured, with a mass of brachiopod and spirifer residues, in the bottom with frequent remains of bryozoans with inclusions of pyrite (reference horizon – "Medium-spirifer limestone"). Its thickness is from 2 to 4 meters [4-5].

2. **Correlating core analysis and well logging**

The accuracy of geological modeling depends on the correct determination of the depth of rocks samples, therefore core analysis data should be correlated with well logging curves.

The study of well sections and refinement of core sampling depths begin with the identification of reference horizons - spatially and lithologically continuous rock strata that have a clear and explicit well logging characteristic – and an accurate correlation of core, collected from these horizons. The reference horizons on the studied deposit are (from top to bottom) "Medium-spirifer limestone", "Lingula clay" and a shaly sand pack of the Ufimian stage, underburden on the main productive strata.

Correlating core analysis and well logging was done by comparing the values of porosity, saturation, permeability, density, carbonate content and other characteristics determined on core samples with well logging curves. As a result, separate continuous formations were identified.

For correlating core analysis and well logging a well logging complex was used, including: GR, NGL, PZ, SP, density gamma-gamma logging. In addition, Caliper, LLD, etc. were used (Fig. 1).

3. **Building of the structural framework**

Building of the structural framework of the three-dimensional model is the next step after the completion of the detailed correlation.

A peculiarity in structural constructions of a shallow deposit is the specificity of processing and interpretation of seismic data in the upper part of the section, which does not allow obtaining sufficiently accurate data about structures in the inter-well space. This is due to the low speed zone and, consequently, the poor resolving capacity of seismic exploration in the upper part of the section (~ 300 m). Exploration wells in the investigated deposit have been drilled with a close spacing, and therefore structural constructions based on well data are more accurate.

When constructing a grid model, a smaller grid should be defined than in traditional models in order to characterize the heterogeneity of a closely drilled reservoir and also to describe the process of steam-assisted gravity drainage (SAGD) in reservoir modeling, if the SAGD method is planned to be used during the deposit development. Rotation of the grid should be chosen from the calculation of its orientation, not only along the strike of the geological structures, but also in the cross of the main mass of the horizontal boreholes.

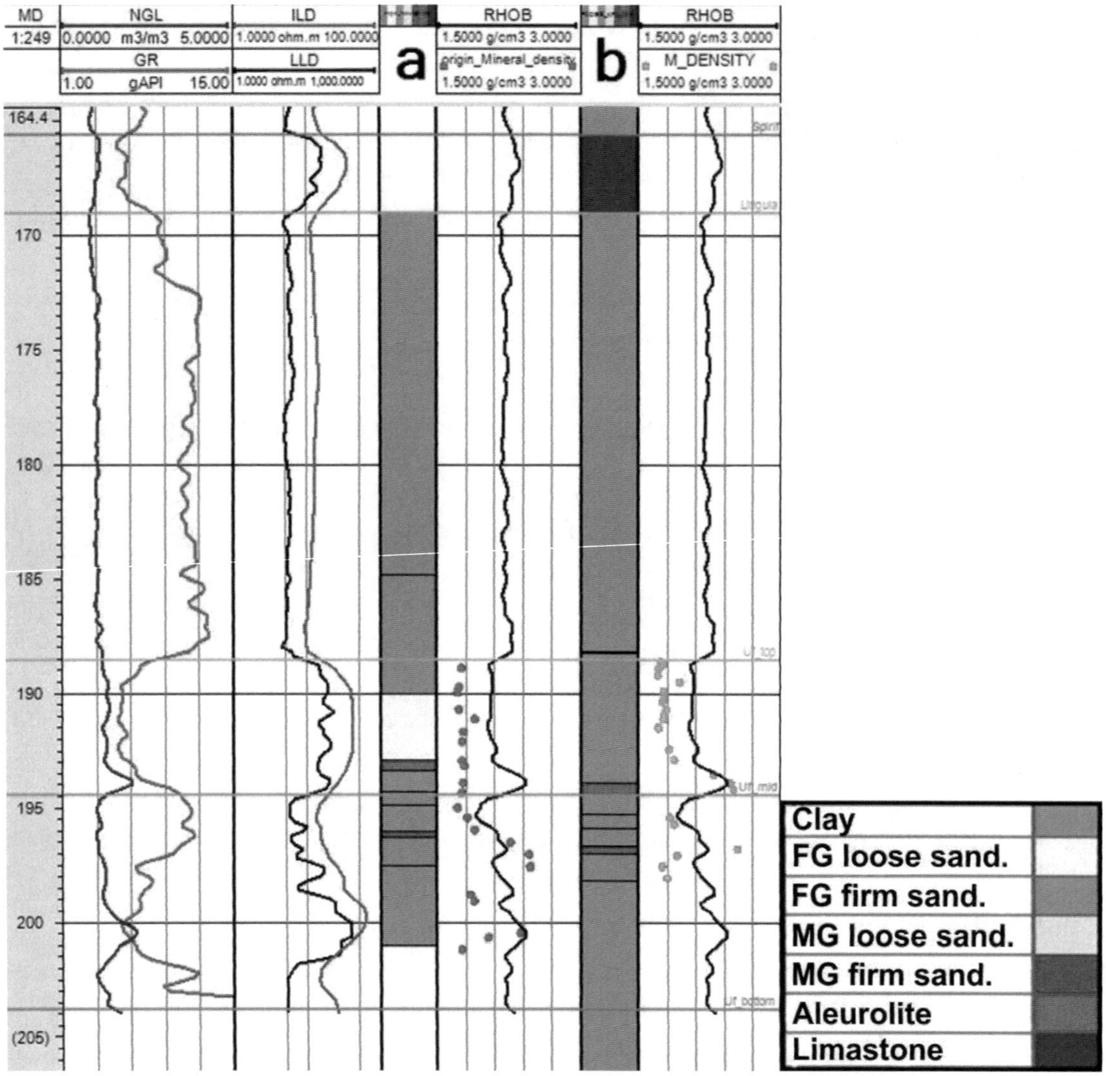

Fig. 1 An example of correlating core analysis and well logging:
a - initial data, b – correlated data, FG - fine grained, MG - medium grained

4. **Geometrization of deposit, sandstone lithotypes classification based on carbonate content and density properties**

While the geometrization of deposits the main difference from the work with traditional model is presence of a large volume of core material, selected from prospecting, exploration, structural wells.

Based on the correlating core analysis and well logging the following sedimentation concept was adopted for further building of the geological model:

- the upper sandstone (productive) part of the deposit has undergone erosion, "Lingula clays" pack overburden this part with unconformity;
- layering of the upper sandstone pack was carried out by follow base method;

- shaly sand pack, lying at the base of the upper sandstone pack, is rather sustained in thickness, and, therefore for this zone were choose proportional layering between top and bottom.

After the layering process of 3D grid, the reservoir/non-reservoir parameter was calculated from normalized GR value. It should be noted that the upper (sandstone) pack of the Ufa stage throughout the section is represented by a reservoir of different quality (depending on carbonate content and density), and the lower (shaly sand) pack is largely represented by a non-reservoir with interlayers of reservoirs.

In the obtained reservoir volume on the basis of carbonate content property (according to core studies) and density property (according to core studies and well logging) lithological types were divided. In total, 2 lithologic types of sandstones were distinguished - loose sandstone and firm sandstone with carbonate cementation. (Fig. 2).

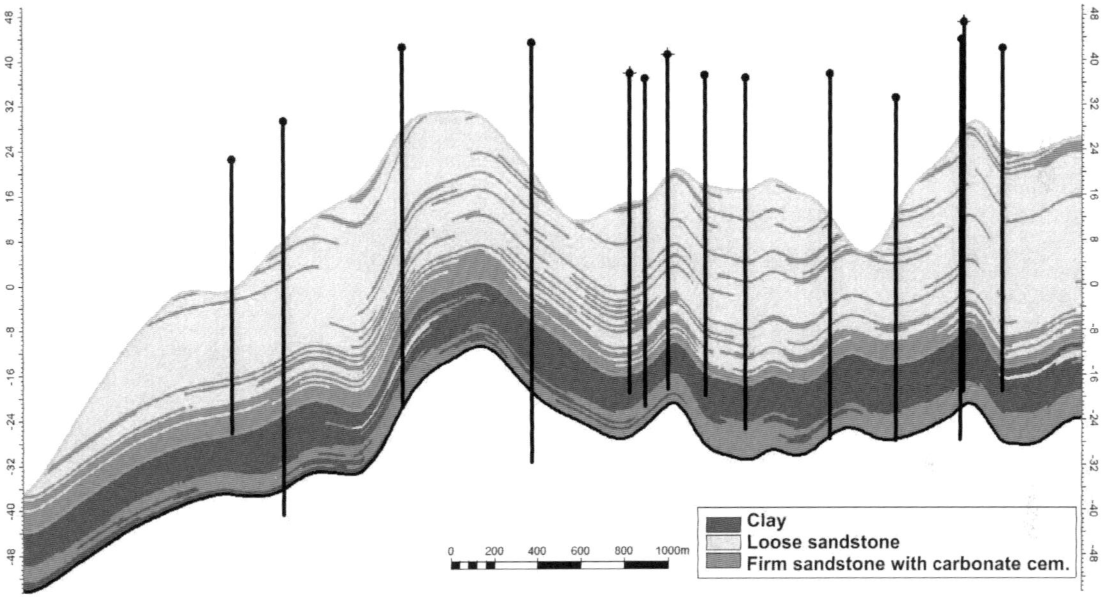

Fig. 2 Lithological types of sandstones slice.

Further, to determine the reservoir saturation behavior, the strata with mass bitumen saturation above and below 4,5% were distinguished (according to oil producing company condition criteria). For this, cells with mass bitumen saturation above conditioned were assigned the value 1, cells with saturation below conditional - 0, after which this parameter was interpolated in the reservoir volume using the Kriging method. The final discrete property of the reservoir saturation behavior (Fig. 3) was calculated using a cutoff of 0.5 (more than 0.5 — an interval with a bitumen saturation higher than the conditioned one).

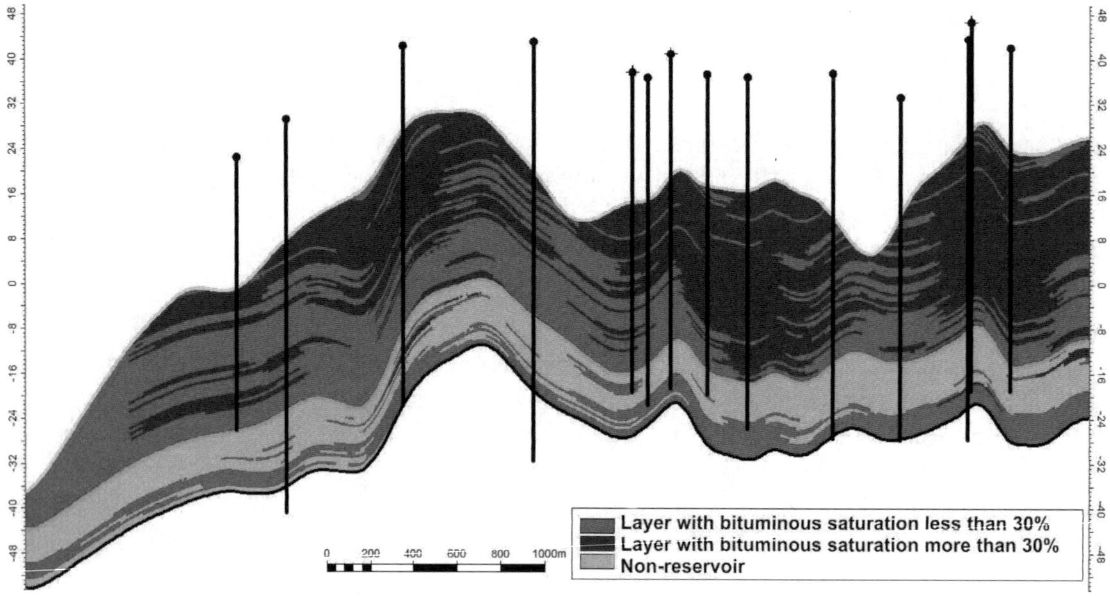

Fig. 3 Reservoir saturation behavior slice.

5. Building a petrophysical model

As a basis for the distribution of porosity in the model core analysis data correlated with well logging were used.

The methods of porosity distribution do not differ from the work with the traditional model. Porosity distribution in the reservoirs was carried out using the Gaussian random function simulation method for each lithological type of sandstone separately.

The permeability property also was distributed on a basis of core analysis data (horizontal absolute permeability). The porosity property was used as a trend in the permeability distribution, since the number of porosity measurements is greater than the permeability measurements.

6. Building a saturation model

For the distribution of bitumen saturation, the dependence of volumetric bitumen saturation on mass is obtained, since the measurements of mass bitumen saturation are more accurate for super-viscous oil. After recalculation, the obtained values of volumetric bitumen saturation were distributed separately in the zones with bitumen saturation above and below 30% (according to oil producing company standards).

7. Geophysical monitoring, optical fiber thermograms analysis

When developing a shallow super-viscous oil deposit by the SAGD method the temperature is periodically measured [6-7]. Using of thermograms in complex with well logging and production data allows us to determine the correct position of the horizontal wellbores in a 3D model and to clarify the surface of the oil water contact which is not horizontal in a shallow super-viscous oil deposits.

Fig. 4 shows an example of a thermogram. The lowest temperature in the toe of the production well is due to the presence of a large amount of produced water. It is worth noting that such an analysis should be carried out in conjunction with the development analysis, if the salinity of the produced water is significantly lower than the reservoir water, then the assumption of the proximity of the conditional oil water contact is not correct - more likely steam has break through from the injection well.

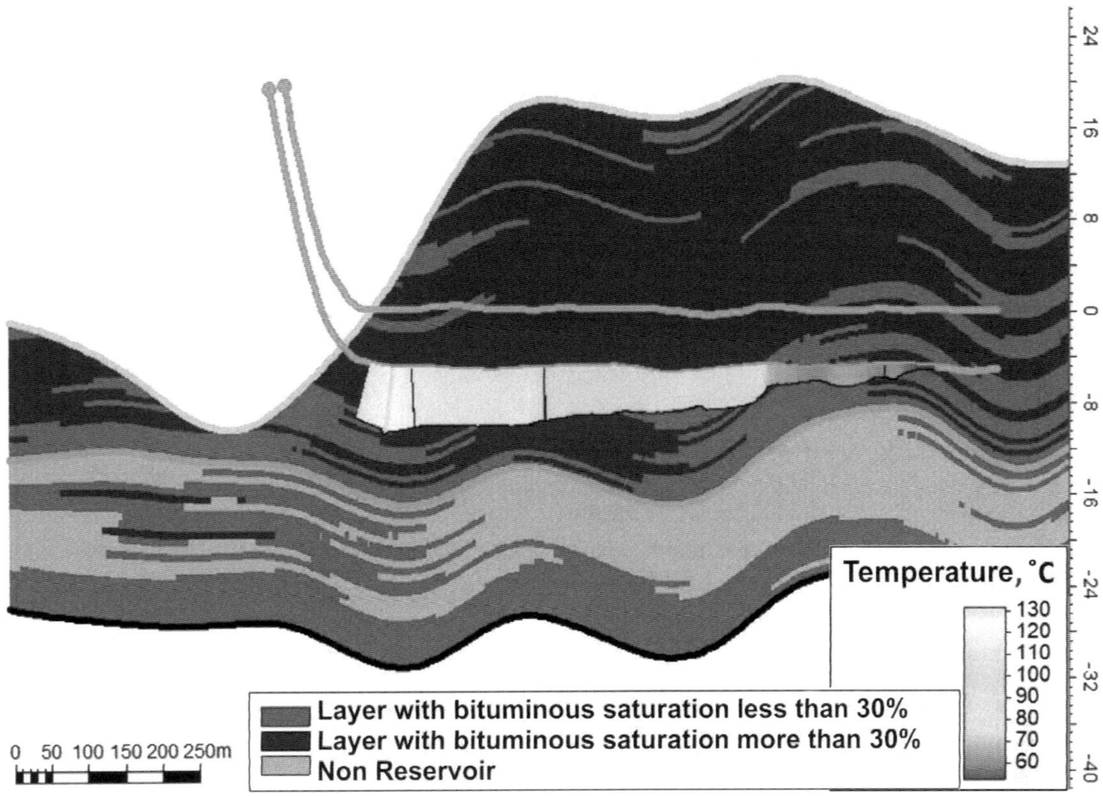

Fig. 4 An example of thermogram analysis.

8. Conclusions

On the basis of correlated core analysis data lithological types of sandstones and the reservoir saturation behavior models was constructed, the following features of the construction of geological models of a shallow super-viscous oil deposit was distinguished:

- The presence of a large volume of core material can serve as a source of data allowing to build detailed geological model core analysis data should be correlated with well logging data.
- The complexity, and often the impossibility of processing and interpreting seismic data in the upper part of the section do not allow us to obtain sufficiently accurate data on structures in the inter-well space. In the case of close spacing of drilled wells, structural constructions for well data will be more accurate.

- When building a bitumen saturation model, it is necessary to allocate conditioned and substandard zones of saturation (according to oil producing company standards), since there may be layers with high porosity and low saturation within the deposit. Measurements of mass bitumen saturation, as a rule, are presented in a larger volume and more reliable than measurements of volumetric bitumen saturation.
- Geophysical monitoring based on analysis of thermograms in production wells allows to clarify the position of the conditional OWC.

Acknowledgment. The work was supported by the Ministry of Education and Science of the Russian Federation (project No. 02.G25.31.0170) and by the subsidy allocated to Kazan Federal University as part of the state program for increasing its competitiveness among the world's leading centers of science and education.

References

[1] Uspenskii B V, Valeeva I F 2008 Geologiya mestorozhdenii prirodnykh bitumov Respubliki Tatarstan [Geology of natural bitumen fields of the Republic of Tatarstan] *Kazan: PF Gart* 347 p (In Russian)

[2] Mullakaev A I, Delev A N, Usmanov S A, Sudakov, V A, Khasanov R R 2018 Tectonic causes of uneven cementation zones distribution in the bituminous sandstones productive part of the Sheshminsky horizon of the South Tatar arch *Neftyanoe Khozyaystvo - Oil Industry* **No.2** pp. 23-25

[3] Mullakaev A I, Khasanov R R, Galiullin B M 2017 Mineralogy of sandstones and localization of oil matter in productive horizons of high-viscosity oil in permian deposits of the Volga-Ural region (Russia) *17th International Multidisciplinary Scientific GeoConference (SGEM2017) Proceedings* **Vol.17 Issue 11** pp. 353-358

[4] Novikova S P, Nurgaliev D K, Sudakov V A, Delev A N, Nurgalieva N G, Lyabipov M R, Akhmadullin R R 2017 The Main Features of the Geological Modeling Process of a Shallow Deposit of Super-Viscous Oil in Aspect of Development Strategy Planning with the Use of Steam-Assisted Gravity Drainage Method *Georesursy = Georesources* **V.19 No.4 Part 1** pp. 331-340

[5] Novikova S P, Usmanov S A, Delev A N, Nurgalieva N G, Lyabipov M R 2017 Geological modeling of super-viscous oil reservoir of the Permian sedimentary complex on the Western slope of the South Tatar arch. Some features of the geological model building of the super-viscous oil shallow deposits *SGEM2017 Vienna GREEN Conference Proceedings* **Vol.17 Issue 15** pp. 219-226

[6] Sudakov V A, Amerkhanov M I, Nurgaliev D K, Khasanov D I, Stepanov A V, Kosarev V E, Usmanov S A 2016 Geophysical monitoring of the steam chamber during the development of super-viscous oil deposits on the example of the PJSC «Tatneft» field *Innovatsii v razvedke i razrabotke neftyanykh i gazovykh mestorozhdenii: materialy mezhd. nauch-prakt. konferentsii [Innovations in the exploration and development of oil and gas fields: Proc. Int. Scientific-Practical Conference]* **V.2.** pp. 220-222 (In Russian)

[7] Sudakov V, Nurgaliev D, Khasanov D, Stepanov A, Khamidullina G, Kosarev V, Galukhin A, Usmanov S, Zaripov A, Amerkhanov M 2016 Technology of Integrated Monitoring of Steam Chamber Evolution During the Oil Production from the Shallow Deposits of Super-Viscous Oil *SPE Russian Petroleum Technology Conference and Exhibition* SPE-182000-MS

Regulating temperature of oil saturation with paraffins to avoid asphaltene, resin and paraffin substances deposition during oil production

D Z Valiev[1], R A Kemalov[1] and A F Kemalov[1]

[1]Institute of Geology and Petroleum Technologies, Kazan Federal University, Kazan, RUSSIA

E-mail: valievdz@gmail.com

Abstract. In order to regulate temperature of oil saturation with paraffins to avoid asphaltene, resin and paraffin substances (ARP-S) deposition using thermal method as well as chemical reagents and additives implementation was considered. A particular attention is paid to the chemical composition of the produced hydrocarbons, as well as to the effect of individual high-molecular components of hydrocarbons during the extraction and further transportation at low temperatures and with a temperature gradient difference. The paper presents the influence of hydrocarbon composition on the development mechanism, the composition and the properties of ARP-S. The methods for the removal of deposits associated with the use of various additives and reagents, as well as the methods for the prevention of these sediments development are considered. A brief description of the main classes of chemicals used for the solution of ARP-S prevention and removal problem is given. It was shown that in order to select the most effective ways of organic substance sediment prevention and removal from the physical-chemical point of view, it is necessary to obtain an idea about the composition and the properties of the initial oil and the developed deposits.

1. Introduction

In the process of oil producing well exploitation with some temperature and pressure decrease accompanied by oil degassing, a sharp decrease of paraffin, asphaltene and resinous substance solubility in it takes place which leads to their precipitation on the surface of the mining equipment [1]. During oil pumping and storing the internal walls of tanks, pipes can also experience the development of ARP-S. Therefore, it is necessary to clean the technological equipment, pipes and tanks from time to time [2].

Nowadays, the problem of asphalt-resinous and paraffin deposits is very significant, because these deposits are contained in the oil of almost all oil-producing regions of Russian Federation. The chemical composition depends on the properties of an extracted oil, the thermal and hydrodynamic conditions of a reservoir, the geological conditions and the method of a field development and its operation [3].

During oil production, they encounter one of the problems causing the complication in a well robot, oilfield equipment and further transportation through pipelines. These deposits are accumulated in the flowing parts of an oilfield equipment and on the inner surface of pipes, which leads to the decrease of oil production and transportation system productivity.

ARP-S increase the wear of equipment, energy costs and pressure in a system. Nowadays, this problem is a very urgent task for the production and transportation of hydrocarbons. The following factors influence the development of ARP-S:

• Pressure reduction,
• Intensive gas release,
• Temperature drop,
• The composition of hydrocarbons,
• Temperature gradient change.

The main factors for reducing the temperature of the fluid: earth geotherma in the area of the well; thermal resistance of the well structure; the dynamic level of oil in the annular space.

2. Methodology

The main method of ARP-S removal is the prevention of deposit development on an oilfield equipment, but if the deposits developed, then the secondary method is the removal of these sediments [4].

ARP-S deposits are formed in a bottom hole zone of a well, on the walls of a wellbore and oilfield equipment. The following methods of prevention and disposal of these sediments are proposed:

• The injection of reagents, acids, coextractants, suppressing the development of paraffin bunches during the extraction and the transportation of hydrocarbons into a system,

• The covering of the inside of oil field equipment with a layer reducing the adhesion of ARP-S to the pipe walls,

• The injection of biodegradable composition into an extracting medium, which does not allow the development of ARP-S,

• The installation of electric heaters at a well bottom designed to maintain the temperature above the melting point of paraffin in a well,

• The increase of asphalt-resinous compound percentage,

• The maintaining of constant pressure in the system of oil production equipment at the well bottom,

• The use of exothermic reactions, and thus the increase of produced hydrocarbon temperatures at the bottom of a well and the prevention of asphalt-resinous and paraffin deposit development,

• Steam heating. The use of steam generated by steam-mobile units like steam field installations PPUA-1200/100 with a temperature of up to 310 °C and pressure up to 10 MPa is only effective up to a depth of 300-400 m for borehole countering purposes. Washing with hot oil is energy-intensive; the inter-treatment interval is usually 90 days.

It should be noted that during the selection of one or more methods for ARP-S prevention or removal, the component composition and the properties of hydrocarbons, the percentage ratio of asphaltenes, resins and paraffin should be taken into account (Table 1.2).

Table 1. Physical and chemical properties and fractional composition of degassed oil

Name		The number of studied		Change range	Average value
		wells	sample		
1		2	3	4	5
Dynamic viscosity, mPa s at 20 °C at 50 °C		 50 54	 73 79	 11.0 – 67.6 6.1 – 26.7	 31.1 11.3
Freezing point, ^{0}C		48	72	(-35.0) – 6.0	-13.9
Melting point of paraffin, ^{0}C		52	77	49.0 – 64.0	54.9
Initial boiling point, ^{0}C		54	79	60.0 – 145.0	101.1
Wt., %	sulphur	54	79	0.70 – 2.10	1.29
	silica gel resins	54	79	3.70 – 19.20	2.94
	asphaltenes	54	79	0.30 – 17.60	8.33
	paraffins	54	79	1.20 – 3.60	2.42
	water	39	56	0.00 – 11.00	0.92
Volumetric output of fractions, %	up to 100 ^{0}C	32	37	0.0 – 5.0	0.6
	up to 150 ^{0}C	54	79	1.0 – 14.0	7.6
	up to 200 ^{0}C	54	79	7.0 – 24.0	16.4
	up to 250 ^{0}C	54	79	18.0 – 34.0	25.9
	up to 300 ^{0}C	54	79	29.0 – 51.0	38.1
	up to 350 ^{0}C	2	2	54.0 – 80.0	67.0
Oil classification		II $T_2П_2$			

Table 2. Component composition of oil gas, degassed and reservoir oil

Name	With a single degassing of reservoir oil under standard conditions				At the differential degassing of reservoir oil under operating conditions				Base oil	
	Extracted gas		oil		Extracted gas		oil			
	% wt.	% moles	% wt.	% moles	% wt.	% moles	% wt.	% moles	% wt.	% moles
Hydrogen sulfide	отс.	отс.	-	-	отс.	отс.	-	-	отс.	отс.
Carbon dioxide	2,23	1,41	-	-	2,77	1,56	-	-	0,16	0,54
Nitrogen + rare	0,86	0,85	-	-	1,01	0,89	-	-	0,06	0,31
Including helium	0	0	-	-	0	0	-	-	0	0
methane	35,65	61,94	0,02	0,25	44,02	68,05	0	0,05	2,45	23,33
ethane	11,07	10,26	0,04	0,30	13,24	10,91	0,06	0,46	0,79	4,03
propane	21,35	13,49	0,31	1,59	20,78	11,68	0,62	3,10	1,74	6,04
isobutane	4,11	1,97	0,17	0,66	3,14	1,34	0,28	1,06	0,44	1,15

N butane	11,29	5,41	0,67	2,63	7,81	3,33	1,02	3,85	1,39	3,67
isopentane	3,06	1,18	0,48	1,53	1,72	0,59	0,60	1,82	0,66	1,40
N-pentane	3,81	1,47	0,80	2,54	2,10	0,72	0,95	2,88	1,01	2,14
hexanes	4,67	1,51	3,06	8,09	2,40	0,69	3,22	8,21	3,17	5,63
heptanes	1,37	0,38	2,42	5,51	0,73	0,18	2,45	5,37	2,35	3,59
octanes	0,53	0,13	2,65	5,29	0,28	0,06	2,64	5,07	2,51	3,36
Residue (C9+highest)	0	0	89,38	71,61	0	0	88,16	68,13	83,27	44,81
Molecular mass	27,86		227,8		24,79		219,7		153,0	
Molecular weight of the residue	-		284,4		-		284,3		279,8	
Density - gas, kg/m3	1,1582		-		1,0304		-		-	
- gas relative (by air), share of un.	0,9611		-		0,8551		-		-	
- oil, kg/m3	-		874,0		-		869,2		788,0	

Proceeding from the mentioned above, it should be noted that the rough surface of pipe walls, contributing to the release of gas from the oil and its cooling. Thus, the solubility of ARP-S in oil deteriorates and the rate of deposit formation increases. However, the increase of a gas-liquid mixture flow rate can somewhat slow the growth of paraffin deposits. It was [5,6] found that the solubility of paraffin decreases in heavy oils. Therefore, the intensity of paraffin deposition in these oils is increased and implies the use of it. Thus the concentration of paraffin compounds in oil. The higher this value, the more intense the deposition of paraffin on the walls of these pipes [7].

3. Results and Discussions

ARP-S development factors. The following results were obtained: the thickness of ARP-S deposits depends on the depth.

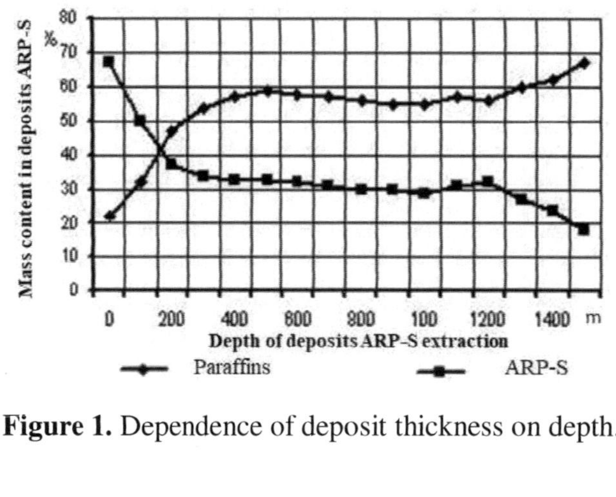

Figure 1. Dependence of deposit thickness on depth.

Figure 2. Deposits of asphaltene, resin and paraffin substances.

Analyzing Fig. 1 and Fig. 2 it should be noted that the development of ARP-S on the borehole wall is associated with a temperature drop, thereby making it difficult to extract hydrocarbons from the formation and no dynamics growth occur in oil extraction industry.

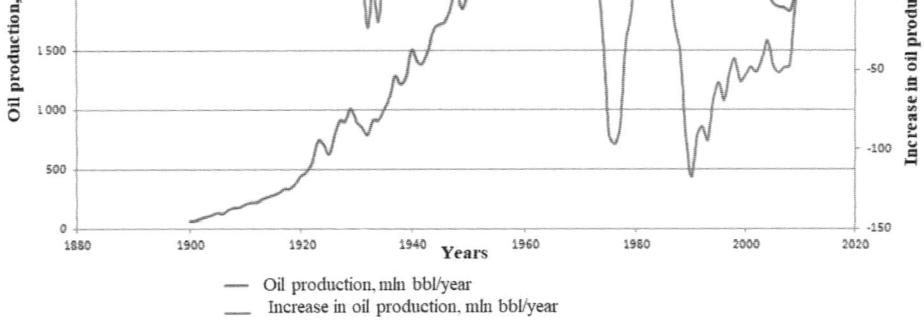

Figure 3. Graph of oil production growth dependence on produced oil.

Analyzing Fig. 3, it should be noted that the increase of oil production increased with the discovery of new fields and their putting into operation, but the volume of produced oil has not changed, which is very surprising nowadays [8]. ARP-S makes it very difficult to extract the production of oil and its further transportation.

The most effective method of heat treatment is to heat the well bore with an electric heater (flexible skin-heater).

The main objective of such systems is to maintain the temperature of the moving fluid above the wax deposition temperature (Fig. 4).

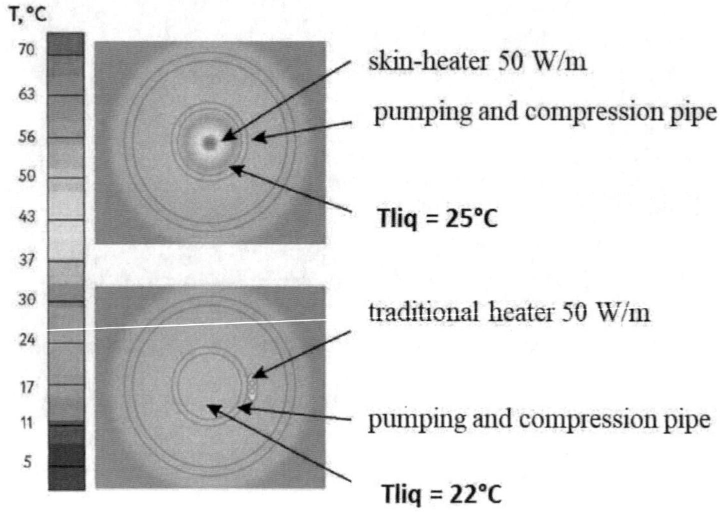

Figure 4. Comparison of fluid heating efficiency using flexible skin-heater and a resistive cable.

The optimal solution from the point of view of energy consumption is when the heating system works only in the zone where the fluid temperature in normal conditions drops below the wax deposition temperature and the heating element has a variable heat generation along the depth of the well. Moreover, the power of such a heater should vary smoothly in a wide range: the linear power of its lower part will be close to zero, whereas in the near-surface part the power can reach 70 W/m.

The heater maintains operability: with external pressure up to 150 atmospheres and the temperature of the oil and gas medium up to 70 ° C.

4. Conclusions

With any production method, an internal section of a tubing is narrowed gradually due to the development of asphalt-resinous and paraffin deposits, which leads to the decrease of a well productivity and oil production volume, and to the failure of underground equipment.

The problem of ARP-S development becomes more serious due to the transition of many ields to the late stage of development. Solving this problem one should be guided by common approaches, namely, to find out the causes of these deposits. The process of ARP-S development is determined by many factors, among which there are the operating conditions of the process equipment during extraction, transportation and the storage of hydrocarbons, the properties of the hydrocarbons themselves. High-molecular components of oil, namely, their composition, structure and percentage ratio have a particular importance in a number of factors that determine the propensity of hydrocarbons to the development of asphalt-resinous and paraffinic deposits. Knowing the properties and the composition of hydrocarbons, their thermal and hydrodynamic conditions in a productive reservoir and selecting the methods of fighting with ARP-S properly, it is possible to avoid the wear of oilfield equipment and pipelines. At one time, this is an economically advantageous condition for oil and gas industry, which in turn contributes to the reduction of overhaul number and a reduced downtime of wells, allows to extend the life of equipment, provides rhythmic work of industry and reduces the cost of oil production.

5. Summary

Adsorption processes are based on the action of acids, reagents and inhibitors of paraffin deposits. When they use chemical methods based on dosing in obtained products of chemical compounds that reduce, and sometimes completely remove or prevent the formation of asphalt-resinous and paraffin deposits.

The most effective method of heat treatment is to heat the well bore with an electric heater (flexible skin heater).

While maintaining a constant pressure in a well, asphalt-resinous and paraffin deposits will not be deposited in oilfield equipment.

Acknowledgements

This work was funded by the subsidy allocated to Kazan Federal University for the state assignment in the sphere of scientific activities (10.7636.2017/7.8).

References

[1] Glushchenko V et al 2008 *Oilfield chemistry: Complications in a reservoir well system - UPPN / MAX Press* p 328
[2] Kitov V, Kemalov R 2017 Methods of asphalt-resinous and paraffinic deposit control *Revista Publicando,* vol 4 no 13 (2) pp 92-99
[3] Ibragimov N et al 2003 *Complications in oil production* (Ufa) p 302
[4] Fink J 2003 *Oil field chemical* (A. – T.: GPP) p 495
[5] Kemalov A, Kemalov R and Valiev D 2013 Study of the structure of complex structural units of heavy oil from Zyuzeevskaya field by NMR relaxometry and rheological studies *Oil Industry* no 2 pp 63-65
[6] Acevedo S, Castro A et al 2007 Relations between asfaltene structure and their physical and chemical properties: the rosary-type structure. *Energy & Fuels* vol 21 pp 2165 – 2175
[7] Kemalov A, Kemalov R, Abdrafikova I et al 2015 Quantum energy calculations of technological parameters of electromagnetic impact on heavy hydrocarbons *Modern Applied Science* vol 9 Issue 4 pp 312-318
[8] Antonova E et al. 2003 Fundamentals of oil and gas business (Textbook for universities - M6 LLC Nedra) p 307

IOP Conf. Series: Earth and Environmental Science **282** (2019) 012024 doi:10.1088/1755-1315/282/1/012024

Thermal properties of the bituminous sandy rocks of the Volga-Ural oil and gas province.

A I Mullakaev[1] and R R Khasanov[1]

[1] Institute of Geology and Petroleum Technologies, Kazan Federal University, Kazan, Russia

E-mail: almazmullakaev@gmail.com, rinat.khassanov@kpfu.ru

Abstract

Natural accumulations of bitumen in the Volga-Ural oil and gas province are unconventional sources of hydrocarbons. They are mainly concentrated in the Republic of Tatarstan and are confined to psammitic (sandy) rocks of lower Permian age (Sheshmian Horizon of Ufimian Stage). Sands and sandstones belong to the greywacke group, so rock-forming minerals are represented by quartz, feldspars and effusive fragments. The internal structure of the bitumen-containing reservoir is characterized by an extreme degree of lithological and mineralogical heterogeneity, which is the result of the imposition of a whole complex of processes. Since the beginning of the development of bitumen deposits by thermal methods, the issues of studying the internal heterogeneity and its influence on the production process become particularly acute. Internal reservoir heterogeneity may prevent uniform heating and the formation of a full-fledged steam chamber. Sands and sandstones of the productive part are divided into three rock types (bituminous sands, bituminous sandstones and residual bituminous sandstones). Thermal studies using methods of thermogravimetry (TG) and differential scanning calorimetry (DSC) showed differences in rock types in terms of the decomposition of mineral and organic components, as well as the removal of gaseous components. As a result of thermal studies, it was found that not only loose, bitumen-saturated sands, but also denser, cemented rock types can serve as a reservoir for hydrocarbons. This is due to the conservation of hydrocarbon fractions in the intergranular space during the formation of carbonate cement.

1. Introduction.

The development of oil fields in the Volga-Ural oil and gas province has been underway for more than 60 years. For the most part they are exhausted and in the final stages of development. The increase of the raw material base occurs due to the intensification of the involvement in the operation of new sources of hydrocarbon raw materials, which include accumulations of natural bitumen in the territory of the Republic of Tatarstan. They are confined to sandy rocks of the Permian age (Sheshmian Horizon of the Ufimian Stage) [1]. The bituminous psammitic deposits of the Sheshmian Horizon occur in the east of the East European Platform within the Volga-Ural oil and gas province. They are tectonically confined to the western slope of the South Tatar arch, which is one of the largest positive structures in the region. The bodies of the tar sands form a system of linearly located elongated elevations, which are morphologically distinctly expressed on the surface of the sheshmzan Horizon.

IOP Conf. Series: Earth and Environmental Science **282** (2019) 012024 doi:10.1088/1755-1315/282/1/012024

Discussions about the nature of sand bodies have been going on for quite a long time, but there is no consensus on their nature. Various assumptions have been made: deltaic, coastal-sea-bar, alluvial, etc. [1, 2]. According to recent studies [3], sand deposits may have a complex genesis involving aeolian processes, which has led to their two-tier structure. The detrital material was originally carried by river water flows to the flat areas from the zone of the collapsing mountain system of the ancient Urals [4]. After that, the upper part of the sandy mass was above sea level and groundwater level and underwent aeolian processing, which led to the formation of linear dunes. These processes explain the two-tiered structure of the Horizon with the division into the lower sand-clay layer and the upper sand layer. The sandy layer is devoid of clay components, which was removed as a result of wind activity. Further, the studied strata was covered with a thick layer of clays and became an excellent reservoir for future penetration and accumulation of hydrocarbons (HC) [3].

Sands and sandstones belong to the greywacke group, rock-forming minerals are mainly represented by quartz, feldspars and effusive fragments [5]. Their mineral composition is determined by the composition of the feeding provinces, paleogeographic and facial features of sedimentation [4, 6]. The reservoir rocks in the productive Horizons of bitumen deposits are sands and sandstones of varying degrees of strength. The reason for the high porosity of sands and sandstones is that detrital grains loosely contact and form communicating pores [6]. The use of thermal technologies for the development of superviscous oil fields [7] dictates the need to control the circulation of high-temperature fluids [8]. In this regard, the requirements for studying the internal structure of bitumen-containing sediments are increasing. It is characterized by an extreme degree of lithological and mineralogical heterogeneity. The nature of the heterogeneity is complex and is the result of the imposition of a whole complex of sedimentological, postsedimentological and diagenetic processes [9]. Tectonic movements also make a major contribution to the development of internal heterogeneity [10]. The technology of steam assisted gravity drainage involves the creation of a steam chamber in the reservoir, within which a viscous oil-bituminous substance is mobilize and acquires flowable properties. Internal reservoir heterogeneity, which complicates the circulation of fluids, may interfere with the uniform heating and formation of the steam chamber. According to mineralogical and lithological features, various rock types are distinguished among the sands and sandstones of the productive part of the Sheshmian Horizon of the Ufimian Stage, which are characterized by different fluid permeability. The main purpose of this work is to clarify the role of various sandy differences during thermal effects on the reservoir.

2. Methodology

Studies of the composition, structural and textural features of sands and sandstones were carried out using petrographic and mineralogical methods. At the first stage, an optical-microscopic study of rocks in transparent thin sections was carried out. The samples were examined in transmitted light using a Carl Zeiss Axio Imager A2 polarization microscope. Mineralogical studies of the fine components of sands and sandstones were studied using an FEI XL-30 ESEM scanning electron microscope with an energy dispersive spectrometer. The method allows to identify fine-sized mineral phases and to establish their chemical composition. As a result of the studies, a typification of the studied sands and sandstones was carried out based on the degree of carbonate cementation and saturation with oil-bitumen fractions. At the second stage, the analysis of rock types of sands and sandstones was carried out using thermogravimetry (TG) and differential scanning calorimetry (DSC). Thermal analysis of sandy rocks and hydrocarbons containing them was carried out on a synchronous thermal analysis device with highly sensitive sensors - NETZSCH STA 449 F3 Jupiter. The analysis interval was set in the temperature range from 30 to 1000 °C with a heating step of 10 °C per minute. The analysis was performed in air and using powder preparations. Data processing and calculations were carried out using the software package NETZSCH. The choice of TG and DSC as a laboratory method of research is due to the fact that the initial typification of samples into three main rock types was made according to the degree of bitumen saturation and cementation. The TG and DSC methods were used to determine the content of carbonate cement and hydrocarbon substances, as well as its fractional composition in various sand and sandstone rock types [11].

3. Results and Discussions

The whole variety of Sheshmian sands and sandstones can be divided into three main rock types [12]. The allocation of rock types is based on two main macro signs: cementation and bituminous saturation. According to these signs, *bituminous sands, bituminous sandstones and residual bituminous sandstones* are distinguished. Below is their detailed description.

Bituminous sands are black loose sands, the color of which is due to the presence of a bituminous substance. The most bituminous intervals of the Horizon are characterized by the lowest consolidation of rocks. The mineral composition is represented by rock-forming grains of quartz, feldspar and effusive rock fragments. They differ from cemented differences by the almost complete absence of carbonate cement, which is found only in rare strata and is not the main cement for the rock-forming minerals of this rock type. The role of cement in them is carried out by oil-bituminous substance. Weakly consolidated fragments of bituminous sands easily collapsing with little pressure on them. This rock type can be called the most important among all since it is the main object of industrial development.

Bituminous sandstones are represented by quite strong black and dark brown sandstones. The mineral composition is represented by the same rock-forming detrital minerals, which are cemented by carbonate minerals. The cement is predominantly calcite with a small proportion of authigenic dolomite. Dolomitization of calcite occurred under the influence of incoming groundwater from the lower sulfate-carbonate Horizons. Cement mainly pore and contact. Bitumen saturation is determined by the developed porous-permeable properties. This rock type has lower bitumen saturation values relative to bituminous sands, which is explained by the presence of carbonate cement, which displaces and prevents the penetration of oil-bituminous substances.

Residual-bituminous sandstones are the most consolidated types of rocks. They are painted in colors ranging from gray and greenish gray to light brown. The mineral composition is composed of the same rock-forming detrital minerals as other rock types. Carbonate cement, less often clay-carbonate, according to the nature of filling is basal, pore. The minerals of carbonate cement are calcite and dolomite. Dolomitization is more developed than in bituminous sands. With an increase in the depth of occurrence, the proportion of dolomite increases. This is due to the fact that the magnesium-containing water necessary for the formation of authigenic dolomite, penetrated from the underlying Lower Permian sulphate-carbonate sediments [10]. This rock type is characterized by the highest content of carbonate minerals in its composition and, visually, the lowest content of bituminous substance. For this reason, the rocks are usually light, but the areas contain signs of residual oil saturation as traces of saturation and characteristic odor. Signs of oil saturation increased in samples of strong cemented differences during sample preparation and heating, which suggests the presence of fractions of the light, non-dense, dark colored hydrocarbons sealed in the cement material. To confirm these findings samples of the above-described rock types were analyzed by thermal methods (TG and DSC). In the course of the experiment, samples were heated to high temperatures (up to 1000 ° C) necessary for the decomposition of mineral and organic components. At the same time, the change in the sample mass and the change in the magnitude of the thermal effect (enthalpy) of the reaction were recorded. Comparison of the curves of change in mass and the magnitude of the thermal effect in certain temperature ranges allows us to draw accurate conclusions about the material composition of rocks in terms of the destruction temperatures of minerals and hydrocarbon fractions. It is noteworthy that all selected rock types have a characteristic behavior of the curves of mass loss and enthalpy.

Typical thermograms of the bituminous sands rock type are shown in Fig.1. This rock type has a high HC content and low carbonate content. Oil saturation is measured according to the mass loss curve (TG line) and exothermic reaction effects (DSC line) in the temperature range from 200-250 °C to 600 °C. The decomposition of carbonates takes place at about 650–750 °C and in this rock type is characterized by a slight decrease in the sample mass and a negative value of the reaction enthalpy.

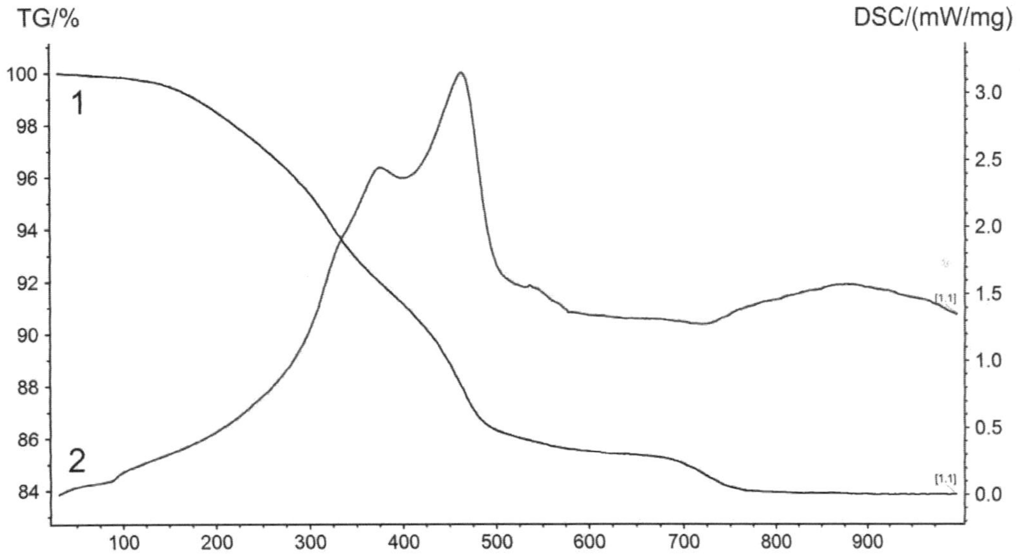

Pic.1 Typical thermogram of the bituminous sands rock type: 1 – weight loss curve, 2 - reaction enthalpy curve

This is due to the fact that almost all carbonates decompose at the same temperature, and is accompanied by heat absorption (ie, the reaction is endothermic), while the combustion process of the HC is stretched in the temperature range, the mass loss is smooth, and the reaction occurs with the release of heat (exothermic reaction). Smooth mass loss is due to the fact that the hydrocarbons in the studied samples are characterized by heterogeneity of the fractional composition. Conventionally, they can be divided into light and heavy hydrocarbons. Light hydrocarbons begin to burn at the beginning of the temperature range of 200-600 °C, while heavy ones react later. The fractional heterogeneity of the HC also explains the rather wide reaction interval of almost 400 ° C. In this case, the oil saturation varies between 8-12% by weight. Carbonate level 1-2%.

The thermogram of the bituminous sandstone rock type is shown in Fig.2. From thermograms, it is clearly seen how the rock type "bituminous sands" differs from the rock type "bituminous sandstones". The proportion of carbonate material increases significantly, while oil saturation decreases. For the samples, the oil saturation varies within 3-5%, and carbonate content - 8-10%.

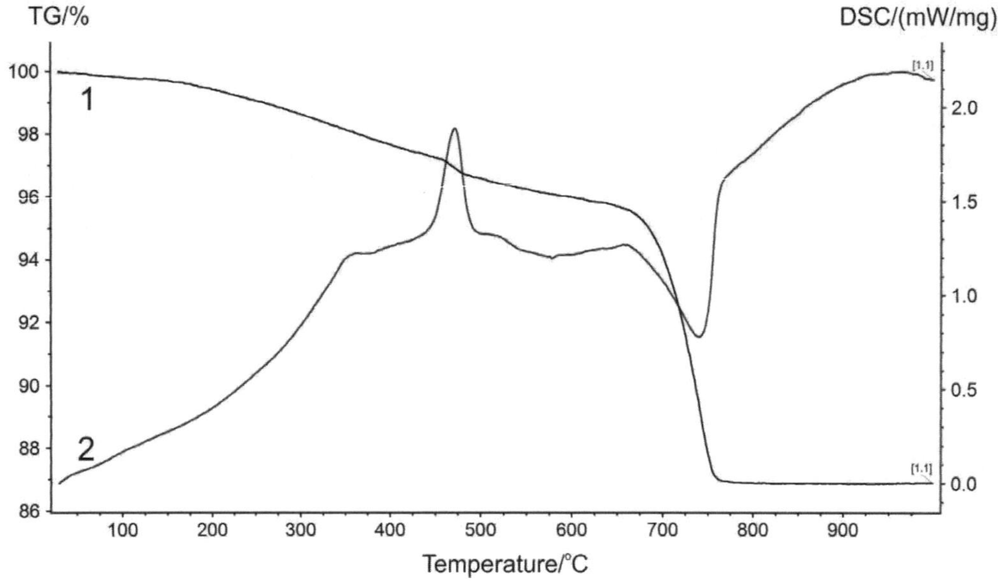

Pic.2 Typical thermogram of the bituminous sandstone rock type: 1 – weight loss curve, 2 - reaction enthalpy curve

The thermograms of the residual-bituminous sandstone rock type are shown in Figure 3. This rock type is characterized by a significant increase in the proportion of carbonates associated with the cement of rocks (18-20%). At the same time, the sample is characterized by a high hydrocarbon content of up to 4-5%. This rock type is characterized by the greatest density and consolidation. However, the external signs of oil saturation in it appear much weaker than in other rock types. High hydrocarbon contents in it apparently explained by their conservation in the pore and intergranular space of the mineral substance with cement

It should be noted that the studied rock types of sands and sandstones differ in fractional diversity of the hydrocarbon substance. In rock type "bituminous sandstone", a curve of the magnitude of the exothermic reaction with a small amplitude is observed, which is typical for heavy hydrocarbons. In the littotype "residual-bituminous sandstones", bright exothermic effects appear, which may serve as a sign of the presence of not only heavy hydrocarbons, but also much more active light ones.

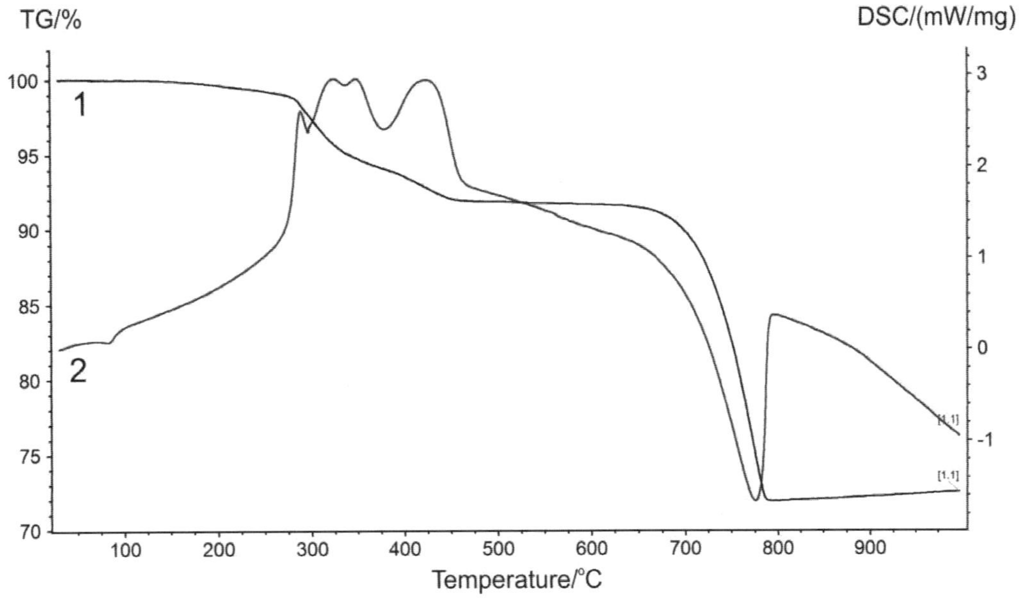

Pic.3. Typical thermogram of the residual-bituminous sandstone rock type: 1 – weight loss curve, 2 - reaction enthalpy curve

4. Conclusions

As a result of the research the following conclusions were made. All selected rock types of sands and sandstones have a similar mineral composition of detrital material and differ from each other only in the content of cement carbonate minerals. The content and composition of hydrocarbon fractions is controlled by the substance of cement, which is caused by epigenetic processes proceeding under the influence of many factors. The main among these factors is the chemistry of the environment. The introduction of hydrocarbon fluid and changes in the chemical composition of the contacting water are the main cause of variations in the chemical environment, which results in different degrees of cementation and oil saturation. It has been established that part of the hydrocarbon fractions can be conserved in the mineral matrix. Due to isolation from hypergenic destruction, preserved hydrocarbons retained the original fractional diversity. The reservoir for hydrocarbons can serve not only loose bituminous sands, but also denser, cemented rock types. The preservation of hydrocarbon fractions in the intergranular space of sandstones may occur during the formation of carbonate cement. The results obtained can be used in the calculation of reserves and involvement in the development of dense sandstone differences.

This study was funded by the subsidy allocated to Kazan Federal University as part of the state program for increasing its competitiveness among the world's leading centers of science and education.

References

[1] Troyepolskiy V.I., Lebedev N.P. Produktivnyye bituminoznyye tolshchi permskikh otlozheniy Melekesskoy vpadiny i Tatarskogo svoda. Kazan: Izdatelstvo Kazanskogo universiteta, 1982. – 104. (In Russian)

[2] Uspenskiy B.V., Valeyeva I.F. Geologiya mestorozhdeniy prirodnykh bitumov Respubliki Tatarstan. Kazan: Izdatelstvo OOO "PF "Gart", 2008. - 348. (In Russian)
[3] Mullakaev Almaz, Khasanov Rinat. The Aeolian factor in the formation of the Sheshmian Horizon sands and sandstones in the Permian of the South-Tatar Arch / In book: Barclay, M., Nikolaeva, S., Silantiev, V. (Eds.) Kazan Golovkinsky Stratigraphic Meeting, 2017: Advances in Devonian, Carboniferous and Permian Research: Stratigraphy, Environments, Climate and Resources. Filodiritto International Proceedings, Bologna, Italy, pp. 381-385.
[4] Khasanov R.R., Mullakaev A.I. Paleogeographic factors of the formation of Permian reservoir rocks of bitumen deposits in the east of the Russian plate (Russia) /International Multidisciplinary Scientific GeoConference Surveying Geology and Mining Ecology Management, SGEM Volume 1, 2016, 469-474 pp. DOI: 10.5593/SGEM2016/B11/S01.059
[5] Khasanov R.R., Mullakaev A.I., Dusmanov E.N. The structure of sandstones in productive horizons of the Permian bituminous deposits of Tatarstan (Russia). Uchenye Zapiski Kazanskogo Universiteta. Seriya Estestvennye Nauki, 2017, vol. 159, no. 1, pp. 164–173. (In Russian)
[6] Pettijohn F.J., Potter P., Siever R. Sand and Sandstone. New York, Springer, 1972. 618 p.
[7] Muslimov R.Kh., Romanov G.V., Kayukova G.P., et al. Integrated Development of Heavy Oils and Natural Bitumens of the Permian System of the Republic of Tatarstan. Kazan, Fen, 2012. 396 p. (In Russian)
[8] Korolev, E.A., Khramchenkov, M.G., Khramchenkov, E.M., Eskin A.A., Gabdelvalieva R.R., Garaeva A.N. Modeling of the suffusion cavities development in bitumen saturated sandstones of Ashalchinskoye deposit under SAGD technology application // Neftyanoe Khozyaystvo – Oil Industry. – 2018. – 1. – P. 55-57.
[9] Korolev, E.A., Bakhtin A.I. Eskin A.A., Hanipova R.R. Diagenetic changes of sandstone reservoir of Ashalchinskoye bitumen deposit // Neftyanoe Khozyaystvo – Oil Industry. – 2016. – 10. – P. 26-28.
[10] Mullakaev A.I., Delev A.N., Usmanov S.A., Sudakov, V.A., Khasanov R.R. Tectonic causes of uneven cementation zones distribution in the bituminous sandstones productive part of the Sheshminsky Horizon of the South Tatar arch//Neftyanoe Khozyaystvo - Oil Industry. - 2018. - Vol., Is.2. - P.23-25. DOI: 10.24887/0028-2448-2018-2-23-25
[11] Yusupova T.N. Identifikatsiya nefti po dannym termicheskogo analiza / Yusupova T.N., Petrova L.M., Ganeyeva YU.M., Lifanova Ye.V., Romanov G.V. // Neftekhimiya -1999. -№ 4. -S. 254-259. (In Russian)
[12] Mullakaev, A.I, Khasanov, R.R., Galiullin, B.M. Mineralogy of sandstones and localization of oil matter in productive Horizons of high-viscosity oil in permian deposits of the Volga-Ural region (Russia)/ 17th International Multidisciplinary Scientific GeoConference SGEM 2017, SGEM2017 Conference Proceedings, 29 June - 5 July 2017, Vol. 17, Issue 11, 353-358 pp. DOI:10.5593/sgem2017/11.

Paleotemperature research of coals in the Volga-Ural oil and gas province (Russian Federation) in connection with forecasting of non-traditional hydrocarbon sources

R R Khasanov[1], Sh Z Gafurov[1], A I Rakhimzyanov[1], A I Mullakaev[1]

[1]Institute of Geology and Petroleum Technologies, Kazan Federal University, Kazan, Russia

e-mail: Rinat.Khassanov@kpfu.ru

Abstract

Using of thermal methods of influence on the formation allows engaging in the extraction unconventional sources of hydrocarbons. One of the most important unconventional oil and gas objects are highly carbonaceous deposits. On the territory of the Volga-Ural oil and gas province, such objects is the Domanik Formation, which are clayey-siliceous-carbonate shale rocks with a high content of organic matter (OM). To identify and clarify the mechanisms of formation of oil and gas fields, as well as forecast and search for promising areas in the area of distribution of shale strata, the method of measuring the reflectivity of vitrinite is used. However, the absence of residues of plant organic matter in the Domanikites creates difficulties in applying the method. To assess the prospects of individual sections of the Domanik Formation, it is possible to use the paleotemperatures of the field of adjacent and overlying sediments, where coal seams and inclusions of plant organic matter are present. The basis of this approach is the paragenetic association of oil, coal and shale sediments. The formation of shale and coal-bearing formations is due to the primary conditions of sedimentation, and the formation of oil and gas deposits is associated with deep thermal effects during periods of tectonic activation on the territory. The structure of modern thermal fields, due to changes in temperature values, is controlled by the location of zones of deep faults in the crystalline basement of the platforms. Paleotemperatures are determined quite accurately by the degree of catagenetic transformation of organic matter (OM), which can be used to reconstruct the structure of ancient thermal fields in sedimentary formations and identify areas of maximum warming. Regional patterns of change in the degree of coal metamorphism can be used as a basis for forecasting and identifying promising areas of oil and gas shale raw materials.

1. Introduction

Using of thermal methods of influence on the reservoir allows engaging in the exploitation of unconventional sources of hydrocarbons, which include highly carbonaceous deposits. Within the Volga-Ural oil and gas province, the most promising as shale sources of hydrocarbon raw materials are rocks of the Domanik Formaition, confined to the Devonian sediments. The area of potentially oil-bearing deposits of the Devonian Domanik Formation covers an area of about 2 million km^2 [1, 2, 3]. The thickness of the Domanik Formation varies from 13 to 72 meters. In the section of the Devonian deposits of the Volga-Ural region, it covers the range from the Semilukian (Domanikovian) Horizon of the Frasnian Stage to the very tops of Famennian Stage [1, 2, 3]. Domanikites are dark clay-siliceous-carbonate shaly rocks with a high content of organic matter (OM) up to 1-5% and above. The rocks of the Semilukian (Domanikovian) and Rechitsian (Mendymian) Horizons of the Upper Devonian possess the greatest oil-bearing potential and perspective. In petroleum geology for forecasting and exploration, the degree of catagenetic transformation of organic matter is used as an indicator of the processes of oil and gas formation [4, 5]. Determining the degree of metamorphism of organic matter is produced by the coal-petrographic method, which is based on the determination of the vitrinite reflectivity index (Ro). However, in the pre-Carboniferous sediments, the application of the method is difficult due to the absence of microcomponents of plant organic matter. To assess the prospects of individual sections of the Domanik Formation can be used the variability of paleotemperature fields in adjacent or overlying rock complexes with inclusions of coal or coal seams. The temperature distribution of modern thermal fields is associated with zones of deep faults in the crystalline basement of the platform and clearly correlates with the degree of catagenetic transformation of organic matter (OM) in sedimentary formations. The basis of this approach is a paragenetic relationship of oil, coal and shale sediments [6]. The paragenesis of coal and oil deposits is global. At the planetary level is observed a distinct geochronological regularity in their location. Coal deposits of various sizes are contained within the post-Devonian oil and gas provinces around the world. This is due to the wide distribution of terrestrial vegetation on land, beginning with the Carboniferous period. The formation of shale and coal-bearing formations is due to the primary conditions of sedimentation, and the formation of oil and gas deposits is associated with deep thermal effects during periods of tectonic activation of the territories. On the territory of the Volga-Ural oil and gas province, early Carboniferous coal-bearing deposits are located directly on Devonian sediments. They are associated with the Visean deposits of the early Carboniferous, which are also oil-bearing. The Visean terrigenous stratum fills in erosion-karst incisions within the northern and southern domes of the Tatar arch and in the inter-domed depressions [7]. The number of coal seams in incisions is usually from 1 to 3, the middle of which is the thickest and reaches 10-30 meters with a depth of 900-1400 m. By their nature, coals are humus. Coals are mostly metamorphosed to the stone-coal stage, but the sections retain the properties of brown coal. Regional patterns of change in the degree of coal metamorphism can be used as the basis for forecasting and delineating promising areas of oil and gas shale HC raw materials.

2. Methodology

To clarify the structure of the ancient thermal field and clarify the sources of heat fluxes, paleotemperatures were determined by the degree of metamorphism of the Visean coal beds. The main method was the coal-petrographic method, based on the measurement of the reflectivity of vitrinite (% Ro). Vitrinite is one of the microcomponents (maceral) of coal, which has a high reflectivity. During the immersion of the coal-bearing strata to a depth, a consistent transformation of the organic part of the coal occurs as a result of an increase in pressure and temperature. At the same time, a gradual and regular increase in the reflectivity of vitrinite occurs, associated with a change in its chemical composition, physical properties and intramolecular structure. The reflectivity of vitrinite in oil immersion (Ro) consistently increases from 0.5–0.64% to 2–2.5%. In coal geology, this indicator is used as an optical criterion for the degree of coal metamorphism. It can also be used to determine the paleotemperatures of the formation of coal and coal-bearing strata. In petroleum geology, the vitrinite reflection index is also used to establish the stage of post-sedimentation transformation of a sedimentary complex containing organic matter [4, 5]. Comparison of reflectivity values with other indicators of the degree of maturity and distribution over the section of oil and gas accumulations allows setting the values of Ro, corresponding to the boundaries of oil and gas formation. Catagenesis of source rocks is analyzed by measurements of the reflectivity of vitrinite (Ro). The lowest reflectivity value corresponding to the generation of oil is 0.45%, and Ro = 0.6% is considered to correspond to the beginning of the formation of industrial oil deposits. The end of oil generation occurs approximately at Ro = 1.3%, gas condensate - at Ro = 2% and methane at Ro = 3.5% [4, 5]. Coal-petrographic studies were conducted in reflected light using a Carl Zeiss Axio Imager A2m polarization microscope. Measurements of the reflectivity of vitrinite were made on an MSF-30U microscope-spectrophotometer, which is designed to record the reflection spectra of minerals and to measure the reflection coefficients of coal and dispersed organic substances.

3. Results and Discussions

Our studies of the metamorphism of coal in terms of the reflection of vitrinite showed an uneven distribution of this indicator within the coal-bearing area. Ro values within the territory under consideration vary in a fairly wide range from 0.44 to 0.73% [8]. They show a distinct tendency to increase in the direction from north to south. In the northern part of the territory, minimal values of the vitrinite reflectance are observed (0.44–0.51%, average 0.47%). The highest values of Ro are established in the southern part of Tatarstan, where the greatest depths of immersion of coal seams are noted up to 1300 1400 meters. In these areas of the earth's crust, the reflection index of vitrinite in coals reaches 0.73% in individual samples. Average values in various deposits vary from 0.58% to 0.66%. The maximum coal metamorphism is fixed in those parts of the coal basin where coal seams lie at maximum depths (1300-1400 meters). The Ro values in the coal in these areas, which coincide with the zones of deep faults in the crystalline basement, correspond to paleotemperatures of 50-70 ° C. These areas are also characterized by elevated values of modern temperatures in the strata. According to field research [9], produced by oil wells, the maximum temperature difference at the absolute level of -1000 meters in the Volga region reaches 30 ° C. Temperatures vary in the range of 14.25-48.2 °C. The structure of the modern thermal field is closely related with tectonic structure of the territory and location of deep faults. Thus, the reason for the increasing of the coalification degree of organic matter is the increasing in temperature in these areas of the earth's crust. However, the effect of the geothermal gradient, which consists in increasing the temperature by an average of 3 ° C every hundred meters, is not sufficient to ensure the observed degree of coal metamorphism. The depth difference of the Visean coal seams does not exceed 300 meters. Devonian sediments (Domanik Formation) occur at a depth of 1.6-2 km. This indicates that the thermal regime in ancient geological epochs was characterized by higher temperatures than at present. The formation of oil should be associated with rocks of the Domanik Formation, where organic matter is about 1-3% or more. Apparently, in the past they have been subjected to higher thermal effects due to tectonic activation of the territory. This led to the generation of oil from the Domanik Formation. The possibility of heat and mass transfer is also confirmed by the results of deep drilling. Within the Melekes Depression, where Domanik thick strata lie in the sedimentary cover, fractured rocks with traces of later hydrothermal activity are fixed in the crystalline basement.

An important applied direction may be the use of unconventional sources of hydrocarbons for additional energy supply for the development of high-viscosity oil and bitumen deposits [11], which are characterized by high energy consumption and require significant amounts of natural gas to generate steam. The problem of providing energy raw materials for steam generation plants can be solved through the use of local coal - raw materials. Depending on the technology chosen, coal seams can be used to produce combustible underground gasification gas or coal-based methane [12, 13] as independent fuels, which can be used for local needs.

4. Conclusions

The results obtained clarify ideas about the nature of the lithification of early Carboniferous sediments and can be used for predictive researches. Based on the analysis of the tectonic structure of the crystalline basement, the structure of the modern temperature field and the patterns of changes in the degree of coal metamorphism, paleo-temperature zoning of the territory can be made, which will serve as the basis for forecasting and contouring prospective areas.

The work was supported by the Ministry of Science and High Education of the Russian Federation contract No. 14.Y26.31.0029 in the framework of the Resolution No.220 of the Government of the Russian Federation and this study was funded by the subsidy allocated to Kazan Federal University as part of the state program for increasing its competitiveness among the world's leading centers of science and education.

References

[1] Aliev M.M., Batanova, G. P., Khachatryan R. O et al. Devonian deposits of the Volga-Ural oil and gas province. M.: Nedra, 1978. 216 p. (in Russian).

[2] Geology of Tatarstan: Stratigraphy and tectonic. Moscow, GEOS. 2003. 402 p. (In Russan).

[3] Kiryukhina T.A., Fadeyeva N.P., Stoupakova A.V., Poludetkina E.N. and Sautkin R.S. Domanik deposits of Timano-Pechora and Volga-Ural basins. Oil and Gas geology. 2013. (3), pp. 76-87 (In Russan).

[4] Vassoevich N.B. The origin of oil. Bulletin of the Moscow State University. Ser. Geology. 1975. (5), pp. 3-23. (in Russian).

[5] Bertrand, R. Standardization of solid bitumen reflectance to vitrinite in some Paleozoic sequences of Canada. Energy Sources Journal. 1993. V. 15, (2), pp. 269-288.

[6] Khasanov R.R., Gafurov Sh.Z., Mullakaev A.I. The paragenesis of coal and oil deposits in paleozoic sediments in the east of the Eastern European platform / 16th International Multidisciplinary Scientific GeoConference SGEM 2016, SGEM2016 Conference Proceedings, Book 1. Vol.1, 627-631pp. DOI: 10.5593/SGEM2016/B11/S01.079

[7] Khasanov, R.R., Larochkina, I.A. Prospects for the joint development of oil and coal deposits in previsean depressions of Volga-Ural oil and gas province. Neftyanoe Khozyaystvo - Oil Industry (1), pp. 36 -39, 2013. (in Russian)

[8] Khasanov, R.R., Gafurov, Sh.Z., Rakhimzyanov, A.I. The degree of the epigenetic transformation of an organic matter in the Early Carboniferous sediments of the central part of the Volga-Ural oil and gas province. 2016. Neftyanoe Khozyaystvo - Oil Industry. 2016. (10), pp. 29-31. (In Russan).

[9] Khristoforova N.N., Neprimerov N.N., Khristoforov A.V. et al., Thermal conditions and evaluation of oil and gas potential of the Volga region, Georesources, 2004. (1), pp. 24–27. (In Russan).

[10] Khasanov, R.R., Mullakaev A.I., Galiullin , B.M., Khayrtdinova L.R. The influence of hydrothermal processes in the crystalline basement on the oil-bearing capacity of the sedimentary cover of the Volga-Ural region (Russia) / 17th International Multi-disciplinary Scientific GeoConference SGEM 2017, SGEM2017 Conference Proceedings, 29 June - 5 July 2017, Vol. 17, Issue 11, 631-636 pp. DOI:10.5593/sgem2017/11 DOI: 10.5593/sgem2017/11/S01.079

[11] Bartis, James T.; LaTourrette, Tom; Dixon, Lloyd; Peterson, D.J.; Cecchine, Gary. Oil shale development in the United States: prospects and policy issues. The RAND Corporation. 2005. 68 p.

[12] Kreinin E.V. Nonconventional hydrocarbon sources. New technologies for their development. Moscow: Publisher: Prospekt. 208 p., 2016. (in Russian)

[13] Moore T.A. Coalbed methane: A review, International Journal of Coal Geology, vol. 101, pp 36-81, 2012

Features of the natural bitumen distribution in the Kazan deposits on the territory of the Melekess depression in Tatarstan

R A Mudarisova[1], Yu V Volkov[1], S E Valeeva[1], E I Fakhrutdinov[1]
[1]Institute of Geology and Petroleum Technologies, Kazan Federal University, Kazan, Russia

E-mail: rayshania@mail.ru

Abstract. This article presents the distribution of natural bitumen in the sediments of the Kazan tier of the middle Permian. Natural bitumens, concentrated in the Permian sediments of the Republic of Tatarstan, occur at shallow depths and are heavy, highly viscous, hypergenically converted oils. In the tectonic relation, the area of maximum Kazan bitumen occurrence is confined to the Melekess depression and the adjacent slope of the South Tatar arch.

1. Introduction

The study of the Permian deposits occupies a special place in the history of the development of oil fields of the Republic of Tatarstan. The discovery of industrial deposits of oil and gas was preceded by a nearly two hundred year study of the surface manifestations of bitumen.

In assessing the actual conditions of occurrence of bitumen and their resources, due to the ambiguity of the factual material interpretation, two diametrically opposite approaches have emerged.

Supporters of the first approach (I.M. Akishev and a group of TatNIPIneft employees) were based on ideas about the continuous development of bituminous in Kazan and other Perm complexes. The second approach to the study of bituminous (V.T. Troepolsky and others) developed earlier ideas about the presence of local accumulations of bitumen in Permian sediments and was based on the analysis of new data on the structure and spatial relationships of bitumen-containing complexes.

Analysis of the structural drilling carried out in the 70s of the last century showed that the bituminous territories are heterogeneous. There are distributed areas of a large number of small deposits, surrounded by a dense background of various intensity bitumen occurrences. At a number of sites, in the presence of favorable prerequisites (reservoir, tires, positive structural forms, etc.), bitumen is localized into larger deposits and their associations. [2]

2. Methodology

Kazan deposits are a thick stratum (more than 100 meters) of carbonate and terrigenous rocks with interlayers and inclusions of sulphates (Figure 1).

There is a rhythm of various types of rocks alternation, which is reflected by a periodically repeated transgression and regression of the basin, a change of shallow-water sediments more deeply. [3]

The deposits of bitumen in deposits of the Kazan usually represent a combination of deposits confined to terrigenous and carbonate reservoirs located at different stratigraphic levels.

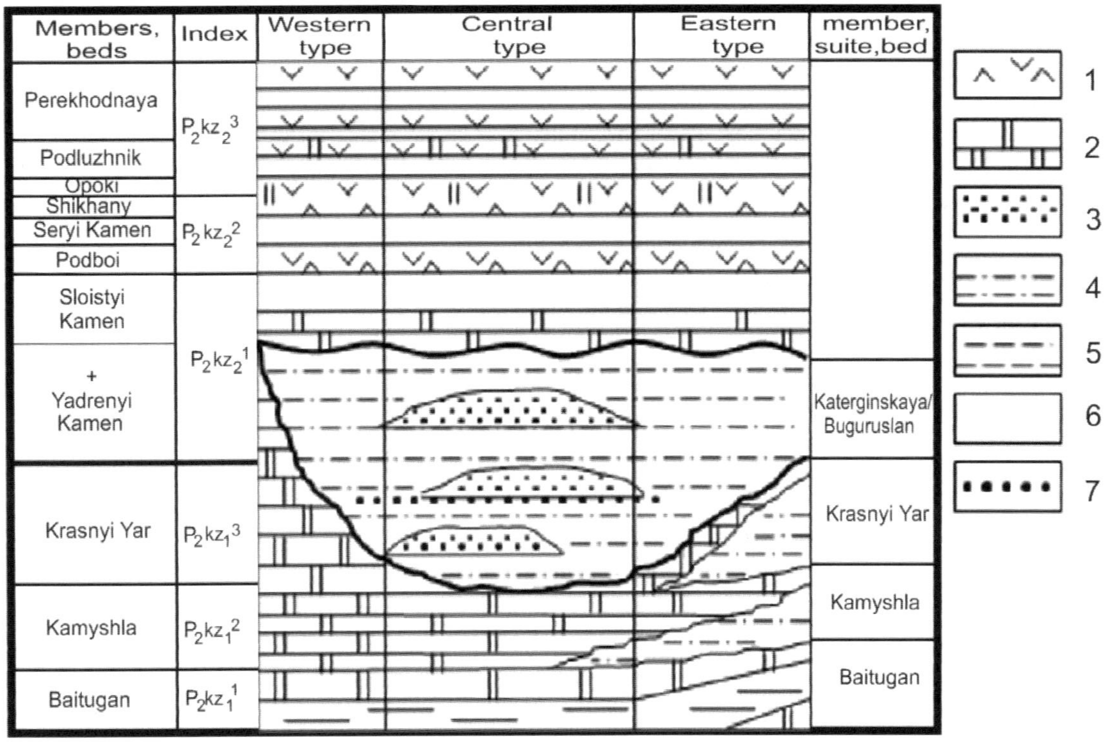

Figure 1. Lithological and stratigraphic section of the Kazan deposits of the Republic of Tatarstan (Ellern, 1984). Symbols of rocks that compose Kazan deposits: 1 - sulphate, 2 - carbonate 3 - sandy, 4 - clayey-silty, 5 - clayey ("lingular clay "), 6 - low-permeable, 7 - the boundary between the lower and upper substages, accepted by geologists of the" TN "(Bogov, Petrov).

The study of the natural bitumen occurrence conditions in the Kazan sediments of the Melekess depression showed that their bulk is concentrated in the lower part, which belongs to the Lower Kazan terrigenous-bitumen complex. The upper boundary of the complex is the sole of a sulphate pack, sufficiently aged along the strike, which is regarded as an analogue of the “slaughter” series of the Prikazansky district. The lower boundary of the complex is held along the roof of the “lingule” clays or in the zone of their absence along the base of the younger Kazan deposits, which overlap directly onto the eroded surface of the lower Permian sediments.

This complex is ambiguous in its lithological composition. Within the eastern part of the Melekess depression, two lithological sequences are distinguished: the lower one is predominantly carbonate and the upper one is mostly terrigenous, containing small interlayers or inclusions of carbonates.

The thickness of the carbonate stratum is variable (from 0 in the northern part to 85 m in the south-west), which is connected both with the conditions of sedimentation and with two erosion: Neogene and Kazan. These changes are manifested in different ways on the territory. Ellern S.S. and Vinohodova G.V. identified three types of cuts.

The western type of the section is geographically distributed in the southwestern part of the eastern side of the Melekess Depression, and covers Akanskaya, Yukhmachinsky, Kondurchinsky, Maximskaya and other areas. This type of section is characterized by the complete absence of terrigenous rocks. The carbonate stratum is stratigraphically represented by deposits of the Baitugan, Kamyshlinsky, and Barbashinsky horizons. Thickness varies from 50 to 70 m.

The central type of the carbonate stratum seam occupies most of the eastern side of the Melekess depression. This type of cut, as well as the West, has a strike from northwest to southeast. The carbonate stratum of this type has a two-layer structure. Its upper part, corresponding to the XI-XIa formations, is absent, which is probably due to erosion; this erosion significantly affected its middle part (XIII-XIIIa layers). The lower part of the stratum (XY-XYII layers) also falls out of the section. The thickness varies in the range of 45-50 m, with the result there are numerous zones of thickness difference. They form carbonate bodies of various configurations along closed isolines of 15, 20, 25, 30 m. Among them, dome-shaped and oval forms of predominantly north-western strike, sometimes almost isometric (1-3x1-7 km) with minimum thickness up to 10 m. In the area of the North-Katerginskoye field, the lowest carbonate thickness was recorded. In some cases, the thickness of the carbonate stratum reaches 45-50 m. Such an increase in its power is observed in the marginal parts of the central zone. North of the latitude of Bilyarsk, the central zone of reduced capacities expands in a northeast direction to the settlement area of Starosheshminsk. Here there are numerous areas of the absence of carbonate stratum or its small thickness of the order of 20–30 m, which is explained by the erosion of its surface in the pre-Akchagyl time. From the analysis of the materials, it follows that the carbonate stratum of the central type of the section corresponds, apparently, only to the Kamyshla horizon. The lithologic carbonate stratum of the central type of the section differs from the western one by the predominant development of organogenic-detrital dolomites, but other species are also found.

The eastern type of the section, which is characterized by the predominance of terrigenous rocks, mainly clays with sandstones and siltstones, is developed in the eastern part of the study area. This type of section is carbonate – terrigenous. The total thickness of the sediments increases to 70-80 m. The heterogeneity of the carbonate stratum of terrigenous rocks has created a kind of cyclical structure of the latter: at the base of each cycle is a stratum of carbonate rock overlapping with terrigenous sediments, with each overlying layer of carbonate sediments lying on the underlying terrigenous unconformity. Towards the central type of the section, the terrigenous deposits pinch out.

The terrigenous stratum of the considered territory of the Kazan deposits, as well as the carbonate stratum, is distinguished by a non-uniform structure. This concerns its lithological composition and sediment thickness. Here intermittency of strata is observed, reflecting its peculiar cyclical structure associated with the transgressive-regressive development of the basin. The thickness of the terrigenous strata is even more variable than the carbonate one. [1]

As a result, the section of the terrigenous sequence was dissected into a series of packs, which are combined into separate elementary cycles. At the base of such a cycle, the most clayey rocks are located, which are replaced upwards along the section by sandy-silt rocks. Four cycles were distinguished by Vinogradova G.V. It should be noted that the cyclical nature of the terrigenous stratum in the direction of both the eastern and western boundaries of its distribution becomes less distinct, which is associated with the formation of the marginal parts of this stratum in various facies. In this case, as a rule, the upper parts of the cycles, represented by sandy-silt rocks, are compacted and pinned out. In this regard, in this area, as well as for carbonate strata, there are three types of terrigenous strata: western, central and eastern.

The most consistently characterized sand-silty "layers" Y and YII. The majority of fixed bitumen occurrences on core are confined to them. The Y and YII strata are most clearly seen in the central type of the section, absent in places in the areas of development of Neogene incisions and local pinch zones. The strip of sandy-silty rocks of each layer in this type of section extends from northwest to southeast. The length of the lanes is 80 - 100 km, the width reaches 40 - 50 km. In each of the layers, the development to the same extent of siltstones and sandy rocks with thin interlayers of clays and dolomites is characteristic; sandstones and siltstones are usually calcareous and often plastered. The thickness of the layers Y and YII unevenly distributed. Thickness fluctuations reach 20 m. Also, there is a coincidence of areas of increased thickness of the terrigenous sequence with the areas of distribution of sandstones or coarse-grained siltstones. Such examples have been established at the Katerginskoye, Pionerskoye, Uzeyevskoye and other fields. It is these areas and forms a kind of local "body". In terms of their configuration, these sand “bodies” have a characteristic shape: a flat base and a convex upper surface. In the western (I) and eastern (III) types of section, the sand-siltstones rocks of these layers are represented by thin layers, as if wedged into predominantly clayey rocks. It is characteristic that the western type of the section directly adjoins the facies zone, alternating with carbonate rocks to the west, and the eastern one borders on the area of the development of red sediments.

A typical representative of the central type of the mine is the Uzeevskoe field. It consists of deposits of the Kamyshla and Barbashin horizons and the series “yadrenyi kamen” and “sloistyi kamen”. Other conditions for the extraction of deposits include: the contour of deposits, the presence of unambiguous ideas about the nature of the distribution of bituminous reservoirs, data on well productivity, a set of laboratory core analyzes to substantiate the categories of reserves (C_1 + C_2), at least half of the reserves to the area of deposits, limited by 5m isolines (optional feature), etc.

Pioneer field. Due to the proximity to the Uzeevskoye field, the sections of the deposits are identical. The most intensive bitumen maintenance is confined to the undifferentiated thicker sediments of the series “yadrenyi kamen” and “sloistyi kamen” of the Upper Kazan and Barbashin horizons of the Lower Kazan.

3. Results and Discussions

Two bituminous complexes are confined to the deposits of the Kazan tier: the Lower Kazan terrigenous-bituminous complex and the Upper Kazanian complex.

The Lower Kazan terrigenous-bituminous complex. The facial environment of sedimentation predetermined the differentiated nature of the distribution of bitumen in the Lower Kazan sediments. In the Kamyshla horizon, favorable reservoir beds are associated with organofragmatic and detrital and oolitic carbonate rocks. The areas of their distribution are confined to the submerged part of the eastern side of the Melekess depression. The boundaries of these lithofacies include the Arbuzovsko-Merezenskaya zone, which is distinguished by a high concentration of bitumen.

Within the zone, the dependence between the intensity of bitumen manifestations and the positive structural plan is clearly manifested. Areas with well-saturated layers (2 or more meters thick) are associated with local uplifts complicating shafts and structural terraces (Gorskoe, Merezenskoe, Aksubaevskoe, Arbuzovskoe, Uzeevskoe, Adelyakovskoe, etc.). Other lithofacies are characterized by the absence of aged collectors and insignificant bituminous.

In the Barbashin horizon, bitumen are much more widespread than in the underlying sediments. However, the uneven development of reservoir rocks left a definite imprint on the distribution of bituminous. According to the scale of concentration of bitumen in this horizon, the territory of the Melekess depression is unequal. In the direction from east to west, three main zones of bitumen concentration can be distinguished: the eastern or Arbuzovsko-Merezen, central and western. The intensity of bitumen decreases in a westerly direction. The majority of bitumen deposits are associated with sandy reservoirs of the eastern zone. Many deposits are small in size due to the complex structure of the host rocks. Lithofacies of carbonate rocks is developed within the central zone. The most intense bitumen are concentrated in the area of Akan uplift. No significant concentrations of bitumen were found in the rest of the territory. Within the western zone, quite numerous, but generally small-scale, concentrations of bitumen are established, which is explained by the heterogeneity and local development of carbonate reservoirs, partial erosion of sediments and unfavorable genetic prerequisites.

The Upper-Kazan complex is an extremely complex, facially changeable multilayer reservoir, divided by a screening sulfate pack of the "podboi" series into two parts: the lower (lower bituminous horizon) and upper (middle and upper bituminous horizons). The lithological composition of rocks is characterized by impermanence and naturally changes in the latitudinal direction. The spatial distribution of bitumen is controlled by types of lithofacies. The confinement of the most intense bitumen concentrations to the central litofacial zone, represented by marine terrigenous-carbonate and carbonate-terrigenous sediments, has been established.

4. Conclusions

An increased concentration of bitumen is noted on the eastern side of the Melekess depression. It is caused by the presence of small lithofacies with a well-defined distribution of reservoir and tire layers and their favorable ratio with positive structural forms of various bituminous horizons. The combined areas of their distribution in terms of coincide with the fragments of the previously allocated Arbuzov-Merezen zone in the Lower Kazan complex. The most significant in the concentration of bitumen is the lithofation of sandy rocks of the "yadrenyi kamen" series. Bitumen occurrences in cavernous and fractured dolomites, sandstones and siltstones of the "sloistyi kamen" series are widely developed. In the middle and upper bituminous horizons, the intensity of bitumen is markedly reduced. This is due to the deterioration of the composition of host rocks due to the uneven distribution of reservoirs, the lack of sufficiently pronounced tires, etc. [2,3,4]

A variety of studies cited in past years did not lead to a complete disclosure of the patterns of bitumen placement and agreed conclusions about the size of their resources. For a long period, prospecting and exploration were carried out without a clear understanding of the actual conditions of occurrence of natural bitumen, which entailed unfounded estimates of the prospects for bituminous. The authors propose to reconstruct the paleogeographic and paleotectonic situation in Kazan time for a more complete study of the conditions of formation of deposits of natural bitumen.

References

[1] Vinokhodova G V *et al* 1985 *On the structure of the lower part of the Kazan tier of the east of the Melekess depression and features of the distribution of bitumen* (Kazan: Kazan State University) pp 8-24
[2] Voitovich E D 1997 *Perm bitumens* (Kazan: Report of the TEO OJSC Tatneft) p 198
[3] Troepolsky V I *et al* 1997 *Productive bituminous strata of the Permian sediments of the Melekeess depression and the Tatar arch* (Kazan: Kazan State University) p 104
[4] Fakhrutdinov E I *et al* 2018 *Features of the particle size distribution of the Ufimian-Lower Kazanian sediments of the section of Karkali* (Academic science - problems and achievements XVII.: Proceedings of the Conference. North Charleston, USA) pp 5-9

IOP Conf. Series: Earth and Environmental Science **282** (2019) 012027 doi:10.1088/1755-1315/282/1/012027

Prospect of heat-treat of the reservoir for the active mechanism of oil recovery, due to the study of the geochemical composition of terrigenous rocks of the section of Karkali, South-East of Tatarstan

EI Fakhrutdinov[1], **NG Nurgalieva**[1], **RA Mudarisova**[1]

1Institute of Geology and Petroleum Technologies, Kazan Federal University, Kazan, Russia

E-mail: MKS-1989eduard@yandex.ru

Abstract. The present paper shows the thermal features of the treatment on the reservoir, geochemistry of sediments, and intensity of chemical weathering of terrigenous rocks of the Ufimian-Kazanian natural reservoir of the section of Karkali in the basin of the r. Sheshma and r. Inesh, to predict the efficiency forecast of the application of heat-treat at the object and decipher the genesis of sedimentary formations in the area. Changes in the main elements in rocks are represented in the variation diagram.

1. Key requirements

Karkali section, heat-treat, geochemistry, climate, oxides, Ufimian-Kazanian natural reservoir, South-east of Tatarstan.

Over the years, the geochemical method of the approach has been used to study sedimentary environments in the sedimentation basin. The geochemical method allows to divide and correlate the formation, without the biogenic component of sediment. This method is based on the study of the geochemical characteristics of the layers in the section.

A known quantity, that both carbonate and terrigenous sediments “remember” the s environment and hydrochemical features of the sedimentary basin in the form of content of “indicator elements”, such as Ca, Mn, Fe, Si, Cl, SO_3, Al, Ti, V, etc. That is, the turn in the sedimentation cycles is a change in the hydrochemical condition, which is reflected in the chemistry of rocks. And thus, the variation of chemical elements within each several of the lithotypes form a peculiar tool in the identification of geological boundaries, with the subsequent description of the sedimentary environments in consequence of the content analysis of the "indicators" elements of the paleodynamic condition.

2.Introduction

The sediments of the Permian system in the east of the Russian Plate are the objects and interests of many studies, since there a thought arises in the knowledge and understanding of the processes occurring in this geological time.

Basin of the Volga and Kama rivers is known as an unique stratigraphic and paleogeographic object formed in littoral, transition and continental environments. We considered the Kazanian stage (~ 272 - 266 million years) within the Late regional cycle [1]. The Late regional cycle can be characterized as a cycle of second order corresponding to the supercycle or sequence that began his formation after the fifth global regressive phase of the Late Paleozoic [2]. Investigated Permian cycle acts as a foreland basin supercycle history from its opening until closing. Three megacycles corresponding to the Ufimian, Kazanian and Urzhumian+Severodvinskian + Vyatskian occurred inside this supercycle [1]. According to lithofacies zonation proposed in [3] the Kazanian rocks of Volga - Kama rivers area belong to the zone of carbonate marine and lagoon sediments. The most interesting Permian sections in the South-Eastern Tatarstan (Volga-Kama rivers region) are exposed in the upper reaches of the Sheshma River, near the villages of Shugurovo and Karkali.

3. Objects and research methods

The Karkali section is exposed in a quarry on the right slope of the r. Sheshma and Inesh near the villages of Shugurovo and Karkali.

The description of the section begins at the road-bed between the villages of Karkali and Shugurovo, there are concentrated outcrops of the sandy-clayey red formation of Ufimian rocks and the boundary between the Ufimian and Lower Kazan sediments (outcrops II – beds 1-7). In outcrop III is concentrated thick series of clayey rocks of bed 8. Outcrop IV (fig. 1) is confined to the western part of the Karkali quarry (beds 8–12) along the right slope of the r.Inesh; outcrop V (beds 13-16), outcrop VII (beds 14-23), outcrop VI (beds 16-18), outcrop VIII (beds 16-18) is confined to the eastern part of the Karkali quarry, on the left slope of the r. Inesh (fig. 1). Outcrop I — the bottom of bed 1 (Ufimian sediments) —in the territory of the Shuguro Tar Plant [4].

Along the whole length section is represented by Ufimian, Lower Kazanian and Upper Kazanian sediments. Consisting of Ufimian sediments includes red sandy-clayey continental sediments. The boundary with Lower Kazanian sediments is accompanied with evidence of erosion. Lower Kazanian sediments are represented by marine facies. The bottom is mainly terrigenous with small layers of carbonate material and the top is carbonate-terrigenous. The carbonate material consists mainly of oolitic carbonates. Upper Kazanian substage consists of continental terrigenous sediments with alternation of siltstone and clayey sandstone [5,6].

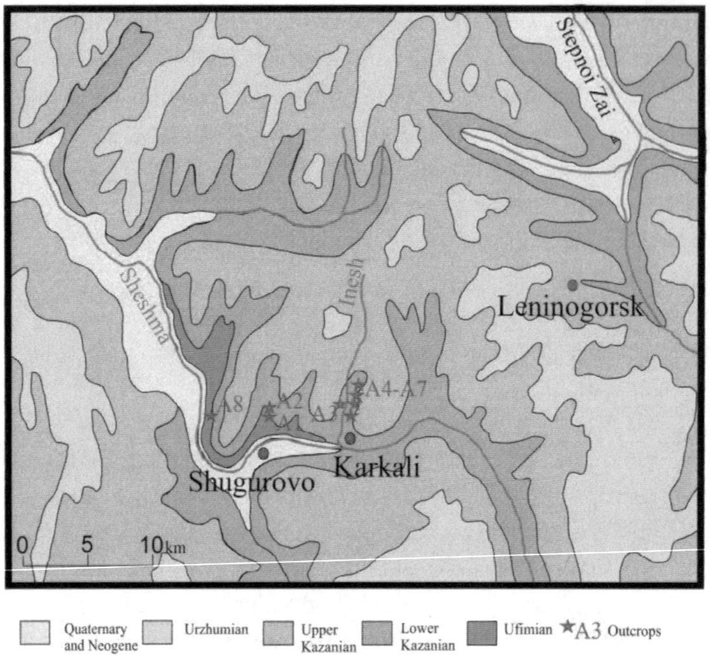

Figure 1. Generalised geological map of the Sheshma river, Republic of Tatarstan. Legend: 1 – Quaternary and Neogene; 2 – Urzhumian; 3 – Upper Kazanian; 4 – Lower Kazanian; 5 – Ufimian (according to Nurgalieva, 2015).

4.Results and Discussions

For indetification the geochemical data, all samples were studied on an X-ray fluorescence spectrometer. Knowing information about precipitation chemistry it is possible of decipher the genesis of sedimentary formations.

The content of rock-forming oxides and trace elements in the terrigenous rocks of the Ufimian and Kazanian ages was studied 116 samples. Such as clay samples -45, sandstone samples -71. Analysis for 43 elements (Si, Ti, Al, Fe, Mn, Ca, Mg, Na, K, P, Sr, S, Cl, V, Cr, Co, Ni, Cu, Zn, As, Br, Rb, Y,Zr,Nb,Cd,Mo,Sn,Sb,Te,Ta,Sm,Nd,Pb,Bi,Se,Cs,Pr,La,Ce,W,Ge,In) performed at the Kazan Federal University using an X-ray fluorescence spectrometer BRUKER S2 Ranger (Bruker, Germany). The obtained results and their mathematical treatment by package STATISTICA allowed to clarify the composition of sediments, as also to get some information about the conditions of sedimentation in the Ural-Biarmian time.

Clay rocks in the section of the Lower Kazanian substage in the basin of the r. Sheshma and r. Inesh were studied according to the data of the modular diagrams by Ya.E.Yudovich and M.P.Ketris: TM(titanium module)-TiO_2 (fig.2) и FM-NKM (femic module-normalized alkaline condition) (fig.3). They allow to classify clay rocks by mineral composition.

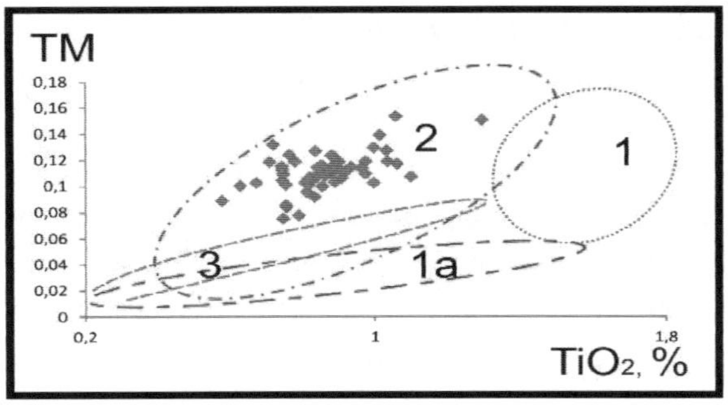

Figure 2. TM-TiO2 diagram for clay rocks in the basin of the r. Sheshma and r. Inesh. 1-kaolinitic clay,1a- low modulus kaolinitic clay – products of catagenetic transformations of montmorillonite and kaolinite substrates, 2- hydromicaceous clay, 3- montmorillonite clay

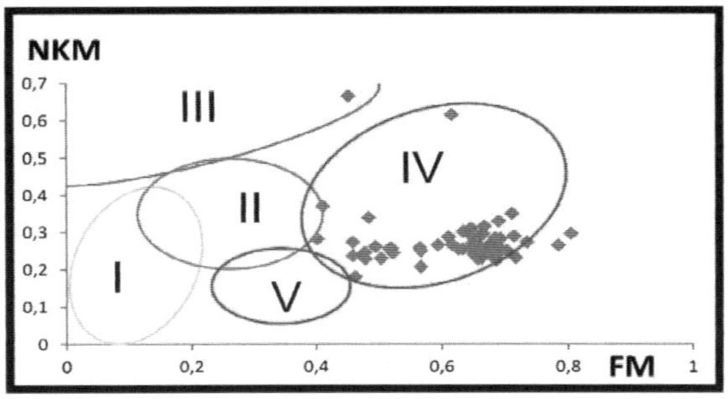

Figure 3. FM-NKM diagram for clay rocks in the basin of the r. Sheshma and r. Inesh. I- kaolinitic clay, II- montmorillonite clay with minor kaolinite and hydromica, III- clay rocks with a composition of chlorite and a small adulteration of Fe-hydromica IV- hydromicaceous clay, V- clay rocks with a composition of three components (chlorite, montmorillonite, hydromica).

According to the TM-TiO_2 and FM-NKM, it can be interpreted that the clay rocks of the Lower Kazanian substage are hydromica in mineralogical composition. The maximum values of TM are typical for tideland, the minimum – for deep marine shelf [6].

For the classification of sandstones, F.J.Pettijohn with coordinates $\log(Na_2O/K_2O)$ – $\log(SiO_2/Al_2O_3)$. A geochemical approach is used, depending on the prevalence of quartz, clay components and feldspar in sandstone. For immature sandstones characteristic $\log(SiO_2/Al_2O_3)$ <1,3 и $\log(Na_2O/K_2O)$>0 [7].

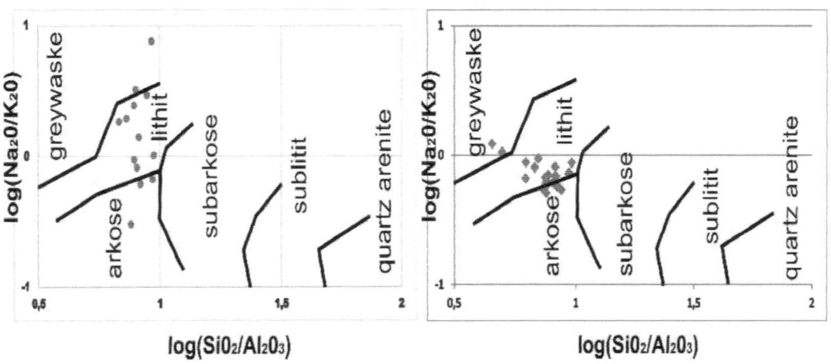

Figure 4. Diagram by Pettijohn for 1-2 beds (A-Ufimian) and 9-10 beds (B-Lower Kazanian) sandstones

The main sandstone samples of beds 1-2 and 9-10 are of lithic sandstones, in the composition of which mainly clay components enriched in Al_2O_3 and depleted alkalis. The samples №5, №8 - bed 1; №100, №101 - bed 10, contain micaceous and chlorite components. Ufimian sandstones multiple re-settle–accumulation occurred far from the provenance area, in them the most pronounced tendency to enrich quartz – mature sandstones. Lower Kazanian sandstones were exposed to significant effects of chemical weathering.

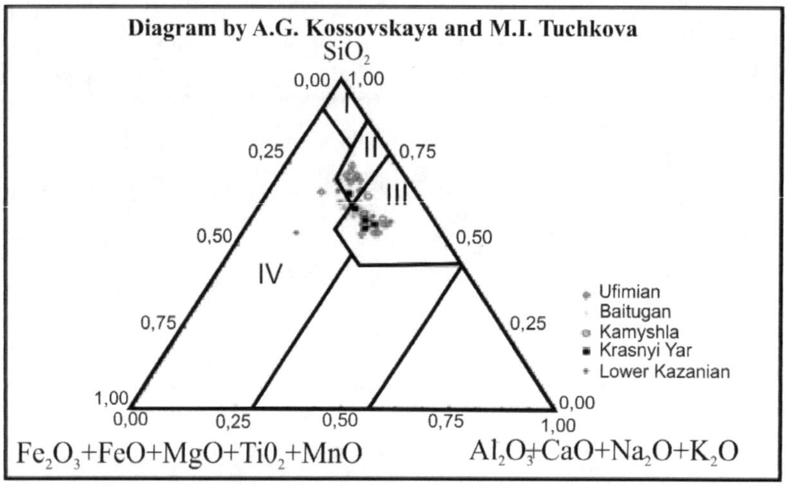

Figure 5. Diagram by A.G. Kossovskaya and M.I. Tuchkova with figurative points of sandstones of Middle Permian sediments

The upper corner in this diagram is either the composition of the rocks in the area of ablation, or the degree of maturity of the fragmentary material. The right corner of the classification triangle reflects the content of feldspars, and the left - the total melanocracy of rocks. The diagram separated four fields: I – quartz sandstone; II – oligomictic sandstone; III – polymictic sandstone; IV - volcaniclastic sandstone. Practically Ufimian sandstones are oligomictic; all Kamyshla sandstones are polymictic.

The most sandy rocks are formed from persilicic or average igneous rock or metamorphic formations in which is present.crystalline quartz.

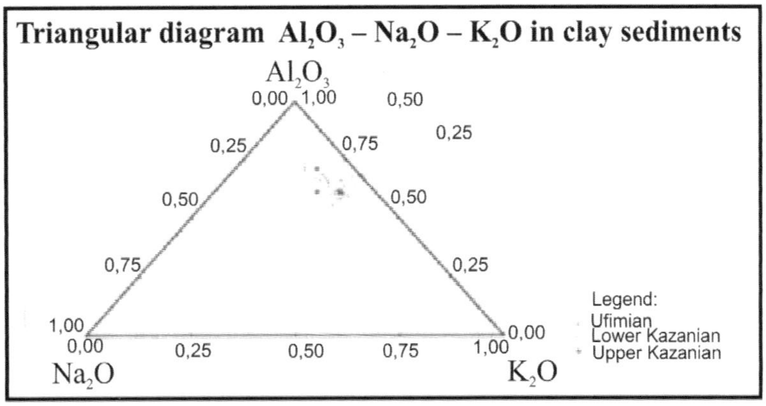

Figure 6. Triangular diagram $Al_2O_3 – Na_2O – K_2O$ in clay sediments

According to the figurative expression of A.G. Kossovskaya and M.I. Tuchkova, sand-and- gravel clastic of abyssal basins and the greater of the mid-ocean ridges, formed by the rocks breaking of the mafics, gradually "melts" and gives rise to a wide range of authigenic minerals and paragenesis of such minerals.

The main triangular diagram is supplemented by two diagrams: AM - [(Al_2O_3/SiO_2-Fe_2O_3+FeO+MgO+$Ti0_2$+MnO)] и AF - [(Al_2O_3/SiO_2)-(CaO+Na_2O+K_2O)]. The ratio Al_2O_3/SiO_2 is there an indicator of the maturity of psammites, CaO+Na_2O+K_2O indicates the presence of feldspar, Fe_2O_3+FeO+MgO+TiO_2+MnO as well as in the main diagram, it reflects the degree of melanocracy. These diagrams show the composition fields of the main types of mafic [7,8].

Clay samples in the section in the basin of the r. Sheshma and r. Inesh on the triangular diagram Al_2O_3 – Na_2O – K_2O (within Al_2O_3 – K_2O) as a result of increased chemical processes of hydrothermal alteration, controlled by warm and humid climatic conditions, lost in its composition alkali metals within closer to Al (fig.6). The upper corner Al in the diagram Al_2O_3 – Na_2O – K_2O does not divide the clay sediments into chlorites, smectites or kaolinites. Samples in the upper parts of the triangular diagram indicate that the maximum loss of alkali metal elements occurred during the weathering process. During intense chemical weathering, elements such as K and other alkaline-earth elements may be identifiers indicating kaolinite as the only residual phase of intense weathering of rocks. In our case, all the samples directed towards the increased value of Al_2O_3, which links the formation of hydromica with potassic feldspar (K-spar) [9].

5. Conclusions

For clays of the Upper Kazanian and Ufimian ages, generally characteristic the continental cold and moderately cold conditions of their formation. In their chemical composition, they carry only minor signs of weathering of the source rocks. These clays formed directly on rocks of provenance area or re-settle into freshwater bodies of water.The compositions of these clays are most similar in composition to igneous rocks, due to which they formed. While the Lower Kazanian clayey sediments in more cases formed in the marine and lagoon conditions. All the clayey sediments in the section in the basin of the r. Sheshma and r. Inesh formed under conditions of for the most part suppressed chemical weathering.

The sandstones in the basin of the r. Sheshma and r. Inesh with minor SiO2 content, on average, 63.77% (in the range of 55.7-76.04). The minimum content of quartz in sandstone in bed 9 (the beginning of the Kamyshla time). The maximum values are characteristic of Ufimian sandstones (beds 1,2). The concentration of TiO_2 is on average 0.814% (in the range 0.386-1.29) —the maximum values in the sandstones of the Upper Kazanian age in bed 25 - minimum are concentrated in bed 10 in Kamyshla time, content of Al_2O_3 is about 8.05% (in the range of 5.76-10.23), Fe_2O_3 the average 6,31% (in the range of 3,68-17,9) and MgO is about 3,84% (in the range of 1,15-5,55). The minimum MgO values are characteristic of Ufimian sandstones, while the maximum values are mainly in the sandstones of the "Lingula" package at the beginning of the formation. Changes in the basic elements of geochemistry in the sandstones in the basin of the r. Sheshma and r. Inesh are presented in a variation diagram of a linear dependance SiO_2 with TiO_2, Al_2O_3, Fe_2O_3, MnO, CaO, MgO, Na_2O, K_2O [1,4,7].

The negative correlation of SiO_2 with other basic elements is associated with a most with silicon dioxide being engrossed in quartz. In these samples, the concentration of TiO_2 increases with Al_2O_3, which indicates that TiO_2 is most likely, associated with phyllosilicates, particularly illites. Such compounds can form only in the presence of water [10].

Geochemical data, as well as lithochemical, allow in many cases a more accurate reconstruction of the facial and tectonic features of sediment accumulation. According to geochemical data, currently information can be obtained on the facies conditions of environment (depth area of sedimentation, mineralization, redox features of the bottom water, etc.)

References

[1] Nurgalieva N G *et al* 2016 T*he Lower Kazanian rocks as shallow marine facies (South-Eastern Tatarstan) on geochemistry* (*ARPN JEAS*, vol 11 **23**) pp 13462-13471

[2] Ignatiev V I 1976 *The formation of the Volga-Urals Anteclise in the Permian Period* (Kazan: Kazan University Press) p 256

[3] Kotlyr G V *et al* 1984 *Osnovnye cherty stratigrafii permskoy sistemy SSSR (Leningrad: The main features of the stratigraphy of the Permian system of the USSR)* p 280

[4] Nurgalieva N G *et al* 2015 Type and reference sections of the Middle and Upper Permian of the Volga and Kama river regions *Karkali section. Ufimian/Kazanian boundary and the Lower Kazanian in shallow marine facies* (Kazan: Kazan University Press) pp 176-192.

[5] Fakhrutdinov E I *et al* 2015 *The Lower Kazanian substage in the key section: Lithologies and Paleoenvironments based on the ESR data* (Uchenye Zapiski Kazanskogo Universiteta. Seriya Estestvennye Nauki vol 157 **3**) pp 87-101

[6] Fakhrutdinov E I et al 2018 *Features of the particle size distribution of the Ufimian-Lower Kazanian sediments of the section of Karkali* (Academic science - problems and achievements XVII.: Proceedings of the Conference. North Charleston, USA) pp 5-9

[7] Yudovich Ya E *et al* 2000 *Basics of lithochemistry* (Saint Petersburg.: Nauka) p 479

[8] Pettijohn F J *et al* 1972 *Sand and Sandstone* (Heidelberg: Springer)

[9] Kay Scheffler 2004 *Reconstruction of sedimentary environment and climate conditions by multi-geochemical investigations of Late Palaeozoic glacial to postglacial sedimentary sequences from SW-Gondwana*
[10] Dabard M P 1990 *Lower Brioverian formations (Upper Proterozoic) of the Armorican Massif: Geodynamic evolution of source areas revealed by sandstone petrography and geochemistry* (France: Sedimentary Geology vol 69 **1-2**) pp 45-58

IOP Conf. Series: Earth and Environmental Science **282** (2019) 012028 doi:10.1088/1755-1315/282/1/012028

An estimation of Ufimian terrigenous reservoir lithology influence on oil recovery factor by steam stimulation basing on tube experiments

A. Rojas[1], I. Idrisov[1], V. Sudakov[1], D. Nurgaliev[1], A. Grachev[2]
[1]Unconventional Reservoirs Modeling Center, Institute of Geology and Petroleum Technologies, Kazan Federal University, Kazan, Russia
[2]Research Laboratory "Rheological and thermochemical investigation", Chemical Institute A. M. Butlerova, Kazan Federal University.
E-mail: allanrojas5@gmail.com

Abstract. The settings, scale and geometry of lab facilities to simulate enhanced recovery methods play a significant role in the approximation of underground conditions of oil fields. The reliability more or less of the best replication of these conditions is a question of the reservoir engineer when forecasts are made for cases when enhanced recovery methods are applied. This paper presents the results of two experiments carried on in a steam flooding tube facility adding clay in one of them to assess the influence in temperature growth and production rates and onset times of a highly viscous oil field of the Tatarstan Republic. The results of the runs show that 4% (by weight) of clay content in the porous media has a strong effect on the overall oil recovery, reducing the temperature growth rate in average 73% when steam stimulation methods are used.

1. Introduction

The steam stimulation process at certain physical and chemical conditions leads to aquathermolysis. It is considered a catalytic process where different reactions take place and consequently upgrading oil fractions [1]. In the reservoir rock of the Ufimian horizon the main lithologies are found dispersed clay sand deposition and clay micro interlayering [2]. The effect of clay minerals contained within the reservoir rock has been tested and found that in dependence on the mineral content, oil production might be affected by nanoparticles migration and consequently throat blockage and modification of relative permeability curves [3,4]. Another study [5] found that the sensitivity on relative permeability curves has a dependence on the type of clay and salinity content of water. These were tested in absence of heat induction. Clay minerals in the reservoir rock is thought to play a role in the overall production reduction effect due to differences in heat capacities. On the other hand, depending on the natural state of reservoir rock, clay minerals or metals might produce a catalytic upgrading effect on high molecular hydrocarbon chains simplifying them into lighter ones at certain physical conditions while steam is reacting with oil [6-8] and therefore relatively improving recovery; however, the latest is not the focus of this paper and any possible catalytic effect will be neglected. The purpose of this research is to assess the degree of influence of the reservoir lithology of the mentioned horizon not on the oil's composition modification but on the total recovery factor under steam stimulation conditions in lab scale, using core steam flooding, as a preliminary stage for the planned future aquathermolysis studies in our research center. Additionally, to provide an overview of our experience on the settings of the facility to simulate steam stimulation methods.

2. Experimental

After having been assessed geological data from cores and well logs it was identified the lithological units and obtained a geological model [2], a cross section is shown in figure 1. The experiment's preparation involves the use of a methodology protocol created in our "Unconventional Reservoirs Modeling Center" and adapted to core rock measurements and fluid samples' analysis. To measure the porosity a total of 8 loose sand samples underwent through a process of bitumen extraction and found an average representation of 30% for the reservoir rock of the Ufimian horizon. It was observed macroscopically the presence of clay minerals in the sand samples and of our interest to investigate the effect in the recovery factor, therefore x-ray analysis for identification and quantification of mineral content was made for several samples of the pay formation.. Oil samples underwent analysis of viscosity with temperature dependence. The oil used for the experiments corresponds to produced oil using thermal recovery with steam injection and after being produced having density of 960 kg/m3 and 8700 cP of viscosity at 8^0 C. The purpose of the experiments carried on do not involve a comprehensive study of the oil fractions' change nor oil catalysis due to aquathermolysis and mineralogy, therefore oil data of the shallow highly viscous oil reservoir of the Ufimian horizon of the Tatarstan Republic will be skipped.

The assemblage of the core into the reactor pipe was made for two experiments: sand + crude oil (case 1) and sand + clay + crude oil (case 2). This was done by previously mixing dry sand with crude oil and stirred at a constant rate for about 5 minutes or until homogeneity is reached by using a sand blender. For the second experiment (with addition of clay), it was mixed before adding crude oil with 4% clay (weight %) to simulate a clay dispersed system. The amount of sand, clay and oil in mixture was weighted and recorded as shown in (table 1). Oil saturation was planned to be 100%, meaning no water was added to the system. Later, the prepared core was added to the flooding facility in three batches, pressurizing to reservoir conditions (4.5 T) each time until the desired pressure stabilizes.

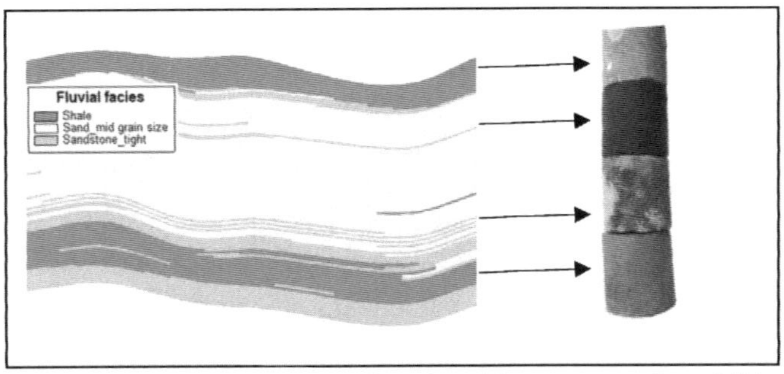

Figure 1. Cross section of field of study with core samples of lithotypes

Case #	Sand mass (%)	Oil mass (%)	Shale mass (%)
1	90.31	9.69	
2	82.67	13.35	3.98

Table 1. Core masses percentages for cases 1 and 2 of experiments under steam stimulation

In order to isolate the core, an epoxy solution at both edges of the pipe was poured and a special temperature resistant silicon-like material was placed in between of the epoxy and the pipe's wall to plug off from leaking fluids. Two wells were set, an injector and a producer. The wells are approximately 5cm long, 6mm external diameter and 3mm internal diameter. Temperature sensors placed all along at the top of the pipe to measure temperature development and steam chamber growth. Within the geometry of the reactor was used a cylindrical pipe with a pair of wells to simulate steam flooding technology as an EOR method. The flooding facility is a cylinder shaped with 40cm long and 10cm diameter. The facility operates under pressures of up to 50 bars and the steam generator heats with temperatures of up to around 400°C. The reservoir's initial temperature is 8°C, however both experiments were run initially at room temperature (25°C). Distillated water was used for the steam injection. An average of 17 kg of water were used per experiment to be injected. Every run was set to inject steam constantly under the same conditions (T:205°C, P:18 bars). The temperature of the reactor was controlled by a temperature controller. Pressure was regulated to keep constant temperature of 205 °C as in figure 9, therefore injection rates might be sensitive to these changes. Injection rate was set constant: 0.6 L/h of steam (distillated water equivalent). The pipe was thermally isolated to prevent heat losses with the environment. Also production line was isolated to reduce the impact of reduction of oil viscosity due to the cooling and be efficiently recovered into the sampling jar. For the dimensions of the pipe was determined an effective run time of average 30 hours when dismissible oil production was observed into the collection jars. In addition a back-pressure valve was installed next to the production line. Due to the characteristics of the facility, of the injector and producer it is considered a steam flooding setting. After the experiment the collected samples were submitted to centrifugation and separated into three phases: oil, water and sand and so determine the recovery factor. Consequently, viscosity of the oil samples were measured.

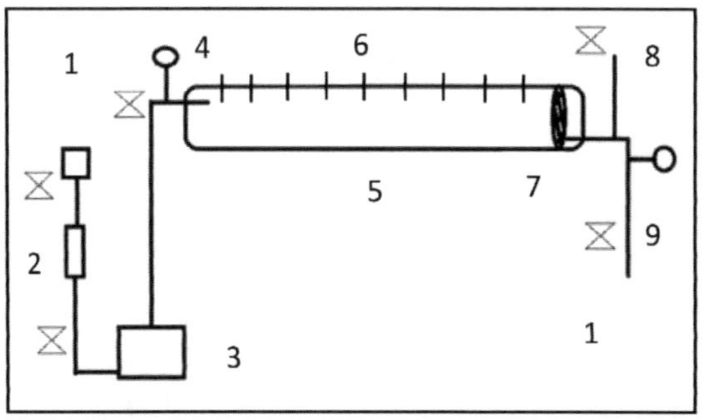

Figure 2. Scheme of steam flooding facility used in the Unconventional Reservoirs Modeling Center. 1) Water cylinder input. 2) Displacement cylinder pressurized with nitrogen. 3) Steam generator. 4) Pressure manometer at the injector well head. 5) Pipe reactor. 6) Thermocouples. 7) Metallic grid filter 8) Back pressure valve. 9) Pressure manometer at the producer well. 10) Sampling point.

3. Results and discussion

3.1. Reservoir parameters

Figures 3- 7. summarize the measurements and analysis of rock samples after bitumen extraction which showed that the mineral with higher abundance is Albite having 41% of its content. It was found that due to mineralogical analysis, the clay embedded in the interlayering of the reservoir sand of the Ufimian horizon, has a weak difference with the one on the top of the reservoir rock "Lingula clay"; however, it shows a fare resemblance to that at the bottom. Moreover, the overall content of clays in the rock samples studied, correspond to 4%. It is worth noting that the reservoir owes carbonate-cemented sandstones distributed in some parts of it and the influence on the overall recovery was not studied in this paper since it was addressed to only loose sand.

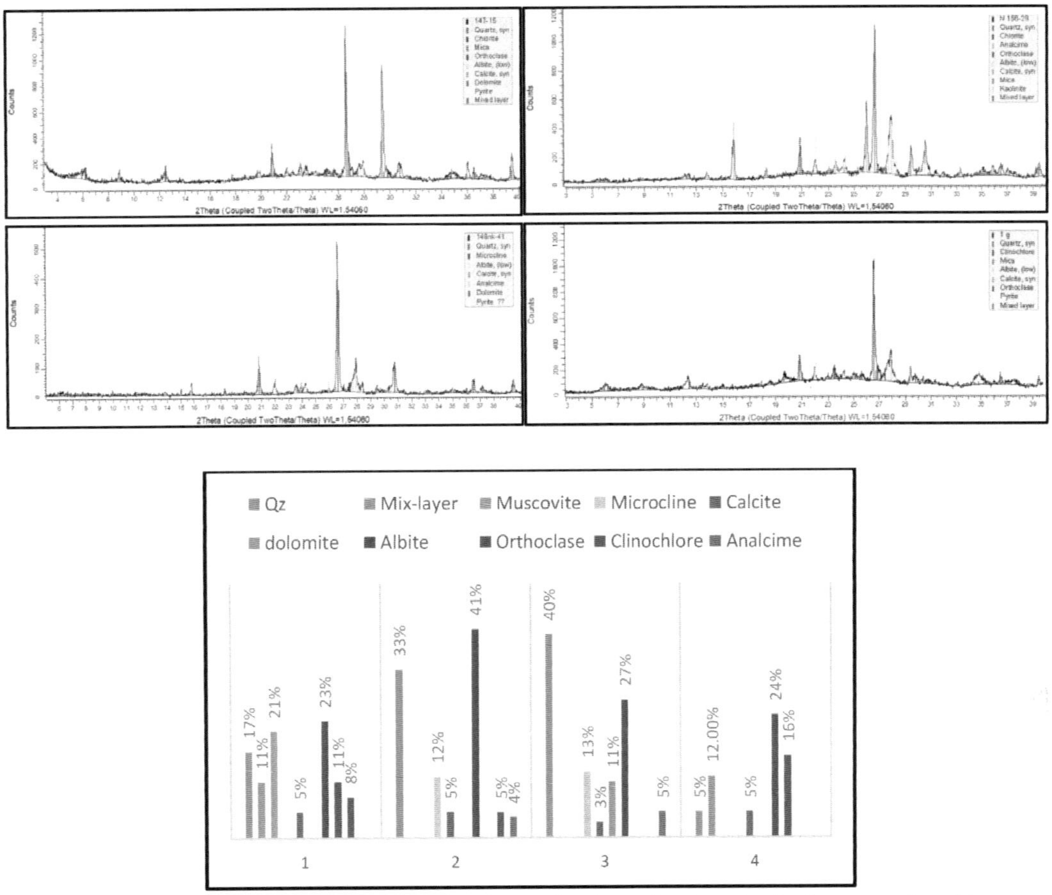

Figures 3-7. X-ray diffraction analysis for determination of mineralogical content in a bitumen-extracted loose sand sample of the Ufimian horizon.

3.2. Material balance

After the experiments were finished, the collected samples were centrifuged and the material balance was obtained. Within the results it was observed almost similar dynamics on the instantaneous oil production and with a few hours delay in the case 2 experiment which is agreed due to the higher heat capacity in comparison with that of sand; however it showed different dynamics on the rates response and on the total recovery factor, and so, being that 81,92 % without addition of clay (case 1) and 50.26 % adding 4% clay (case 2), suggesting that the clay has a strong influence in the overall recovery although it is a small percentage of 4 % as in figures 8. Absorption and adsorption studies of the oil into the clay were not carried on. The total Steam-Oil-Ratio was determined for case 2 being higher than case 2, as expected, due to the presence of clay as in figure 8b. An average of 17 steam pore volumes were injected during both experiments of steam stimulation with the prepared terrigenous core of the Ufimian horizon shown in figure 8a. In figure 8.b, in the fifth hour a peak of SOR in case 2 as a consequence of dismissible oil production in that time step, as in case 1 in time step 2 taking into account that it is an hourly time step basis. It is evident in figure 8.b, how the SOR gives a clue of how the steam injection becomes less economical after 12 hours (7 pore volumes) for case 1. Case 2 shows a delay in the plateau zone which is approximately at the hour 17 (10 pore volumes) which is in agreement with the 5 hours delay of the initial production time step in comparison with case 1.

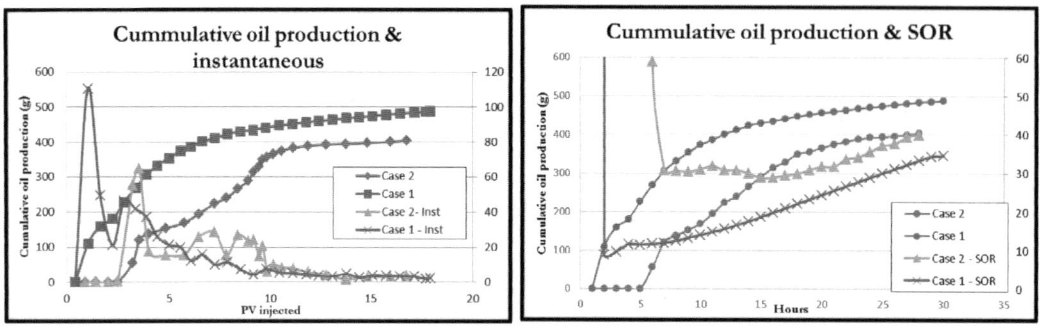

Figure 8. Oil production results comparisons of cases 1 and 2. a) cumulative and instantaneous oil, b) cumulative and SOR.

3.3 Temperature profile analysis

As shown in figure 2, thermocouples were placed along the tube to monitor temperature development during the steam displacement experiments. Case 1 and case 2 show quite different profiles (figures 9 and 10); even though both experiments were set to inject steam under 205° C. As in figure 10 during the first half of the run, temperature was increased slightly and progressively up to 215° C due to no response in oil production within the same period in which case 1 produced (hour 2). Temperature profile of case 2 shows a lean on of the curves. It shows two stages and it is hypnotized that: the first (up to 400 min) the delay is thought to be as a result of clay swelling and change in relative permeability curves. The second stage (400 – 1000 min), as a result of the higher heat capacity of swollen clay.

It is evident from figures 9 and 10 that the required amount of time for the last thermocouple to start temperature increase is by a factor of two, for case 2. For the oil tested it is a highlight to notice that the apparent mobility onset is around 40° C; however, at different operation times in dependence on clay concentration (case 1: 100 min, case 2: 400 min). Additionally, the oil production rate drastically decreases after temperature is normalized to 205° C (case 1: 400 min, case 2: 1000 min) (figures. 8 - 10). It can be said it represents the state of irreducible oil saturation; however, higher temperatures were not tested to prove this wrong. The temperature growth rate can be expressed as:

$$T\ (rate) = \frac{\Delta T}{\Delta time} \qquad (1)$$

Where:

T(rate) – Temperature growth rate (°C/min)

ΔT – Temperature increment (°C)

Δtime – Time from start of temperature increment until 205 °C are reached for last thermocouple (min)

From (1) it can be obtained that T (rate) is 0.56 °C/min and 0.15 °C/min for case 1 and case 2, respectively. It is worth mentioning that these observations were performed for a tube steam displacement facility where thermodynamic conditions are considered isobaric and confined to a volume of $0.00314 m^3$.

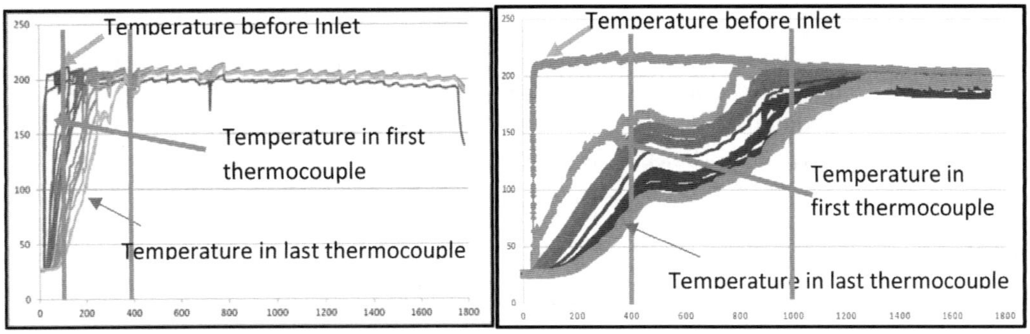

Figures 9 and 10. Temperature profile of experiment for case 1. X-axis: time in minutes, Y-axis: temperature in ^{0}C. Left – case 1. Right – case 2

4. Conclusions

X-ray analysis found that the reservoir clay does not show any determinant difference to those at the top and the bottom; therefore, heat capacities can be assumed to be also closely similar. Two cores were elaborated and mounted on a steam flooding tube facility. Two experiments were launched and run successfully. One of the experiments required the addition of clay (4% by mass) to measure the influence on the overall oil recovery. As expected the cumulative oil production for case 2 was lower than that of the case 1. It was correlated from the temperature profiles of the tests run that the oil's apparent mobility onset is around 40° C. Also temperature growth rate was estimated to be 0.56 °C/min for case 1 and 0.15 °C/min for case 2. Additionally, SOR is shown higher for case 2 all over the experiment. It can be concluded that the presence of clay has a strong influence on the reduction of the overall recovery factor and delay on the production times. Moreover, the contact time of reservoir with steam from the results obtained in the experiments performed could be used as an approach for prediction of production times and scheduling when interpolations of the clay index distribution in the reservoir could be done. Furthermore, for a more efficient timing forecast to applying hybrid enhanced oil recovery methods in the terrigenous reservoir rock of the Ufimian horizon since the highest SOR's were observed after approximately 17 hours (10 pore volumes) for both cases, under the conditions of the reported experiments here, where it shows itself to be non-economical. It is worth mentioning that there is a need for repeatability of the carried on experiments and also to trial the effect on different clay deposition patterns.

Acknowledgements

The work was supported by the Ministry of Education and Science of the Russian Federation (project No. 02.G25.31.0170) and by the subsidy allocated to Kazan Federal University as part of the state program for increasing its competitiveness among the world's leading centers of science and education.

References

[1] Oki Muraza, Ahmad Galadima Aquathermolysis of heavy oil: A review and perspective on catalyst development. Fuel 157 (2015) 219–231

[2] Delev A.N., Nurgaliev D.K., Sidorov S.V., Akhmadullin R.R., Aslyamov N.A. Construction of geological model of a shallow super-viscous oil deposit based on core analyses and geophysical monitoring data.

[3] Hirakawa S. Effects of clay and clay minerals on relative permeability curves. Vol 22, NO. 3

[4] Wang Xingxin. Characteristics of clay minerals and their effects on production capacity of the Cretaceous sandstone reservoirs of Songliao Basin, China. In: Proceedings of the 9th international Clay Conference, Strasbourg, 1989. Vol IV : Clays in sediments. Diagenesis and hydrothermalism. Strasbourg : Institut de Géologie – Université Louis-Pasteur, 1990. pp. 105-114. (Sciences Géologiques. Mémoire, 88)

[5] Baptist O. C. and Sweenrey S. A.. Effect of clays on the permeability of reservoir sands to water of different salinity contents.

[6] S.M. Petrov, D.A. Ibragimova, A.V. Vakhin, I.M. Zaidullin, G.R. Valieva, Z.R. Zakirova. Investigating the structure and composition of heavy oil under thermal-catalytic treatment in presence of carbonaceous minerals.

[7] Abdrafikova, I.M. Conversion of extra-heavy Ashal'chinskoe oil in hydrothermal catalytic system / I.M. Abdrafikova, G.P. Kayukova, S.M. Petrov, A.I. Ramazanova, R.Z. Musin, V.I. Morozov // Petroleum Chemistry. – 2015. – Vol. 55. – Issue 2. – P. 104-111)

[8] Abdrafikova, I.M. Structural-Group Composition of Heavy Oil Conversion Products Using FTIR Spectroscopy / I.M. Abdrafikova, G.P. Kayukova // Helix. – 2018. – Vol. 8(1). – P. 2898-2907. - DOI 10.29042/2018-2898-2907

IOP Conf. Series: Earth and Environmental Science **282** (2019) 012029 doi:10.1088/1755-1315/282/1/012029

Lab rock core studies of the Ufimian horizon as a prior step to displacement experiments with steam

I. Idrisov[1], A. Rojas[1], R. Sagirov[1].
[1] Federal University, Kazan, Russia
E-mail: ilyas_irekovich@mail.ru

Abstract. The author in this paper determined the coefficient of porosity from well log data and carried on a study of core samples using X-ray microtomography and traditional methods. A method was elaborated and later tested for measurement of porosity of samples taken from the production interval of shallow super viscous oil deposits of the Tatarstan's Republic. Also determination of mineralogical composition and characteristics of the reservoir rock for application to future steam flooding experiments. On the basis of the data obtained, was concluded which of the methods gives the most reliable information on this of the reservoir rock under study.

1. Introduction

Definition of reservoir properties and bitumen saturation of productive rocks is one of the minimum input data to enable assessing reserves and planning methods for extracting bitumen. The source of information may be from core studies or from geophysical surveys. However, in the case of bitumen-saturated loose sands of the Tatarstan Republic, the question of estimates of reservoir rock properties, both from core studies and geophysical surveys, acquires significant importance.

The purpose of the work is to study the properties of the terrigenous reservoir rock of the Ufimian horizon of the super-viscous oil field of the Republic of Tatarstan, in order to clarify the parameters of the bulk model when performing steam flooding experiments. The task comprised the determination and comparison of the coefficient of porosity of weakly cemented – unconsolidated sands, using different methods. Furthermore, to develop and test a method for the determination of porosity in unconsolidated samples. The determination of mineralogical composition and characteristics of the reservoir for application of steam flooding experiments and analysis of the results obtained.

2. Methodology

Determining weakly cemented – unconsolidated sands' parameters of the Nizhnoye-Karamalskoye deposit is of considerable challenge in laboratory practice. This fact is due to the core sampling actual practices which modify the original natural structure of the rock and adding some difficulty to the direct measurement of reservoir rock properties. Hence the need to apply fundamentally different methodological approaches to the study of these rocks when special sampling methods are not used.

The main methodological issues of laboratory analysis of weakly cemented rocks – unconsolidated sands were worked out in collaboration with TatNIPIneft and Kazan State University, which resulted in a Standard of the Company for the analysis of uncemented and weakly cemented rocks in laboratory, introduced into Tatneft in 1996. Due to the fact that bitumen is not mobile, meaning no drilling mud filtration and while performing downhole core sampling, bitumen from the core is not displaced by the drilling mud, therefore with some modification of the technique in this case, gives the possibility to get advantage of this feature and obtain additional information about the reservoir. The developed volumetric method for measuring the porosity of uncemented samples made it possible to determine the clay content, and to conduct a granulometric analysis and X-ray phase analysis.

Nizhnoye-Karamalskoye deposit of super-viscous oils of the Cheremshanskoye field is spatially confined to a sedimentation trap, formed as a result of a local increase in the thickness of the sand pack of the Sheshminsky horizon. The deposit is confined to the crest of reservoir and crest position trap parts, catches and refers to the type of Layer-uplifted deposit. Oil sands have high reservoir properties.

2.1 Porosity measurement methods

2.1.1 Porosity measurement based on gamma-gamma density data

Gamma-gamma density log has a high geological efficiency and is applied for the determination of bulk density, porosity, lithotype definition, characterization of geological bodies with anomalously low bulk density. The density gamma-gamma log method is based on measuring the intensity of artificial gamma radiation scattered by rock-forming elements in the process of irradiating them with a stream of gamma rays. From the (1) it follows that to determine KP according to the gamma-gamma density logging data, it is necessary to know the values δ_f and δ_m

$$Kp = \frac{\delta m - \delta}{(\delta m - \delta f)} \quad (1)$$

Where:
Kp: Porosity coefficient (dimensionless);
δ is the bulk density of the rock (g/cc);
δ_m is the density of the matrix (g/cc);
δ_f is the density of the fluid filling the pore space (g/cc).

The density of the matrix = 2.7 g/cc, the density of the reservoir fluid = the density of bitumen 0.98 g/cc.

2.1.2. Porosity measurement with x-ray computed tomography (sandstone)
Core measurements were performed using a 2D inspection of Phoenix v | tome | x s240 in the X-ray computed tomography laboratory of the Institute of Geology and Oil and Gas Technologies of Kazan (Volga Region) Federal University [1] (Fig. 1). The method of non-destructive study of the internal structure of an object is based on measuring the attenuation of X-rays by different parts of the object, with different density, composition, and thickness. For the formation of three-dimensional images of the internal structure of the object, complex computer processing of data sets over a set of its 2D shadow projections is used.

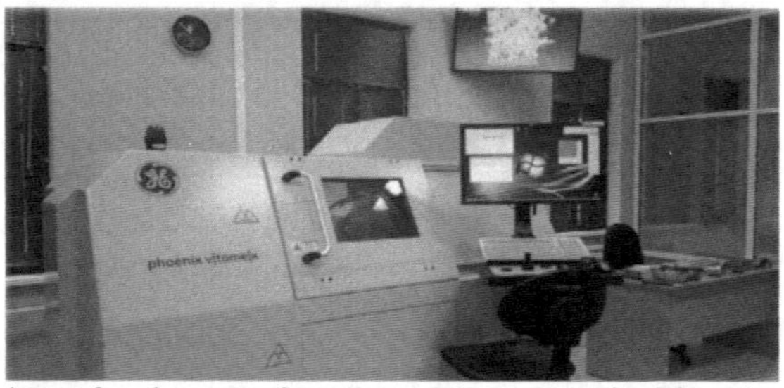

Fig. 1. Micro / nanofocal system for computed tomography Phoenix v|tome|x s240

2.1.3. Study of measurement with traditional method (open porosity with sandstone)
Method for determination of the coefficient of open porosity by liquid saturation. Additionally, bulk density, bulk bitumen saturation. The technological sequence of operations are as follows:

1. The saturation of the rock samples with the fluid (kerosene or a simulated formation water) using the capillary imbibition. [2] Fig.2
2. Weighing saturated sample in the saturating fluid and in air
3. According to the results of weighting are carried out calculations as:
   - the ratio of open porosity,
   - bulk density
   - bulk bitumen saturation
4. Drying samples at 105 ° C for 40 h

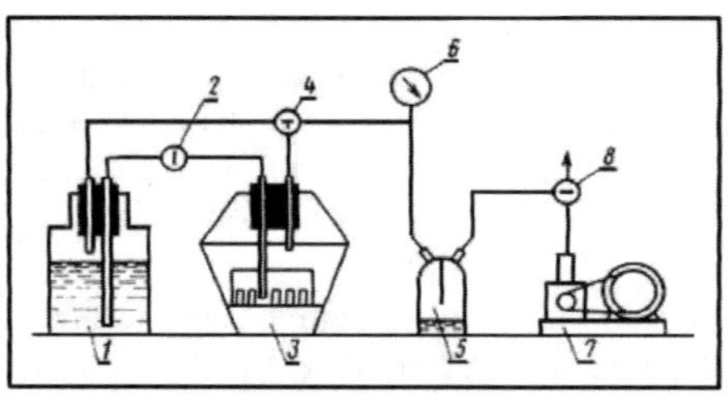

Fig.2 Device for samples' saturation. 1-vessel for saturating fluid; 2- shut-off valve; 3- vessel for saturation of samples; 4- three-way crane; 5-trap for preventing liquid from entering the vacuum pump; 6 - pressure gauge; 7-vacuum pump; 8- shut-off valve of the vacuum pump and filling the system with atmospheric air.

2.1.4. Study of porosity measurement with volumetric method (unconsolidated sand)

Determination of the parameters of the unconsolidated sands of the Nizhnoye-Karamalskoye deposit is of considerable difficulty in laboratory practice. The developed volumetric method for studying porosity allowed to determine the parameters of weakly cemented rocks and unconsolidated sands that are most commonly found in core samples from the Ufimian and Kazanian horizon

The studied samples are weakly cemented – unconsolidated, intensively bitumen-saturated unconsolidated sands. It is relatively weakly compressed by the pressure of the overlying rocks due to the shallowness of the reservoir formation. The highly-viscous oil represents the cementing material. In the volumetric method of determining porosity, we assume that bitumen occupies the entire porous volume of the sample. Thus, immersing the sample in a liquid (benzene), we can determine the total volume of the sample. Cold extraction of bitumen was carried out using two solvents: benzene and chloroform. For the first stage of extraction of bitumen from the sample, benzene was used. The completion of the extraction was determined visually, when the extraction liquid (benzene) showed itself transparent. Chloroform was used to dissolve the residual hydrocarbons, and the end of extraction was also determined visually. After the complete removal of bitumen from the sample immersed in benzene, was determined the volume of bitumen extracted. The difference between the total volume of the sample and the volume of the extracted sample is equal to the pore volume (we assume that bitumen occupies the entire porous volume of the sample). Porosity is defined by:

$$Poro = \frac{Vp}{Vs} \tag{2}$$

Where:
Poro - Porosity (dimensionless);
Vp - pore volume (ml);
Vs - sample volume (ml)

Fig.3 Bitumen saturated, unconsolidated sand of the Ufimian horizon sample. Determination of the sample volume before extraction by immersion in benzene. Determination of the sample volume after extraction by immersion in benzene.

2.2. Determination of the mineralogical composition.

The bitumen extracted samples and remaining sediments on filter paper after the volumetric method for determination of porosity and river sand for bulk model were sent to the laboratory of X-ray phase analysis for determination of its mineralogical composition. X-ray phase analysis identifies the various phases in the mixture based on the analysis of the diffraction pattern given by the sample under study. The determination of a substance in a mixture is carried out by the set of its interplanar distances and relative intensities of the corresponding lines on the radiograph. The sand pack of the Ufimian horizon is composed of polymictic sands representing gray, light gray rocks.

2.3. Granulometrical features of unconsolidated sands.

The granulometric composition of rocks is the quantitative content of particles of various sizes in the rocks. The granulometry characterizes the degree of dispersion of mineral particles composing the rock. Many other reservoir properties of the porous medium depend on the degree of dispersion of minerals:

porosity, permeability, specific surface area, residual water saturation, oil saturation, capillary forces, and others. After cold extraction the samples passed through sieve and determined granulometric composition of rock [3].

2.4. Clay index of unconsolidated sands.

The clay index is a property of rocks to contain particles with a diameter less than 0.01 mm. Clay particles are fragments of clay minerals + impurities. The properties of rocks contain a number of clay particles that occupy a space between larger grains. It is called dispersed system in contrast to the actual thin interlayering system of the rock. On the filter paper after cold extraction and during the determination of porosity by the volumetric method, a remaining sediment was obtained and was sent to X-ray phase analysis. The sediment was confirmed to be a clay. The clay content of sedimentary rocks is characterized by the content in the rock matrix of a rock with an effective diameter of less than 0,01 mm. Clay content is usually set according to the granulometric composition of rocks and is calculated by (3):

$$C = \frac{\text{Mf}}{Mm} \tag{3}$$

Where:
C – Clay index (dimensionless);
Mf - mass fraction with a diameter of <0,01 mm (g);
Mm - mass of the sample (g)

3. Results and discussions.

Studies of bituminous core rocks require a different treatment from the standard approach. The large number of steps of the process along with the number of samples' weighting and corrections for losses in weights lead to a highly time consuming duty in comparison with studies of core samples of conventional reservoirs. The measurement of porosity with the different methods used in this study (Table 1), for weakly cemented sandstones and unconsolidated sands show an average porosity of 30%.

№ Well	X-ray computed tomography Porosity%	Traditional method Porosity %	GGKp) Porosity %
Study of porosity of consolidated sandstone			
149/1	30,11	29,9	29,8
149/2	30,96	28,7	27,8
153/1	27,92	-	25-30
153/2	25,62	-	26
155	32,97	-	31,5
162	26.08	29,8 -33,5	31,5
164	29,9	29,8	29,8
Study of porosity of unconsolidated sand			
		Volumetric method, Porosity %	
150/1	-	30	30
150/2	-	29	30
156	-	29	29
160	37,27%	25	27

Table 1. Results of porosity measurements of core samples of the Ufimian horizon using different methods.

The sand pack of the Ufimian horizon is composed of polymictic sands (Fig.4a) representing gray, light gray rocks. For bulk model we will use common river sand which was also analyzed and showed a more inert nature (Fig.4b). These rocks consist mainly of grains of quartz rounded to different degrees.

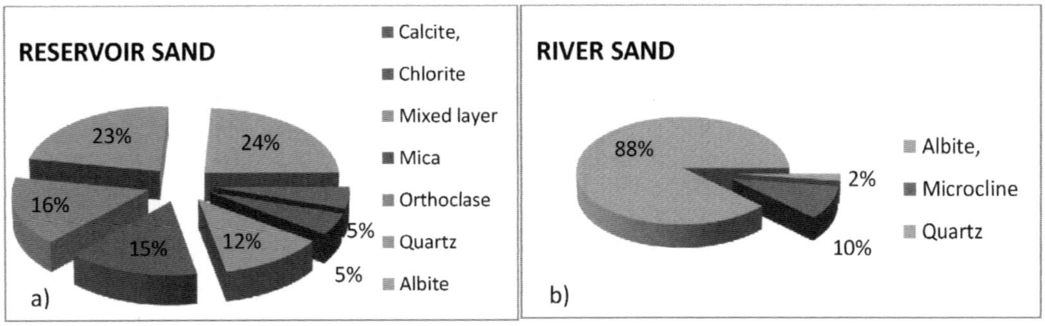

Fig.4 X-ray phase analysis results for reservoir sand and common river sand

The Nizhne-Karmalskoye deposit of the Cheremshanskoye field is represented by the Sheshminsky horizon (Fig. 5), consisting of a lower sand-clay pack and the upper sandstone pack. The overlaying rock, which lies on top of the sand pack is composed of clay, dark gray, with a bluish tinge and is a good cap rock (Lingula clay). X-ray phase analysis results show, from Lingula clays, clays from the pay zone, clays from the lower sandy clay package have the same clay components. The granulometric composition of the mixture of various fractions up to 0.5mm of quartz sand (Table 2). On the filter paper after cold extraction and during the determination of porosity by the volumetric method, a remaining sediment was obtained and was sent to X-ray phase analysis. The sediment was confirmed to be a clay. The clay index of the studied samples was determined (Fig.6). From (4): The clay content is equal to 6%. However, among all the samples analyzed it was obtained and average value of 4% (by weitght) (table 3).

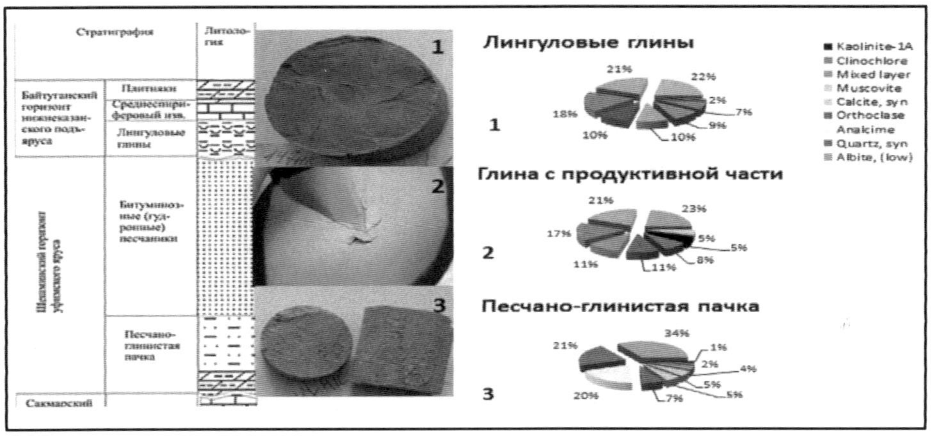

Fig.5 Average well section of the cap rock and pay zone samples of the Ufimian horizon for research in X-ray phase analysis (1 - Lingula clays, 2 - clays from the sand pack pay zone, 3 - clays from a sandy clay package)

Cell size mm	Sand from reservoir, g	River sand, g
1	-	0,85
0.8	-	0,683
0.63	-	1,183
0.5	-	2,007
0.315	0,356	17,78
0.16	7,405	47,205
0.08	2,611	6,345
0.063	0,278	0,186
0.032	0,419	0,076
0.025	0,051	0,004
Sample mass without losses	11,28	76,916

Table 2. Granulometric composition of samples of the Ufimian horizon under study.

$$C = \frac{2.601}{45.028} = 6\% \quad (4)$$

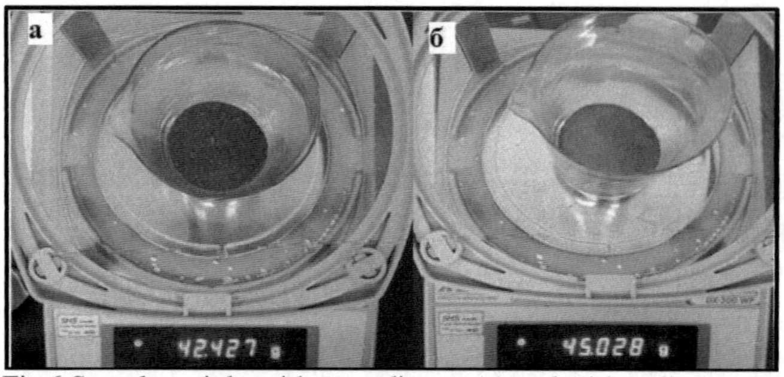

Fig.6 Sample weight without sediment (a) and with sediment (b)

№ Well	Clay content, %
149/1	4
149/2	4
153/1	5
153/2	6
155	3
162	4
164	4
150/1	4
150/2	3
156	5
160	4

Table.3 Results of measurements of Clay content in studied samples

Conclusions

The studies run have shown that the developed volumetric method for measuring porosity can be technologically simple, and is recommended for determinations of porosity of weakly cemented sandstones and unconsolidated sands. It expands the capabilities of their research, allowing measurements on wet and dry core samples, which cannot be studied by the traditional method, when no special downhole sampling methods are used. As a result of the obtained values the reservoir sands and sandstones are characterized by good reservoir properties and have an average porosity of 30%, a mixture of various fractions up to 0.5mm of quartz sand. X-ray phase clay analysis results show that for a bulk model can be used clay from a sandy clay package with average value clay content of 4% (by weight). According to the results obtained, they will be used for estimations of pore volumes of future displacement experiments with steam.

References

[1] https://kpfu.ru/laboratoriya-rkt-70706.html
[2] GOST 26450.1-85 "Method for determination of open porosity by fluid saturation"
[3] GOST 12536 -2014 "Soils. Methods of laboratory granulometric (grain-size) and microaggregate distribution"

IOP Conf. Series: Earth and Environmental Science **282** (2019) 012030 doi:10.1088/1755-1315/282/1/012030

Building a model of the isoprene polymerization process in the presence of microheterogeneous neodimiumcatalytic systems

Eldar Miftakhov[1], Sofya Mustafina[2], Olga Medvedeva[3], Dmitriy Zhavoronkov[4] and Svetlana Mustafina[2]

[1]Ufa State Aviation Technical University, 12 K. Marx str., Ufa, Russia
[2]Bashkir State University, 32 Zaki Validi str., Ufa, Russia
[3]Kazan Federal University, 18 Kremlevskaya str., Kazan, Russia
[4]Open Joint Stock Company "Synthez-Kauchuk", 14 Tech str., Sterlitamak, Russia

E-mail: OAMedvedeva@kpfu.ru

Abstract. In the research is built the mathematical model of the isoprene polymerization process in the presence of microheterogeneous neodimium catalytic systems. The direct problem is solved and is analyzed the molecular-mass distribution of the polymer. An algorithm of solving the inverse problem for search and elaboration of the kinetic parameters of the research process is developed.

1. Introduction

Production of polymer materials takes one of the leading places in the modern chemical and petrochemical industry. A significant share in it brings the production of synthetic polyisoprene, which accounts for more than 700 thousand tons of the product in year. Polyisoprene is a product of solution polymerization of isoprene in the presence of microheterogeneous titanium and neodymium catalytic system type Ziegler-Natta, characterized by high stereo specificity and the possibility of obtaining rubber of stamps SKI-3 and SKI-5 with a complex of necessary consumer properties. Earlier made studies [1] have shown, it's possible to impact on the dispersion of microheterogeneous catalytic system of the Ziegler-Natta due to local hydrodynamic influence on the reaction mixture by changing the technological scheme of the preparation of the titanium catalyst. Building of the mathematical model taking into account these changes allows not only to predict the properties of the product, but also to optimize the production process.

2. The technology of the production process

The process of isoprene polymerization in rubber production is carried out in a continuous method in a battery of sequentially interconnected polymerizers, by way of which are used the vertical cylindrical apparatus with a volume of 16.6 m^3, equipped with four-tier two-vane mixing devices. The power of the electric motor of the mixing device is 30 kW, what provides a mixing speed of 20÷40 rpm. The delivery of the heat carrier to the polymerizers is not carried out, and the temperature of the reaction mixture is regulated by the temperature of the incoming monomer solution and the dosage of the catalyst.

Particularly the polymerization of isoprene in the presence of a neodymium catalyst in industrial production is carried out using a cascade of three bulk reactors. The initial temperature of the isoprene solution in isopentane is 0-minus 5^0C, the mass flow rate is 20 t / h. The consumption of neodymium catalyst is 1 mol. NdCl3/10000 mol. isoprene. In order to regulate the molecular weight of the polymer, into the process line is continuously supplied the diisobutylaluminium hydride (DIBAH) .The scheme of the production process is shown in Fig.1

3. A kinetic scheme of the polymerization process

The kinetic method of modeling polymerization processes is the compilation and numerical solution of kinetic equations for the concentration of all types of particles involved in the process [3]. The construction of a mathematical model is complicated by the kinetic heterogeneity of the neodymium catalyst, which, alongside with other factors, is determined by the amount of isopropyl alcohol (IPA) included in the complex. The results of the experiments and the solution of the inverse problem of the formation of the molecular mass distribution of polyisoprene allows to judge that in the process of polymerization of isoprene on the catalytic system NdCl3·nIPA-TIBA (triisobutylaluminium)-PP is possible the existence of two types of active centers (AC).This leads to a twofold increase of number of considered reactions, which are complicated by the search of kinetic parameters[4,5] characterizing the rate of one or the other reaction. However, the hydrodynamic effect on the reaction mixture of synthesisNdCl3·nIPA synthesis (method 2) leads to an increase in the content in structure of the complex IPA and the formation of a monocenter catalyst, which greatly simplifies the search for kinetic constants.

In addition, by the formulation of the reaction scheme, it was assumed that the rate of the initiation reaction is much greater than the rate of the chain growth reaction, and the initiation reaction is not limiting. Then the kinetic scheme of the isoprene polymerization process in the presence of neodymium-containing catalytic systems can be described as the following stages:

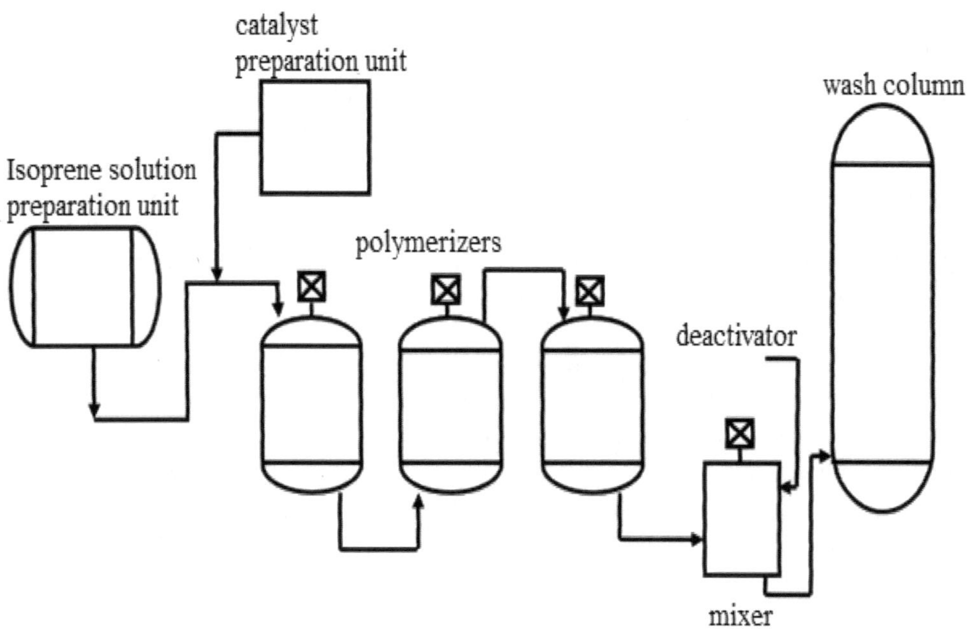

Fig.1. Scheme of the continuous production of rubber SKI-3

1. Chain growth

$$P_i + M \xrightarrow{k_p} P_{i+1},$$

2. Chain transfer to monomer

$$P_i + M \xrightarrow{k_m} Q_i + P_1,$$

3. Chain transfer to organoaluminum connection (OAC)

$$P_i + A \xrightarrow{k_a} Q_i + P_1,$$

4. Death of active centers

$$P_i \xrightarrow{k_d} Q_i,$$

where M – is the monomer, A – the concentration of OAC, P_i – the active chain of the polymer length i; Q_i – inactive chain of the polymer length i, k_p, k_m, k_a, k_d – the rate constants of the reaction of chain growth, transfer to monomer, to the OAC and the death of active centers respectively.

4. The construction of the mathematical model and the solution of the direct problem

The principal difference between polymerization processes from all other chemical processes is that instead of a product with a fixed molecular weight is obtained a certain molecular weight distribution of the polymer (MWD). The static theory of polymerization is based on the possibility of MWD analysis by a special kind average molecular masses $\overline{M}_w, \overline{M}_n$, called the average mass and the average count [6]. For the analysis of MWD are entered concepts of moments of active and inactive chains:

$$\mu_j = \sum_{i=2}^{+\infty} i^j P_i, \quad \eta_j = \sum_{i=2}^{+\infty} i^j Q_i. \tag{1}$$

In addition, it is necessary to find derivative moments of active and inactive chains by the location time

$$\frac{d\mu_j}{dt} = \sum_{i=2}^{+\infty} i^j \frac{dP_i}{dt}, \quad \frac{d\eta_j}{dt} = \sum_{i=2}^{+\infty} i^j \frac{dQ_i}{dt}. \tag{2}$$

On the basis of the described kinetic scheme using the law of active masses, we write a system of differential equations describing the change in the concentrations of all components of the polymerization mass:

$$\begin{cases} \frac{dM}{dt} = -M\sum_{i=0}^{\infty} P_i(k_p + k_m), \\ \frac{dA}{dt} = -k_a A\sum_{i=0}^{\infty} P_i, \\ \frac{dP_1}{dt} = -k_p MP_1 + (k_m M + k_a A)\sum_{i=2}^{\infty} P_i, \\ \frac{dQ_1}{dt} = k_m MP_1 + k_a AP_1, \\ \frac{dP_i}{dt} = k_p M(P_{i-1} - P_i) - k_m MP_i - k_a AP_i - k_d P_i, \\ \frac{dQ_i}{dt} = k_m MP_i + k_a AP_i + k_d P_i. \end{cases} \tag{3}$$

The derived system of equations containing infinite sums we transform by the method of moments to a system of differential equations for statistical moments of the molecular mass distribution of polyisoprene (4), where μ_i, η_i - moments of molecular mass distribution of active and inactive chains respectively.

$$\begin{cases} \frac{dM}{dt} = -MP_1(k_p + k_m) - M\mu_0(k_p + k_m), \\ \frac{dA}{dt} = -k_a AP_1 - k_a A\mu_0, \\ \frac{dP_1}{dt} = -k_p MP_1 + (k_m M + k_a A)\mu_0, \\ \frac{dQ_1}{dt} = k_m MP_1 + k_a AP_1, \\ \frac{d\mu_0}{dt} = k_p MP_1 - (k_m M + k_a A + k_d)\mu_0, \\ \frac{d\eta_0}{dt} = (k_m M + k_a A + k_d)\mu_0, \\ \frac{d\mu_1}{dt} = 2k_p MP_1 + k_p M\mu_0 - (k_m M + k_a A + k_d)\mu_1, \\ \frac{d\eta_1}{dt} = (k_m M + k_a A + k_d)\mu_1, \\ \frac{d\mu_2}{dt} = 4k_p MP_1 + k_p M\mu_0 + 2k_p M\mu_1 - (k_m M + k_a A + k_d)\mu_2, \\ \frac{d\eta_2}{dt} = (k_m M + k_a A + k_d)\mu_2, \\ \frac{d\mu_3}{dt} = 8k_p MP_1 + 3k_p M(\mu_2 + \mu_1) + k_p M\mu_0 - (k_m M + k_a A + k_d)\mu_3, \\ \frac{d\eta_3}{dt} = (k_m M + k_a A + k_d)\mu_3, \end{cases} \tag{4}$$

The initial conditions for the system (4) are as follows:

$$\begin{aligned} &M(0)=M_0, A(0)=A_0, \\ &P_1(0)=P_0,\ Q(0)=0, \\ &\mu_i(0)=\eta_i(0)=0, i=0\ldots3. \end{aligned} \tag{5}$$

The calculation of the average count $(M_n)$and average mass(M_w) molecular masses will be produced on the basis of the obtained values of the moments of active and inactive chains:

$$\begin{aligned} M_n &= m_0 \frac{\mu_2+\eta_2}{\mu_1+\eta_1}, \\ M_w &= m_0 \frac{\mu_3+\eta_3}{\mu_2+\eta_2}, \end{aligned} \tag{6}$$

where m_0 – is the molar mass of isoprene (m_0 = 68.119 g/mol.).

For further calculation of averaged molecular characteristics it is necessary to know the values of all rate constants of elementary reactions of the kinetic scheme of the process. It should be taken into account that the required values of constants in the case of isoprene polymerization on catalytic systems with different mole content of IPA have a different in size character. To find kinetic constants, the inverse problem was solved [7].

5. Solution of the inverse problem

The strategy of finding the kinetic parameters of polymerization consist insearch of the minimum matching function of the experimentally determined parameters and calculated on the basis of a mathematical model describing the polymerization process.

The solution of the inverse problem for this process was carried out in two stages. At the first stage, on the basis of experimental data on the conversion change, was carried out the search of constants of chain growth rates and chain transfer to the monomer by the way of minimizing of the functional following type:

$$H(z)=\sum_{i=1}^{p}\left(K_i^{\text{exp.}}-K_i^{calc.}\right)^2,\ \vec{z}=\left(k_p,k_m\right), \tag{7}$$

where $K_j^{\text{exp.}}$ – value of conversion defined in the course of the experiment, $K_j^{calc.}$ – the calculated values of conversion.

At the second stage, on the basis of experimental data on the change of the average count and average mass molecular masses of polyisoprene was carried out the search for the constants of the chain transfer rates to the OAC and the reaction of the death AC by the way of minimizing two functional types simultaneously:

$$\begin{aligned} F(w) &= \sum_{i=1}^{p}\left(M_n^{\text{exp.}}-M_n^{calc.}\right)^2,\ \vec{w}=\left(k_a,k_d\right), \\ G(w) &= \sum_{i=1}^{p}\left(M_w^{\text{exp.}}-M_w^{calc.}\right)^2,\ \vec{w}=\left(k_a,k_d\right). \end{aligned} \tag{8}$$

For the refinement of the kinetic parameters by the way of minimizing the functional (7)–(8) was used the numerical method of multidimensional minimization-the method of configurations (Hook - Jeeves) [8].

The results of the calculations for the solution of the inverse problem: the rate constant for the reaction of chain growth $k_p = 48 l/(mol \cdot s)$, the rate constant for the reaction of chain transfer to the OAC $k_a = 240 l/(mol \cdot s)$, the rate constant for the reaction of chain transfer to the monomer and the rate constant for the reaction of the death of AC k_d got value close to 0.

6. Computational experiment

To check the adequacy of the constructed mathematical model and the calculated values of the kinetic parameters was carried out the solution of the direct problem. For the numerical solution of the obtained system of differential equations was used an implicit Adams-Moulton method of the 4th order [9].

The process of isoprene polymerization on NdCl3·nIPA-TIBA-PP catalytic system was carried out under the following conditions:

- concentration of chloride neodymium in toluene solution-0.035 mol/l;
- ratio NdCl3/Al(i-C4H9)3 / piperylene = 1/12/2 mol.
- consumption of neodymium catalyst - 1 mol NdCl3 / 10000 mol isoprene;
- isoprene concentration in isopentane is 15% by weight.
- the temperature of the process 20°C.

The moment values obtained from the solution of the system (4)-(5)were used to determine the dependence M_n, M_w and conversion of the polymerization time (Fig. 2). From the given dependences it can be seen that the values of conversion and averaged molecular masses satisfactorily describe the data of the results of the experiment conducted in the laboratory, what confirms the adequacy of the model we use.

7. Results

The built mathematical model of the process of polymerization of isoprene in the catalytic system NdCl3·nIPA-TIBA-PP allows to make calculations of average molecular characteristics, to obtain the dependence of the consumption of reagents, conversion dependence and polydispersity with time. From the obtained graphs it is seen that the conversion of 70% is achieved in one hour of the process, and the polydispersity of the resulting product after an hour is at the level of 4.2-4.3, which corresponds to the experimental data.

By solving the inverse problem was carried out the identification of unknown kinetic parameters. Analyzing the results of the numerical experiment it can be noted that the reaction of the chain transfer to the monomer and the reaction of the decay of the active centers have less influence on the MMP of the product and its conversion dependence in the case of a controlled polymerization process. Computational experiment for industrially significant process shows satisfactory agreement with the experimental data obtained in the laboratory.

IOP Conf. Series: Earth and Environmental Science **282** (2019) 012030 doi:10.1088/1755-1315/282/1/012030

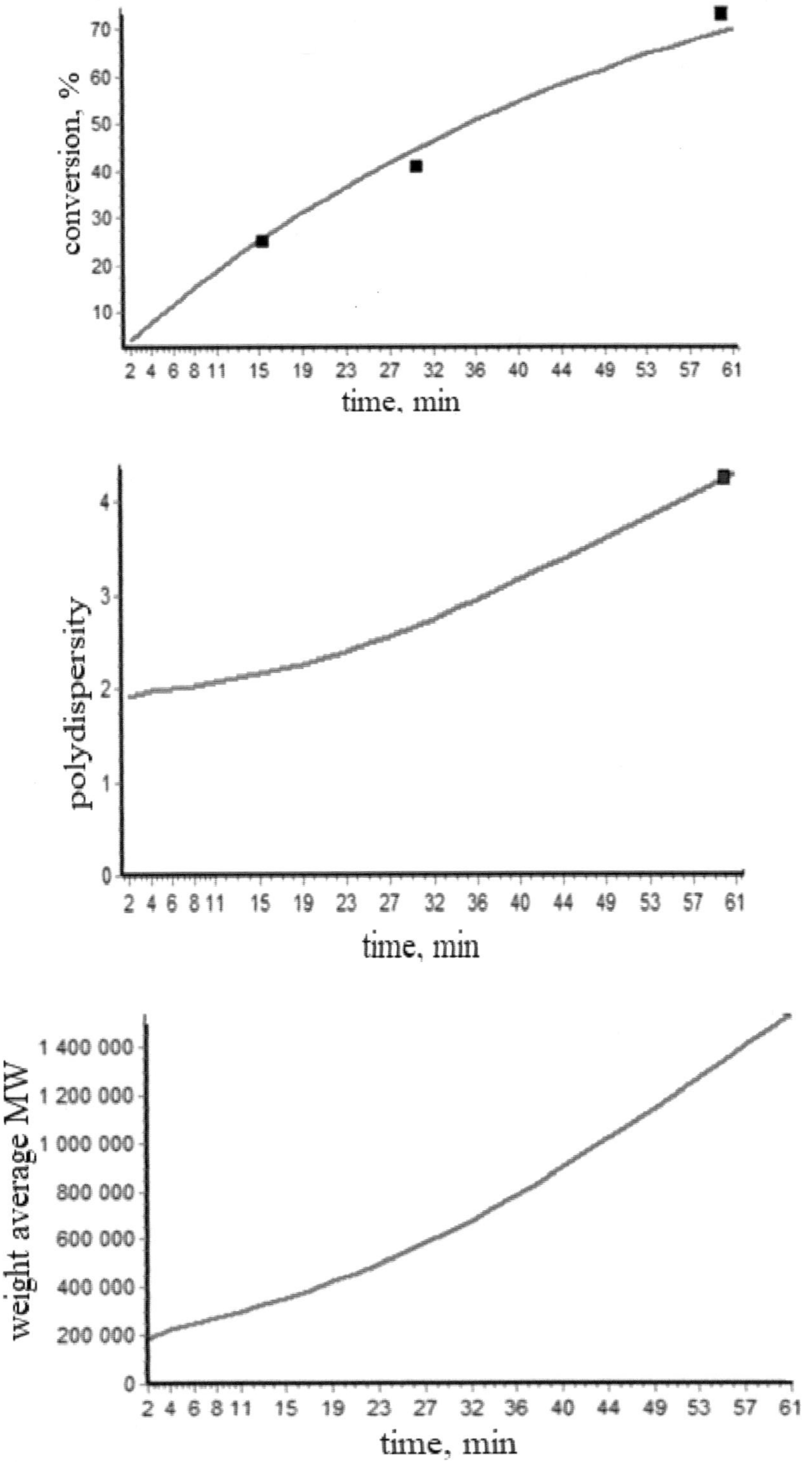

Fig. 2. Dependence of experimental (points) and calculated (solid line) values of conversion, polydispersity and average mass molecular weight on time.

8. Acknowledgement
The reported study is performed according to the Russian Government Program of Competitive Growth of Kazan Federal University and was funded by RFBR and Republic of Bashkortostan according to the research project № 17-47-020068.

9. References
[1] Zakharov V.P., Mingaleev V.Z, Berlin A. A., Nasyrov I.Sh., Zhavoronkov D.A., Zakharova E.M. Kinetic inhomogeneity of titanium and neodymium catalysts of production of 1,4-CIS-polyisoprene. Chemical physics. 2015. Vol. 34, №3. P. 69-75.
[2] Nasyrov I.Sh., Faizova V.Y. A., Zhavoronkov D.A., Zakharov V.P., Zakharova E.M. Comparison of laboratory and industrial data on monomer conversion by isoprene polymerization. Caoutchouc and rubber. 2017. No. 1. P. 6-10.
[3] Kafarov V.V., Dorokhov I.N., Dranishnikov L.V. System analysis of chemical technology processes. - M.: Science, - 1991. - 350 p.
[4] Ulitin N.V., Shirokyh, E.B., Tereshchenko K.A., Peruhin Y.V., Zakharov V.P. effect of the catalyst composition, Ndcl3·n(i-C3H7OH)-Al(i-C4H9) 3 on its activity and molecular weight of the synthesized on it polyisoprene // Vestnik Kazanskogo tekhnologicheskogo universiteta. 2014. No. 18.
[5] Aminova G.A., Gusaerov A.M., Vasenin I.V., Garifullina, E.V., Ignashina T.V., Ismagilova A.I., ManuikoG.V., Bronskaya V.V. Identification of the constants of the process of synthesis of isoprene rubber on neodymium catalytic system // Vestnik Kazanskogo tekhnologicheskogo universiteta. 2012. No. 13.
[6] Podvalny S.L. Modeling of industrial polymerization processes. - M.: Chemistry, - 1979. - 256 p.
[7] Usmanov T.S., Spivak S.I., Usmanov S.M. Inverse problem of the molecular-mass Back-formation distribution – M.: Chemistry, in 2004. - 252 p.
[8] Miftakhov E.N., Nasyrov I.Sh., Mustafina S.A., Spivak S.I. Solution of direct and inverse kinetic problems for the process of emulsion copolymerization of butadiene with styrene // Vestnik Bashkirii. UN-TA. 2011. No. 2.
[9] Mustafina S., Miftakhov E. and Mikhailova T. Solving the direct problem of butadiene-styrene copolymerization. International Journal of Chemical Sciences 2014; 12 (2): 564-572.
[10] Tatiana Mikhailova, Svetlana Mustafina,Vladimir Mikhailov, Sofia Mustafina Automation of Control's Process of Macro of Microsoft Excel File for Data Processing of Chemical Experiments // Entomology And Applied Science Letters, 2018, T.5, №3, pp.21-26.

IOP Conf. Series: Earth and Environmental Science **282** (2019) 012031 doi:10.1088/1755-1315/282/1/012031

Numerical study of polymerization processes on various catalytic systems

Igor Grigoryev[1], Svetlana Mustafina[1] , Olga Medvedeva[2], Sofya Mustafina[1] and Oleg Larin[3]

[1]Bashkir State University, 32 Zaki Validi str., Ufa, Russia
[2]Kazan Federal University, 18 Kremlevskaya str., Kazan, Russia
[3]South-West State University, 50 years of October str., 94, Kursk, Russia
E-mail: Mustafina_SA@mail.ru

Abstract. A set of algorithms for determining the averaged molecular characteristics of the polymer and finding the optimal initial concentrations was built on the basis of the mathematical software for solving rigid and non-rigid systems of ordinary differential equations describing various polymerization processes. With the help of the developed mathematical and software computational experiments were carried out for various technological parameters and production recipes for the considered processes. The dependencies of the molecular mass characteristics of the copolymer on the time of polymerization or conversion of monomers were built, which showed agreement with the data obtained as a result of experiments in the central factory laboratory of Synthez-Kauchuk OJSC (Sterlitamak, Republic of Bashkortostan) and the laboratory of physico-chemical studies of polymers of BashSU.

1. Introduction

Polymerization processes are the most popular research processes due to their widespread use and great opportunities. The use of synthetic rubbers has found itself in various branches of science of technology: they are not limited to the tire industry, they are used in medicine, and in the production of household appliances, and in other areas. One of the most important and urgent tasks of studying polymerization processes is the study of the kinetics of the process, the relationships between the reagents, and their influence on each other. Knowledge of these mechanisms allows you to build models of processes and further determine process parameters [1].

The basis of the formation of the polymer chain is the active center involved in the growth stages of the kinetic and molecular chains. Depending on the nature of the active center, radical, anionic, cationic, and coordination polymerization processes are distinguished [2]. In the industrial production of synthetic rubbers, methods of mathematical modeling of technological processes are increasingly being used to solve problems of forecasting and optimizing production. Then the constructed mathematical model will characterize the changes occurring within the system. Since these changes are reflected in the form of differential equations, then in the aggregate the whole process is represented by a system of differential equations [3]. It should be noted that the resulting systems have a large dimension, which must be limited.

The fundamentals of modeling polymerization processes are given by foreign researchers by Goldfinger G. and Alfrey T. [4], as well as Lewis F. and Mayo F.[5] who proposed independently from each other the simplest model of a limit link. The study of kinetic schemes continued in the works of foreign researchers Mertz E., Willing C and domestic scientists Semchikova Yu., Berlin A. A., Wolfson S. A. [6]. The modular principle for constructing a mathematical model is proposed in the works of Podvalniy S. L. and Kafarov V.V.[7], according to which the kinetic description of the process should be supplemented with a macrokinetic module that will take into account the hydrodynamic and energy patterns of the process under consideration. The study of processes and devices on the scale and conditions of industrial production is complex and time-consuming, therefore, the task of automating the sequence of stages of constructing a model becomes more urgent. More precisely, we are talking about creating automated programs for modeling processes and reactors, which would have great flexibility and versatility and could be used with minimal revision to study complex chemical-technological processes.

2. Methods

Radical polymerization is one of the main industrial methods that receive more than half of the polymers produced in the world: a number of synthetic rubbers, polyethylene, polyvinyl chloride, polyacrylates, water-soluble carbochain polyelectrolytes, polystyrene and many others.

Radical polymerization involves several stages: initiation stage, chain growth, price transfer, and chain termination stage [8].

The initiation of radical polymerization consists in the nucleation of free radicals in the reaction systems, which are capable of starting reaction chains. This stage occurs due to the thermal decomposition into two fragments of the initiator molecules; each fragment in turn is the primary free radical and can initiate the growth of the polymer chain:

$$I \xrightarrow{k_i} 2R, \tag{1}$$

where k_i – the rate constant of the initiator decay, characteristic of this type of initiator, R – free radical, I – initiator.

Then the primary radical attaches to a molecule of the monomer:

$$R + M \xrightarrow{k_{i1}} P_1, \tag{2}$$

where k_{i1} – initiation rate constant, R – free radical, P_1 – an active ("growing") polymer chain of a length equal to one, containing one M monomer unit.

Chain growth consists in the sequential attachment of monomer molecules to a growing radical. Chain growth occurs by the scheme:

$$P_1 + M \xrightarrow{k_{p1}} P_2, \; P_i + M \xrightarrow{k_{pn}} P_{i+1}, \tag{3}$$

where $k_{p1}, k_{p2}, \ldots, k_{pn}$ – chain growth rate constants, P_n – active polymer chains of length n containing n monomer M links.

It is usually assumed that the reaction rate constants for active polymerization chains with an arbitrary length of units are constant, i.e. $k_{p1} = k_{p2} = \ldots = k_{pn} = k.$

Chain termination occurs as a result of deactivation of active particles. There are three options for termination the chain, the first two of which are the most likely:

1) by recombination

$$P_n + P_m \xrightarrow{k_{rec}} Q_{n+m}, \tag{4}$$

2) by disproportionation

$$P_n + P_m \xrightarrow{k_{dis}} Q_n + Q_m, \tag{5}$$

3) spontaneous (random) way

$$P_n \xrightarrow{k_{sp}} Q_n, \tag{6}$$

where k_{rec}, k_{dis}, k_{sp} – chain termination rate constants by recombination, disproportionation, and spontaneous path, respectively, P_n, Q_n – active and inactive polymer chains of length n, respectively, containing n units of monomer M.

The stage of chain transfer is that the regulator that is introduced into the system breaks off the growing chain, but at the same time the regulator itself becomes a free radical and starts a new kinetic chain of polymerization. Chain transfer can occur on the following agents:

– to the solvent

$$P_n + S \xrightarrow{k_{fS}} Q_n + P_1^0, \tag{7}$$

– to the monomer

$$P_n + M \xrightarrow{k_{fM}} Q_n + P_1, \tag{8}$$

– to the polymer

$$P_n + Q_m \xrightarrow{k_{fP}} Q_n + M - 1 + P_1, \tag{9}$$

where k_{fS}, k_{fM}, k_{fP} – the transmission constants of the chains respectively, S – agent, P_n, Q_n – active and inactive polymer chains of length n, respectively, containing n units of monomer M.

Thus, the generalized kinetic radical polymerization scheme comprises the following elementary steps:

1. Initiator decomposition

$$I \xrightarrow{k_i} 2R,$$

2. Chain growth

$$R + M \xrightarrow{k_{i1}} P_1,$$

3. Continuation of the chain

$$P_1 + M \xrightarrow{k_p} P_2, \; P_i + M \xrightarrow{k_p} P_{i+1},$$

4. Death of active centers

$$P_n \xrightarrow{k_{sp}} Q_n,$$

5. Recombination of active chains

$$P_n + P_m \xrightarrow{k_{rec}} Q_{n+m},$$

6. Disproportioning of active circuits

$$P_n + P_m \xrightarrow{k_{dis}} Q_n + Q_m,$$

where I – initiator, R – free radical, M – monomer, P_n, Q_n – active ("growing") and inactive ("dead") chain lengths of the copolymer, respectively, containing n links with monomer M, $k_i, k_{i1}, k_p, k_r, k_{rec}, k_{dis}$ – constants of elementary stages of initiation, growth, and chain termination stages respectively

To describe the kinetics of the mechanism, it suffices to write out the velocity equations for the elementary stages of the process. They will determine the behavior of the catalytic system itself.

Thus, the speed of the chain growth stage will be the product of the concentrations of monomer and active centers, taken with a coefficient in the form of the rate constant of the same stage:

$$W_p = k_p C_a \text{M}. \tag{10}$$

The speed of the growth stage will be determined by the monomer concentration, which decreases with time, since the concentration of active centers remains constant during the process. However, the rate of decrease in monomer concentration decreases with increasing process time. This means that with deep degrees of conversion the values of growth rates will be close to each other. This fact testifies to the so-called conversion "plateau", i.e. reaching a certain degree of conversion, the monomer in the system is practically not spent.

The total chain transfer rate can be written by combining two types of chain transfer stages present in the kinetic scheme:

$$W_o = k_M M C_a + k_A A C_a. \tag{11}$$

The same formula can be rewritten as:

$$W_o = W_M + W_A. \tag{12}$$

Analysis of the dependence (11) suggests that the speed of chain transmissions decreases with the process time. This behavior of the velocity affects the values of the average numerical molecular characteristics of the polymer, which, like the conversion curve, are characterized by the appearance of a "plateau" at deep conversion degrees.

Making a matrix of the stoichiometric coefficients of reactions of the general kinetic scheme, having performed the operations of transposition and multiplication by a column vector of reaction rates, we obtain a system of differential equations for the concentrations of substances:

$$\begin{aligned}
&\frac{d[I]}{dt} = -k_i[I],\\
&\frac{d[R]}{dt} = 2k_i[I] - k_{i1}[M][R] - k_r[P_1][R],\\
&\frac{d[M]}{dt} = -[M]k_p\sum_{i=1}^{\infty}[P_i] - [M]k_{i1}[R],\\
&\frac{d[P_1]}{dt} = k_{i1}[M][R] - k_p[M][P_1] - k_r[R][P_1] - k_{rec}[P_1]\sum_{j=1}^{\infty}[P_j] - k_{dis}[P_1]\sum_{j=1}^{\infty}[P_j],\\
&\frac{d[Q_1]}{dt} = k_r[R][P_1] + \frac{1}{2}k_{rec}[P_1][R] + k_{dis}[P_1]\sum_{j=1}^{\infty}[P_j],\\
&\frac{d[P_i]}{dt} = k_p[M][P_{i-1}] - k_p[M][P_i] - k_r[R][P_i] - k_{rec}[P_i]\sum_{j=1}^{\infty}[P_j] -\\
&- k_{dis}[P_i]\sum_{j=1}^{\infty}[P_j],\ 2 \le i \le \infty,\\
&\frac{d[Q_i]}{dt} = k_r[R][P_i] + \frac{1}{2}k_{rec}\sum_{j=1}^{i-1}[P_j][P_{i-j}] + k_{dis}[P_i]\sum_{j=1}^{\infty}[P_j],\ 2 \le i \le \infty,
\end{aligned} \tag{13}$$

where $[\ldots]$ – concentration of the relevant substances ($[I]$ – initiator, $[R]$ – free radical, $[M]$ – monomer, $[P_n],[Q_n]$ – active ("growing") and inactive ("dead") chain lengths of the copolymer, respectively, containing n links with monomer M), f – efficiency of initiation,

$k_i, k_{i1}, k_p, k_r, k_{rec}, k_{dis}$ – constants of elementary stages of initiation, growth, and chain termination stages respectively.

The resulting system of differential equations (13) for a generalized kinetic scheme allows us to formulate the problem of determining its solution based on the initial data:

$$\begin{gathered} \left[I^{(0)}\right]=[I(0)],\left[M^{(0)}\right]=[M(0)], \\ \left[R^{(0)}\right]=0,\left[P_1^{(0)}\right]=0,\left[Q_1^{(0)}\right]=0, \\ \left[P_i^{(0)}\right]=0,\left[Q_i^{(0)}\right]=0,\ i\geq 2. \end{gathered} \tag{14}$$

Thus, system (13) is a mathematical model of the process, where concentrations of reagents play the role of functions. It is seen that system (13) is autonomous. As can be seen from system (13), it is infinite, since it is assumed that the process goes on until the monomer is completely exhausted, which may mean the formation of polymer chains of a sufficiently long length. Therefore, generalizing the assumption made, system (13) has an infinite number of equations. Each of them is an ordinary differential equation of the first order. Analysis of the right sides of the equations shows that the maximum degree of the addendum does not exceed two. This characterizes the molecular nature of the reaction, since according to the general kinetic scheme; most of the reactions are bimolecular. It can also be seen that the monomer participates in the corresponding reactions as the initial substance no more than times, which indicates the first order of the rate of reactions according to the monomer.

Solving the direct kinetic problem means finding a solution to the system of ordinary differential equations (13) under the given conditions (14) and the known values of the kinetic rate constants of elementary reactions. If it is required to find a set of kinetic constants such that the behavior of a number of functions of system (13) is known in advance, this leads to finding the solution of the inverse kinetic problem.

Analyzing the general kinetic scheme, it can be noted that disproportionation and recombination reactions are included in it as chain termination reactions, which belong to the class of bimolecular chain termination reactions. The change of the right parts of the equations is a distinctive feature of the system of ordinary differential equations characterizing the processes of continuous polymerization. The inclusion of these stages of breakage affects the behavior of inactive polymer chains, that is, we can say that the likelihood of the appearance of such chains increases. In other words, the number of stages in which these chains may appear actually increases.

3. Results and discussion

One of the most important processes in the industrial production of synthetic rubber is the process of isoprene polymerization. The kinetic scheme of polymerization of isoprene comprises the following elementary steps:

1. Initiator decomposition:

$$I \xrightarrow{k_i} P_0.$$

2. Chain growth:

$$P_i + M \xrightarrow{k_p} P_{i+1}\, 0 \leq i \leq \infty.$$

3. Chain transfer to the monomer M:

$$P_i + M \xrightarrow{k_m} Q_i + P_1,\ 2 \leq i \leq \infty.$$

4. Transfer of the chain to an organoaluminum compound A:

$$P_i + A \xrightarrow{k_a} Q_i + P_1,\ 2 \leq i \leq \infty.$$

5. Death of active centers:

$$P_i \xrightarrow{k_d} Q_i,\ 2 \le i \le \infty,$$

where M – monomer, A – organoaluminum compound, I – initiator, P_n, Q_n – active ("growing") and inactive ("dead") chain lengths of the copolymer, respectively, containing n links with monomer k_i, k_p, k_m, k_a, k_d – constants of elementary stages of initiation, growth, transfer to monomer, organoaluminum compound and deactivation of active centers, respectively.

This scheme differs from the previous one in that it lacks the chain termination stages by recombination and disproportionation.

Similarly, we obtain an infinite system of ordinary differential equations for the concentrations of substances:

$$\begin{aligned}
&\frac{d[I]}{dt} = -k_i[I],\\
&\frac{d[P_0]}{dt} = fk_i[P_0] - k_p[M][P_0],\\
&\frac{d[P_1]}{dt} = -k_p[M][P_1] + k_m[M]\sum_{j=2}^{\infty}[P_j] + k_a[A]\sum_{j=2}^{\infty}[P_j] - k_d[P_1],\\
&\frac{d[M]}{dt} = -(k_p + k_m)[M]\sum_{j=1}^{\infty}[P_j],\\
&\frac{d[Q_1]}{dt} = k_m[M][P_1] + k_a[A][P_1] + k_d[P_1],\\
&\frac{d[A]}{dt} = -k_a[A]\sum_{j=1}^{\infty}[P_j],\\
&\frac{d[P_i]}{dt} = k_p[M][P_{i-1}] - k_p[M][P_i] - k_m[M][P_i] - k_a[A][P_i] - k_d[P_i], 2 \le i \le \infty,\\
&\frac{d[Q_i]}{dt} = k_m[M][P_i] + k_a[A][P_i] + k_d[P_i],\ 2 \le i \le \infty,
\end{aligned} \tag{15}$$

where $[\ldots]$ – concentration of the relevant substances ($[M]$ – monomer, $[A]$ – organoaluminum compound, $[R]$ – free radical, $[I]$ – initiator, $[P_n],[Q_n]$ – active ("growing") and inactive ("dead") chain lengths of the copolymer, respectively, containing links with monomer), f – efficiency of initiation, k_i, k_p, k_m, k_a, k_d – constants of elementary stages of initiation, growth, transfer to monomer, organoaluminum compound and deactivation of active centers, respectively.

Initial data in this case can be represented as:

$$\begin{aligned}
&[I^{(0)}] = [I(0)], [M^{(0)}] = [M(0)],\\
&[R^{(0)}] = 0, [P_1^{(0)}] = 0, [Q_1^{(0)}] = 0,\\
&[P_i^{(0)}] = 0, [Q_i^{(0)}] = 0,\ i \ge 2.
\end{aligned} \tag{16}$$

The resulting system (14) with the initial conditions (15) contains an infinite number of differential equations, since the formation of polymer macromolecules goes to infinite chain length.

There are a large number of methods that allow finding a numerical solution of systems of differential equations of this type. The application of some is based on the principle of quasistationary of Semenov-Bodenstein. This principle is that the rate of change in the concentration of radicals is too small compared to the rate of change in the concentration of monomer. This assumption greatly simplifies the calculations, since allows you to equate to zero the kinetic equations for the concentrations of "short-lived" components, as well as being one of the possible methods for solving rigid systems of differential equations. But the application of these methods is limited only by the initial stage, when polymers have a short chain length.

In this work, the so-called method of moments was used. The application of the method of moments is justified in cases when it is necessary to determine the quantitative characteristics of the polymer and on the basis of them to build the molecular mass distribution. The law of its description may be different: the distribution of Beasley, the distribution of Flory and several others. Another advantage of the method is to convert the source system to the final form without imposing additional restrictions [9].

The construction of mathematical models in the framework of the method is based on the expression of the moments of the distribution of active and inactive circuits. The statistical theory of polymerization is based on the assumption of the possibility of analyzing molecular mass distributions (MWD) by average molecular masses M_n , M_ω and polydispersity coefficient K_D. The concepts of moments are used for their analysis, which are used in statistics and probability theory to estimate the distribution of random variables.

For the moments of any order, basically, we have [10]:

$$m_j = \mu_j + \eta_j, \tag{17}$$

where η_j, μ_j is j - th moments of active and inactive polymer chains calculated in accord with the following formulas:

$$\mu_j = \sum_{i=2}^{\infty} i^j [P_i], \tag{18}$$

$$\eta_j = \sum_{i=2}^{\infty} i^j [Q_i], \tag{19}$$

For the substitution of moments in system (15), in addition to the expressions (18) - (19) themselves, it is necessary to define derivatives μ_j and η_j with respect to time, which can be found by the following rule

$$\frac{d\mu_j}{dt} = \sum_{i=2}^{\infty} i^j \frac{d[P_i]}{dt},$$

$$\frac{d\eta_j}{dt} = \sum_{i=2}^{\infty} i^j \frac{d[Q_i]}{dt}. \tag{21}$$

(20)

The found moment values are substituted into the formula for the average molecular weight M_n, M_w and the polydispersity index K_D (22) - (23).

The value of M_n determines the average weight of the polymer macromolecules called average molecular weight. It is calculated by the following formula [11]:

$$M_n(t) = m \frac{\mu_1(t) + \eta_1(t)}{\mu_0(t) + \eta_0(t)}, \tag{22}$$

where m is the molecular weight of the monomer.

If parameter M_n is characterized as a low molecular weight part of the MWD, then parameter M_w is the average of the distribution of molecular weight and is calculated by the formula [12]:

$$M_w(t) = m \frac{\mu_2(t) + \eta_2(t)}{\mu_1(t) + \eta_1(t)}, \tag{23}$$

where m is the molecular weight of the monomer.

To estimate the width of the MMD, a parameter called the polydispersity coefficient is usually used. The coefficient of polydispersity is close in meaning to the dispersion (spread) of molecular weight and is calculated by the following formula:

$$K_D = \frac{M_\omega}{M_n}. \tag{24}$$

Let's perform the necessary replacements and substitutes, and system (15) can be written in the closed form. This system has a finite number of equations and is soluble in regard with the following moments[13]:

$$\begin{aligned}
&\frac{d[I]}{dt} = -k_i[I],\\
&\frac{d[P_0]}{dt} = fk_i[P_0] - k_p[M][P_0],\\
&\frac{d[P_1]}{dt} = -k_p[M][P_1] + k_m[M]\mu_0 + k_a[A]\mu_0 - k_d[P_1],\\
&\frac{d[M]}{dt} = -(k_p + k_m)[M][P_1] - (k_p + k_m)[M]\mu_0,\\
&\frac{d[Q_1]}{dt} = k_m[M][P_1] + k_a[A][P_1] + k_d[P_1],\\
&\frac{d[A]}{dt} = -k_a[A][P_1] - k_a[A]\mu_0,\\
&\frac{d\mu_0}{dt} = \sum_{i=2}^{\infty}\frac{d[P_i]}{dt} = k_p[M][P_1] - (k_m[M] + k_a[A] + k_d)\mu_0,\\
&\frac{d\mu_1}{dt} = \sum_{i=2}^{\infty} i\frac{d[P_i]}{dt} = (1+\mu_0)k_p[M][P_1] - (k_m[M] + k_a[A] + k_d)\mu_1,\\
&\frac{d\mu_2}{dt} = (\mu_2 + 2\mu_1 + \mu_0)k_p[M][P_1] - \left(k_p[M] - k_m[M] - k_a[A] - k_d\right)\mu_2,\\
&\frac{d\eta_0}{dt} = \sum_{i=2}^{\infty}\frac{d[Q_i]}{dt} = (k_m[M] + k_a[A] + k_d)\mu_0.\\
&\frac{d\eta_1}{dt} = \sum_{i=2}^{\infty} i\frac{d[Q_i]}{dt} = (k_m[M] + k_a[A] + k_d)\mu_1.\\
&\frac{d\eta_2}{dt} = \sum_{i=2}^{\infty} i^2\frac{d[Q_i]}{dt} = (k_m[M] + k_a[A] + k_d)\mu_2.
\end{aligned} \tag{25}$$

Initial data in this case can be represented as:

$$\begin{aligned}
&[I^{(0)}] = [I(0)],\ [M^{(0)}] = [M(0)],\\
&[R^{(0)}] = 0,\ [P_1^{(0)}] = 0,\ [Q_1^{(0)}] = 0,\\
&\mu_k(0) = 0, \eta_k(0) = 0,\ k = 0,1,2.
\end{aligned} \tag{26}$$

The result of solving the direct kinetic problem (25) - (26) is the dependence of the change in the molecular characteristics of the resulting polymer on the polymerization time: the average M_n and weight average M_w molecular weights [14].

The constructed model was tested on experimental data obtained in the factory laboratory.

4. Summary

For the catalytic system VOCl3-Al(i-C4H9)3 in the laboratory, an experiment was carried out with the following conditions: $T = 25°C,\ C_M^{(0)} = 1$ mol / l, $C_V^{(0)} = 5 \cdot 10^{-3}$ mol / l, $C_A^{(0)} = 10.8 \cdot 10^{-5}$ mol / l.

Pic. 1 and pic. 2 show that average M_n and weight-average M_w molecular weights are increasing functions versus time. We note that after 60 minutes of polymerization both characteristics take a constant value.

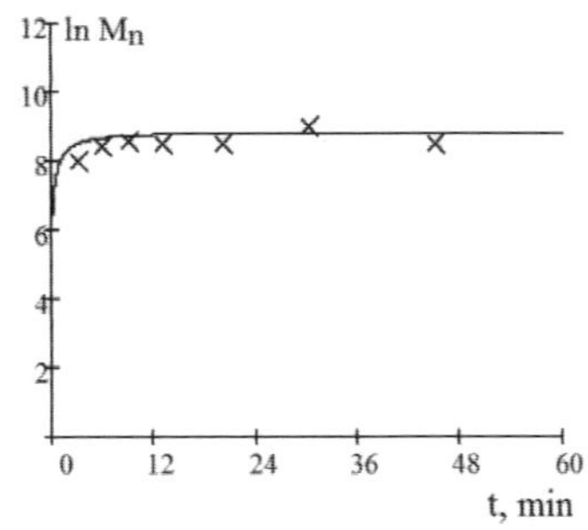

Pic.1 The average molecular weight M_n versus time.

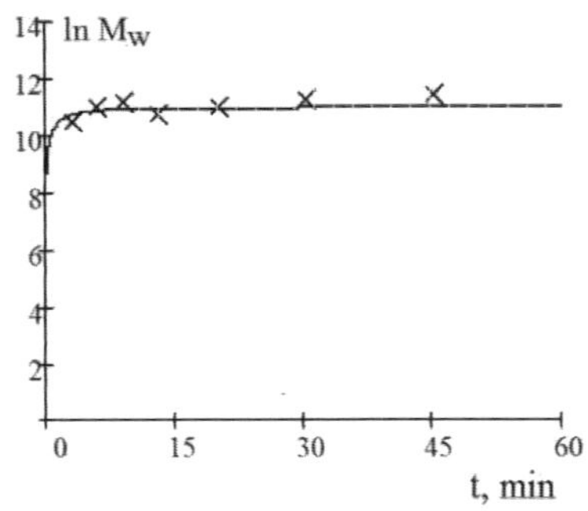

Pic.2 The average of the distribution of molecular weight M_w versus time.

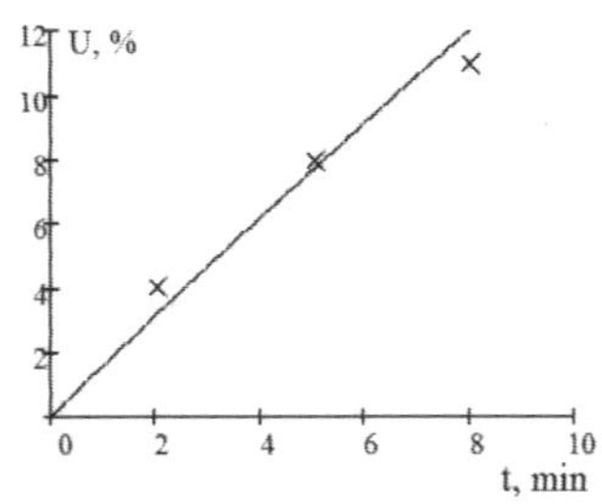

Pic.3 The dependence of monomer conversion U of the duration of the polymerization.

5. Conclusions

Thus, the polymerization conditions have been selected in the work. On the basis of the mathematical model, the values of the monomer concentrations versus the polymerization time have been constructed, and the values of number average and weight average molecular weights have been found. The simulation and experimental results show a satisfactory agreement.

6. Acknowledgements

The reported study is performed according to the Russian Government Program of Competitive Growth of Kazan Federal University and was funded by RFBR and Republic of Bashkortostan according to the research project № 17-47-020068 and №18-38-00255

7. References

[1] Olga A. Medvedeva, Sofya I. Mustafina, Igor V. Grigoryev, Svetlana A. Mustafina Direct Modeling of Chemical Processes in Conditions of Uncertainty of Initial Data//International Journal of Engineering & Technology. – 2018. – Vol. 7. – №3.27. – Pp. 556-560. doi: 10.14419/ijet.v7i3.27.18493

[2] Svetlana Mustafina, Vladimir Vaytiev, Igor Grigoryev The method of research of the direct problem to the variation of the kinetic parameters within a given range // International Journal of Pure and Applied Mathematics. – 2017. – Vol. 112. – №4. – Pp. 805 - 815. doi: 10.12732/ijpam.v112i4.11

[3] I V. Grigoryev and S.A. Mustafina Simulation of the polymerization process of butadiene per neodymium catalytic system //J. Phys.: Conf. Ser. – 2018. – Vol. 1096. – №012152. – Pp. 1-7.

[4] Alfrey T., Goldfinger G. The mechanism of copolymerization // J. Chem. Phys. – 1944. – Vol. 12. – №6. – Pp. 205-209.

[5] Mayo F. R., Lewis F. M. A basis for comparing the behavior of monomers in copolymerization, the copolymerization of styrene and methyl methacrylate // J. Am. Chem. Soc. – 1944. – Vol. 66, – Pp. 1594 – 1601.

[6] Berlin A. A., Wolfson S. A., Enikolopyan N. S. Kinetics of polymerization processes. - M: Chemistry., 1978. - 320 p.

[7] Podvalny S.L. Modeling of industrial polymerization processes. - M .: Chemistry, 1979. 350 p.

[8] Grigoryev I and Mustafina S Mathematical modeling of the copolymerization of α-methylstyrene with maleic anhydride in a heterogeneous environment // Procedia Engineering. – 2017. – Vol. 201. – Pp. 639-644. doi: 10.1016/j.proeng.2017.09.678

[9] Grigoryev I, Mikhailova T and Mustafina S Study of the styrene and maleic anhydride copolymerization process by mathematical modeling methods ARPN Journal of Engineering and Applied Sciences – 2017. – Vol. 12. – №5. – Pp. 1561-1566.

[10] Igor Grigoryev, Eldar Miftakhov, Svetlana Mustafina Mathematical Modelling of the Copolymerization of styrene with Maleic Anhydride in a Homogeneous Environment // International Journal of Chemical Sciences. – 2016. – Vol. 14. – №1. – Pp. 381-386.

[11] Igor Grigoryev, Tatiana Mikhailova, Svetlana Mustafina Modeling of radical copolymerization processes on the basis of the moments method // International Journal of Chemical Sciences. – 2016. – Vol. 14. – №4. – Pp. 2860-2866.

[12] Grigoryev I and Mustafina S 2017 Mathematical modelling of the copolymerization of A-methylstyrene with maleic anhydride in a heterogeneous environment ARPN Journal of Engineering and Applied Sciences. – 2017. – Vol. 12. – №8. – Pp. 2668-2671.

[13] Mustafina S., Miftakhov E. and Mikhailova T. Solving the direct problem of butadiene-styrene copolymerization. International Journal of Chemical Sciences 2014; 12 (2): Pp. 564-572.

[14] Tatiana Mikhailova, Svetlana Mustafina,Vladimir Mikhailov, Sofia Mustafina Automation of Control's Process of Macro of Microsoft Excel File for Data Processing of Chemical Experiments // Entomology And Applied Science Letters, 2018, T.5, №3, pp.21-26.

Natural heavy oil reservoirs porosity and saturation coefficients estimation based on wavelet analysis of X-Ray computed tomography histogram

R Kadyrov[1], A Starovoytov[1], E Utemov[1], E Statsenko[1], R Sagirov[1], M Validov[1]

[1]Institute of Geology and Petroleum Technologies, Kazan Federal University, Kazan, Russia

E-mail: Rail7777@gmail.com

Abstract. With the development of unconventional resources the standard task of evaluation the porosity and oil saturation needs new approach. Heavy oil and bitumen can act as cement in many sand reservoirs, which excludes the use of standard measuring techniques. A new method for estimating the porosity and oil saturation of heavy oil reservoirs, based on wavelet X-ray computed tomography histogram analysis, was proposed. Its advantages are the ability to work with loose sand reservoirs, for which it is not possible to use standard assessment methods. A comparative analysis of the proposed method with a standard method for estimating porosity based on sample saturation with kerosene showed that the values of the porosity coefficient in the tomography resolution range 5-15 μm differed by 3-7% in the direction of the wavelet analysis method. This effect is due to the presence of transient voxels – mixels.

Keywords: heavy oil, reservoir, porosity, saturation, CT, X-ray computed tomography, histogram, wavelet, mixel.

1. Introduction

With the depletion of light and medium oil reserves, heavy oils and natural bitumen become an important source to meet the growing demand for fuel and petrochemicals [1]. The standard tasks of evaluation the porosity and oil saturation of reservoirs become nontrivial in new circumstances. The central problem is that heavy oil and bitumen can act as cement in many sand reservoirs. Thus, the hydrocarbon extraction of such samples leads to the destruction of the rock and the inability to evaluate the filtration and capacitive properties of reservoir using the standard methods. This paper outlines a new approach for estimation the porosity and oil saturation based on analysis of X-Ray computed tomography histogram.

X-Ray computed tomography (CT) is an established and rapidly evolving technology of proven value for geological investigations [2]. The advantage of tomography over other porous media characterization techniques is direct imaging of microstructure, from which we can evaluate statistical and geometrical parameters such as porosity or porous media surface area. Each three-dimensional cell of tomography model (voxel) has the grayscale value of X-Ray attenuation coefficient of material. Standard histogram of X-Ray computed tomography of clean reservoir sample has a form of a double-peaked curve, where the x-ray attenuation coefficient is plotted along the x-axis and the number of voxels with a given grayscale value – along the y-axis. The minimum value of histogram between two peaks usually is a border value between rock matrixes and air in porous media and around the sample (fig. 1a). Heavy oil saturated sample usually has additional little peak after air curve (fig. 1b). The main idea of our approach is using the information from histogram for oil volume classification.

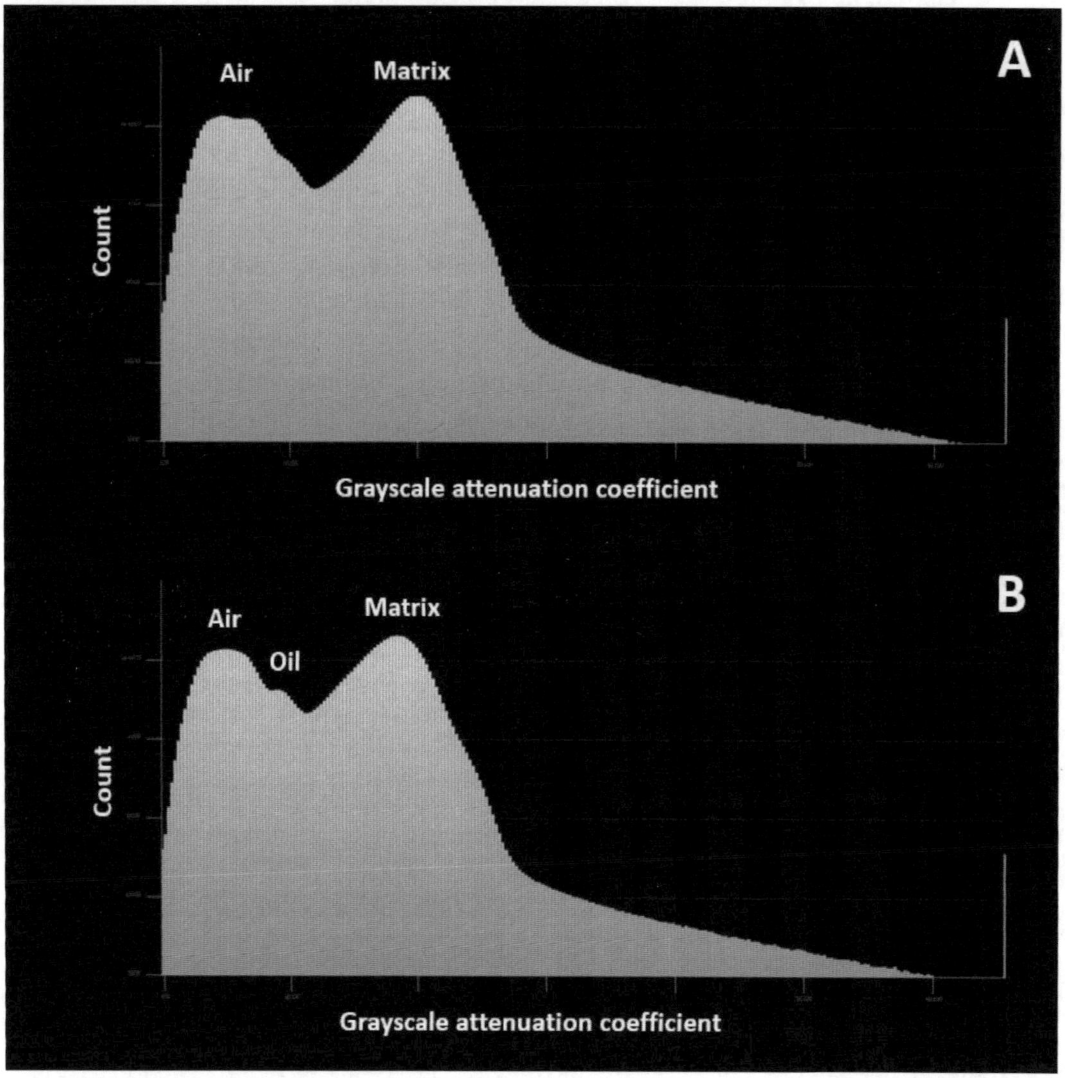

Figure 1. Typical X-Ray computed tomography histogram of the sand reservoir (resolution 36μm): a – after hydrocarbon extraction, b – before hydrocarbon extraction

2. Methodology

We use wavelet analysis to break the main curve to several, which sum of areas is equal to area of main curve. Wavelets are a class of functions used to localize a given function in both space and scaling. The method is based on continuous wavelet transformations with special selected basis in order to solve inverse problem and briefly described in [3].

Data processing starts with cutting the air volume around the sample. This step changes the form of histogram from double picked to one picked curve (fig. 2a). After analysis, the general curve is broken to 5-8 "daughter" wavelets (fig. 2b). Each of them can be interpreted as one material.

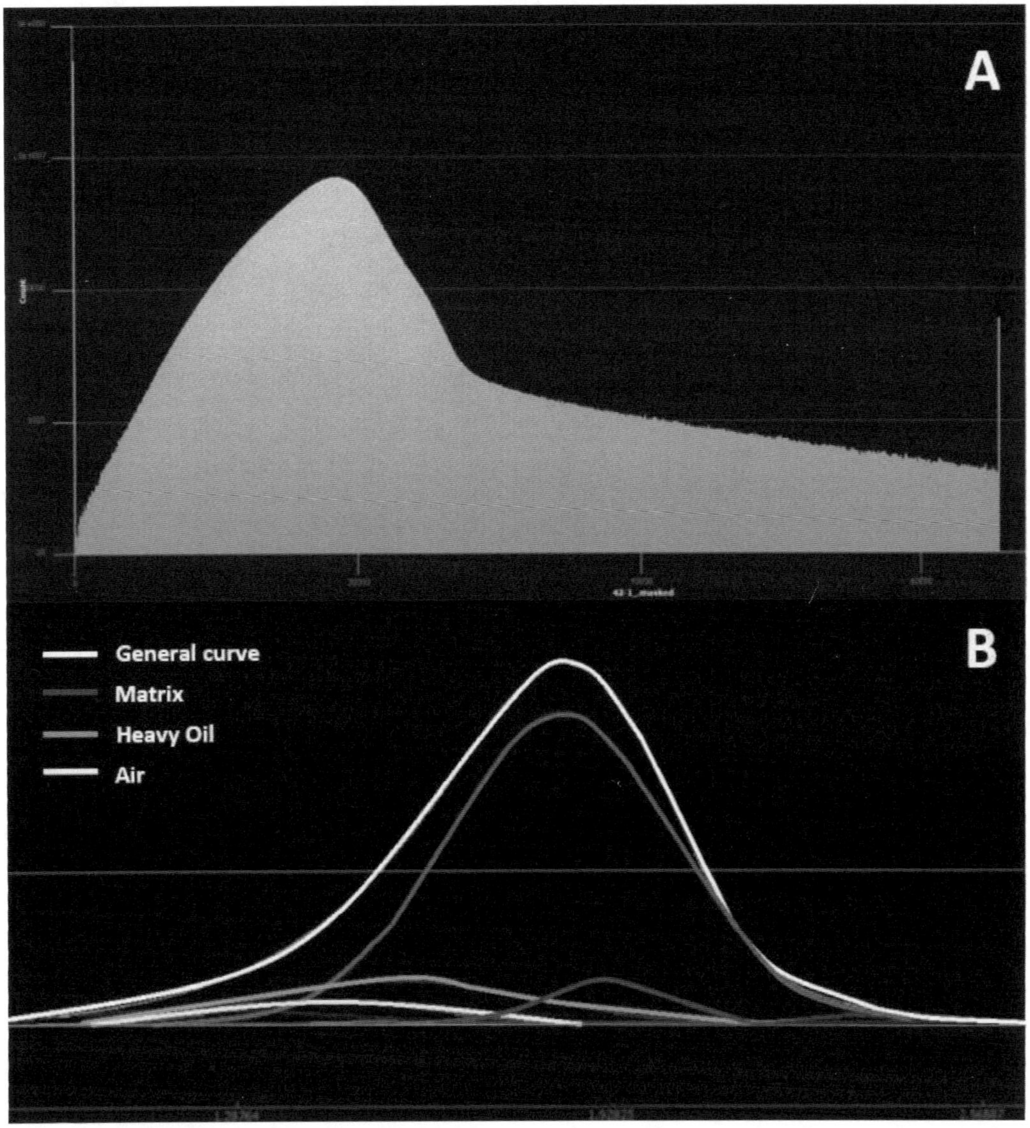

Figure 2. Histogram transformation: a – after cutting the air around the sample, b – after wavelet analysis with defining the materials

3. Results and Discussions

The grayscale values extracted for air and oil from the results of wavelet analysis demonstrate a visual conformity with tomographic slices (fig. 3). Air classified regions correlate with the most black areas and oil classified areas surround them.

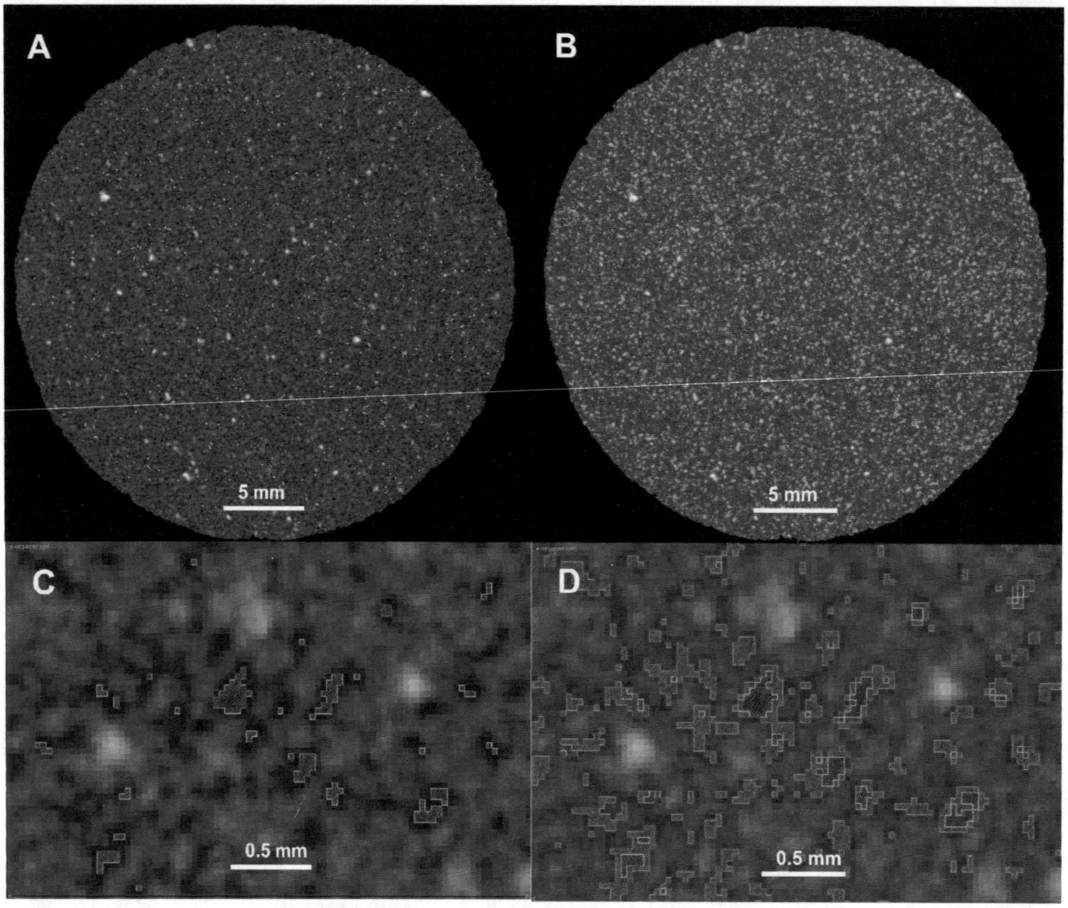

Figure 3. Image classification based on the results of wavelet analysis: yellow – air in pores, green – heavy oil in pores

We compared the porosity coefficients, obtained by standard method of saturating the sample with kerosene and wavelet analysis of tomographic histograms on 7 samples (table 1 and fig. 4). The last one included the sum of porous media filled by air and heavy oil. The results of comparison demonstrate that wavelet analysis data is always higher to 3-7% than standard method in the range of tomography resolution from 5 to 15 μm. This effect takes a place due to wrong classification of the voxels with the same grey value as used for heavy oil. Naturally, this voxels can have transitional gray values between air filled pores and matrix or include voxels with "invisible" pores, which sizes less than tomography resolution [4]. These transitional voxels called "mixels" [5].

Table 1. Comparison between porosities coefficients, obtained by standard method and wavelet analysis for different samples

Sample No.	Tomography resolution, μm	Porosity coefficient, %	
		Standard method*	Wavelet analysis
1	12,696	30,52	33,52
2	5,885	10,73	15,26
3	12,803	24,44	30,11
4	13,303	23,44	29,87
5	12,689	21,28	27,92
6	14,173	22,1	25,62
7	11,578	4,15	8,02

* - standard is method of saturating the sample with kerosene

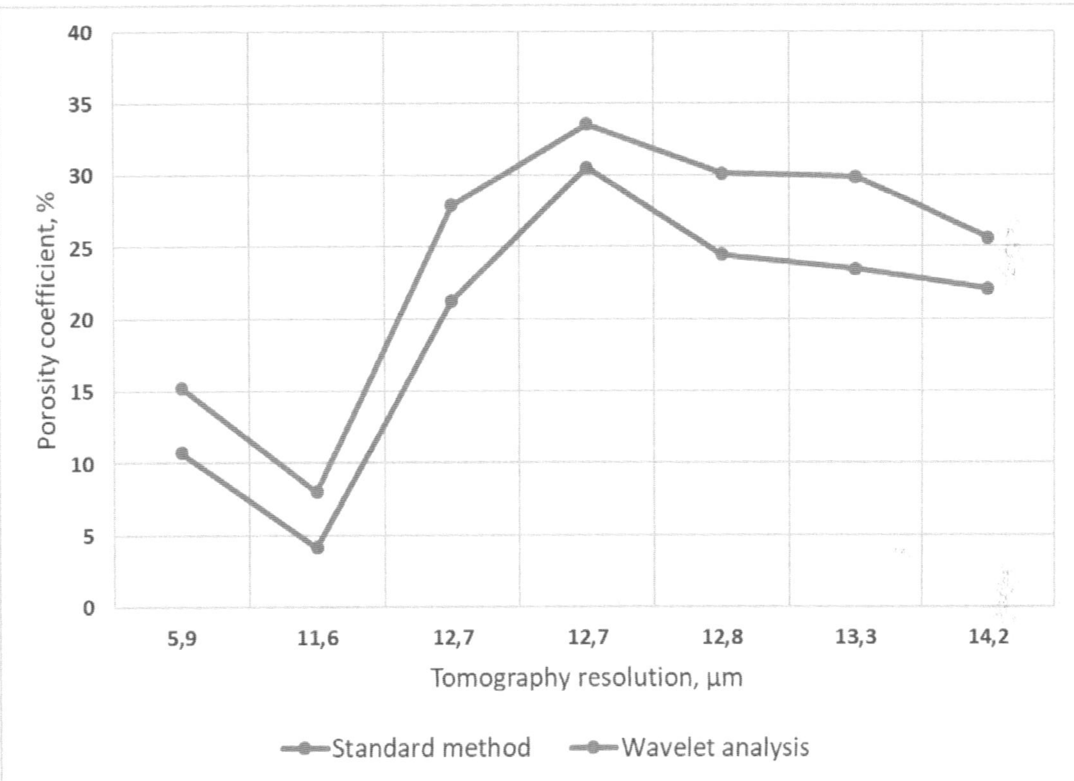

Figure 4. Changing the porosity coefficient versus resolution of sample's tomography

4. Conclusions

Thus, a new method for estimating the porosity and oil saturation of heavy oil reservoirs, based on wavelet X-ray computed tomography histogram analysis, was proposed. Its advantages are the ability to work with loose sand reservoirs, for which it is not possible to use standard assessment methods. A comparative analysis of the proposed method with a standard method for estimating porosity based on sample saturation with kerosene showed that the values of the porosity coefficient in the tomography resolution range 5-15 μm differed by 3-7% in the direction of the wavelet analysis method. The study of the applicability of this method will be continued.

Acknowledgements

The work is performed according to the Russian Government Program of Competitive Growth of Kazan Federal University.

References

[1] Kadyrov R 2018 The porous space structure of Domanik shales in the East of Russian Plate *18th International Multidisciplinary Scientific GeoConference SGEM2018, Science and Technologies in Geology, Exploration and Mining* vol 18 pp 907–14

[2] Ketcham R A and Carlson W D 2001 Acquisition, optimization and interpretation of x-ray computed tomographic imagery: Applications to the geosciences *Comput. Geosci.* **27** 381–400

[3] Utemov E V. and Nurgaliev D K 2005 Natural wavelet transformations of gravity data: Theory and applications *Izv. Phys. Solid Earth* **41** 333–40

[4] Mikhailov D N, Ryzhikov N I and Shako V V 2015 An integrated experimental approach to determination of clogging kinetics of porous media *Neft. khozyaystvo - Oil Ind.* 74–8

[5] Kobayashi Y, Kawasaki S, Kato M, Mukunoki T and Kaneko K 2013 *Evaluation of Porosity in Geomaterials Treated with Biogrout Considering Partial Volume Effect*

The effect of temperature heating on the microstructure of the Domanik oil shale porous space

R Kadyrov[1], E Statsenko[1]

[1]Institute of Geology and Petroleum Technologies, Kazan Federal University, Kazan, Russia

E-mail: Rail7777@gmail.com

Abstract. As the global supply of traditional oil shrinks, the energy industry invests heavily in the unconventional energy resources, such as oil shale. One of big potential resources of oil shale that locates on the territory of Volga-Ural Basin in the East of Russian plate is Domanik. Its increasing significance led to the necessity for deeper understanding of Domanik porous space structure. The main idea of this work is to establish the changing of pore structure during the heating. As result, increasing the porosity to 3 times as a result of heating of the oil shale was established. The distribution of heated to 350° C sample demonstrated decreasing the ratio of pores with sizes 1.5-5 μm and significant increasing the ratio of pores with sizes 5-10 μm. It means that main changing of pore structure during the heating happened with organic matter pores. The increasing of the volumes of pores with sizes 10-25 μm that could be connected with generation of elongated cracks along the shale lamination. The results opens new opportunities for evaluation of maturity of organic matter based on pore structure analysis.

Keywords: Domanik, oil shale, unconventional resources, porosity, heating, eccentricity, pore size distribution, organic matter, CT, X-Ray computed tomography.

1. Introduction

As the global supply of traditional oil shrinks, the energy industry invests heavily in the identification, exploration, production and processing of unconventional energy resources, which are difficult to access or extract [1]. Oil shale is a fine-grained sedimentary rock consisting of organic matter that releases hydrocarbons (shale oil) and combustible gas during pyrolysis [2]. The potential value of a shale deposit mainly depends on its thickness, depth and organic matter content [3]. One of such type of oil shale that locates on the territory of Volga-Ural Basin in the East of Russian plate is Domanik.

The Domanik deposits consist of siliceous-clay-carbonate rocks enriched with organic matter and other components of oil in the dispersed state. They are Middle Frasnian source rocks and have low porosity and permeability. The increasing significance of shale oil and gas has led to the need for deeper understanding of Domanik porous space structure. The total porosity of natural Domanik oil shales is distributed from 0.66 to 1.87 %. A classification of pore space structure in conformity with pores sizes of the Domanik oil shales is proposed in [4]. It includes cracks and cavities with linear sizes more than 0.1 mm that spread in carbonate forms of Domanik rocks, interparticle pores, which have sizes from 10 μm to 100 μm, and organic matter pores with a part of little interparticle pores with sizes from 10 μm and less. The main idea of this work is to establish the changing of pore structure during the heating.

2. Methodology

We took two samples with sizes about 1 mm in diameter and 1 mm in height from the same Domanik core sample of Sargaevian horizon. The information about the sample was briefly described in [4][5]. One of the sample was heated to 350°, another was researched without heating. Then X-ray computed microtomography (microCT) was made by micro- and nanofocus X-ray control system for computed tomography General Electric V|TOME|X S 240. The resolution 1.5 μm was reached for both samples. Virtual models of the porous structure and measurements were performed by Volume Graphics and Avizo Fire software.

3. Results and Discussions

Finally, two microtomographic models of the pore space structure of Domanik oil shale were obtained for the natural sample and heated to 350° C (fig. 1). We note a visual increase in the number of pores in the heated sample. In addition, the newly formed pores predominantly have an elongated shape along the lamination. The results of measurement of the porosity coefficient and eccentricity confirm our findings (table 1). The porosity coefficient of the heated sample is more than 3 times higher than that of the natural one. Eccentricity is a classical elliptic parameter used for the shape of binarized objects. The eccentricity is necessarily between 0 and 1.If it is close to 0, pores predominantly have form of circle (sphere). If it tends to 1, pores have more elongated forms. We can see that after heating of the sample the pores became more elongate than in natural one.

The pore sizes distribution of natural Domanik oil shale shows that the main role in porous structure of the sample plays the pores with equivalent diameter less than 5 μm (fig. 2). They consist of more than 80% of porous volume of the sample. The distribution of heated to 350° C sample demonstrates decreasing the ratio of pores with sizes 1.5-5 μm and significant increasing the ratio of pores with sizes 5-10 μm. This finding suggests the important idea that main changing of pore structure during the heating happens with organic matter pores, which sizes can obtain up to 10 μm [4]. This result opens new perspectives for evaluation of maturity of organic matter based on pore structure analysis. In addition, we marked the increasing of the volumes of pores with sizes 10-25 μm that can be connected with generation of elongated cracks along the shale lamination. Such lithogenetic cracks are briefly occurred in more mature oil shales and their genesis can be connected with temperature separation at the boundary of two media with different physicochemical properties.

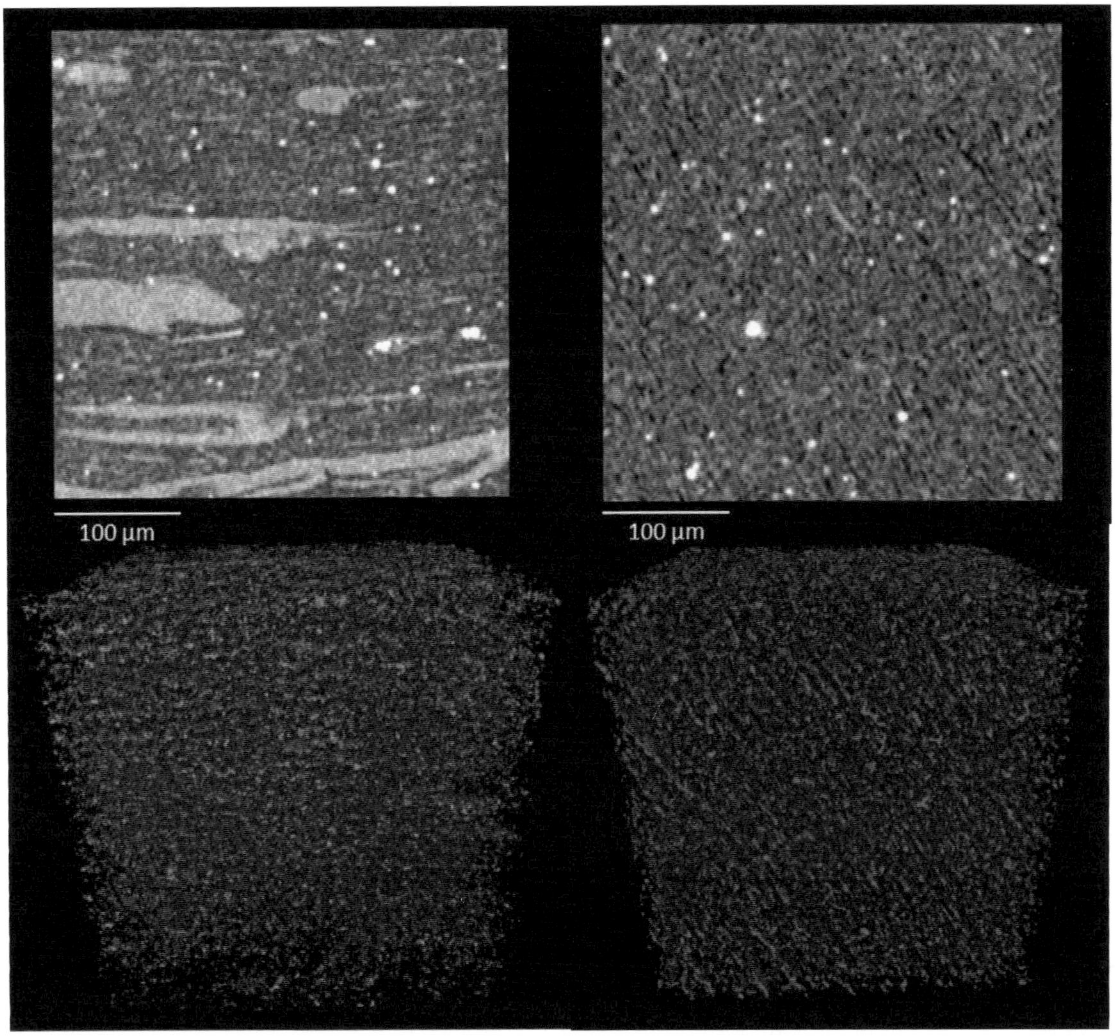

Figure 1. X-ray microtomography slices (on the top) and 3D visualization of pore structure (on the bottom) of natural (on the left) and heated to 350° C (on the right) samples of Domanik oil shale. The resolution of both models 1.5 μm.

Table 1. Comparison between coefficients of porosity and eccentricity for natural and heated to 350° C samples of Domanik oil shales

Sample	Porosity coefficient, %	Eccentricity of porous media
Natural	1.89	0.39
Heated 350°C	6.14	0.79

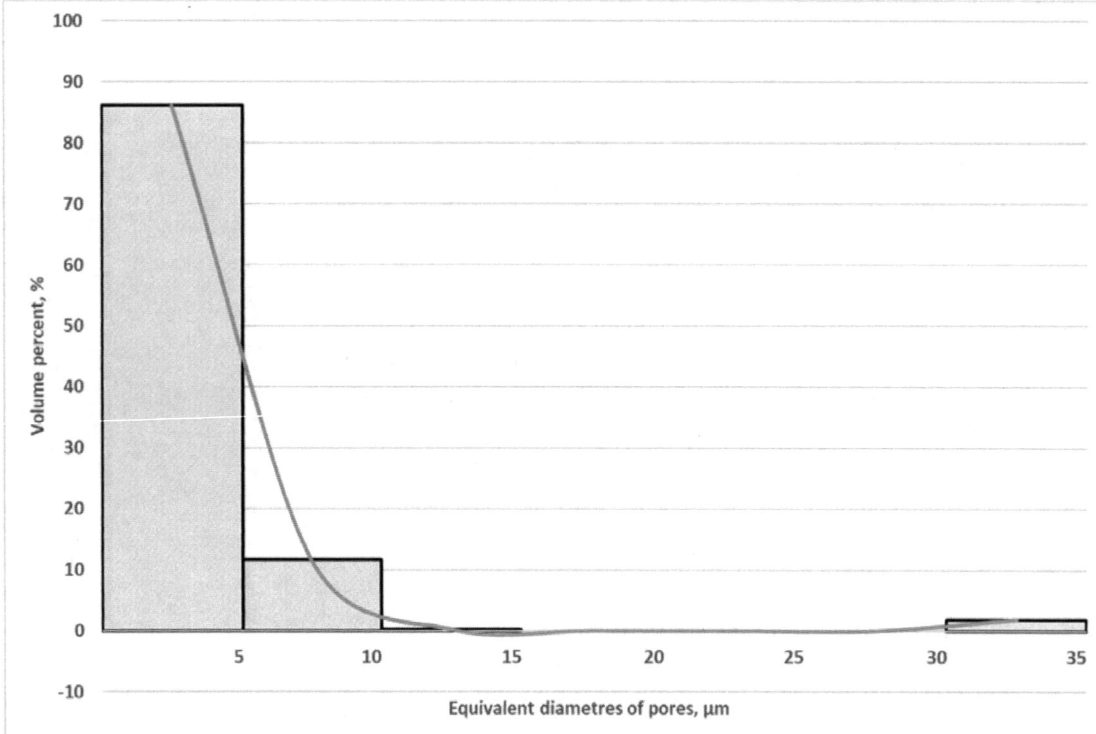

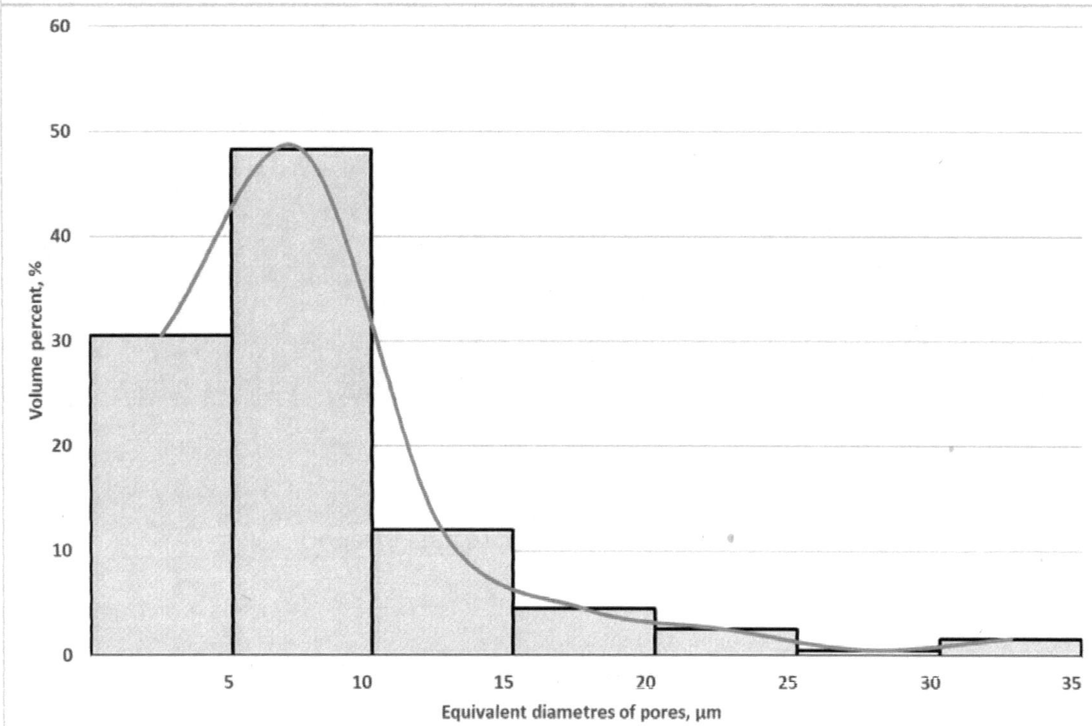

Figure 2. Pore sizes distribution for natural (on the top) and heated to 350° C samples of Domanik oil shale.

4. Conclusions

Thus, there was established increasing the porosity to 3 times as a result of heating of the oil shale. The distribution of heated to 350° C sample demonstrated decreasing the ratio of pores with sizes 1.5-5 μm and significant increasing the ratio of pores with sizes 5-10 μm. It means that main changing of pore structure during the heating happened with organic matter pores. We also marked the increasing of the volumes of pores with sizes 10-25 μm that could be connected with generation of elongated cracks along the shale lamination. Obtained results opens new possibilities for evaluation of maturity of organic matter based on pore structure analysis.

Acknowledgements

The work is performed according to the Russian Government Program of Competitive Growth of Kazan Federal University.

References

[1] Saif T, Lin Q, Gao Y, Al-Khulaifi Y, Marone F, Hollis D, Blunt M J and Bijeljic B 2019 4D in situ synchrotron X-ray tomographic microscopy and laser-based heating study of oil shale pyrolysis *Appl. Energy* 1468–75
[2] Lee S 2019 *Oil Shale Technology* (Taylor & Francis Group)
[3] Dyni J R 2003 Geology and resources of some world oil-shale deposits *Oil Shale*
[4] Kadyrov R 2018 The porous space structure of Domanik shales in the East of Russian plate *18th International Multidisciplinary Scientific GeoConference SGEM2018, Science and Technologies in Geology, Exploration and Mining* vol 18 pp 907–14
[5] Kayukova G P, Kiyamova A M, Mikhailova A N, Kosachev I P, Petrov S M, Romanov G V, Sitdikova L M, Plotnikova I N and Vakhin A V 2016 Generation of Hydrocarbons by Hydrothermal Transformation of Organic Matter of Domanik Rocks *Chem. Technol. Fuels Oils* **52** 149–61

IOP Conf. Series: Earth and Environmental Science **282** (2019) 012034 doi:10.1088/1755-1315/282/1/012034

Measurement of Damköhler number on the X-Ray computed tomography basis in the process of heavy oil carbonate reservoir acidizing

R Kadyrov

Institute of Geology and Petroleum Technologies, Kazan Federal University, Kazan, Russia

E-mail: Rail7777@gmail.com

Abstract. Nowadays, cases of the joint use of thermal and chemical methods of enhanced oil recovery have become more frequent. In this regard, acid treatment has an important role, which allows increasing the productivity of wells with low-permeability carbonate reservoirs. At the experimental level, the presence of an optimal injection mode has been proven, that depends on the dimensionless of Damköhler number and leads to the formation of "wormholes" of dissolution with increasing reservoir permeability. In this work, a new optimized method for calculating the Damköhler number based on X-ray computed tomography data is proposed. A comparison of the calculation of values by the classical and the new technique for different wormholes is presented.

Keywords: EOR, acid treatment, carbonate acidization, wormhole, Damköhler number, microCT, X-ray computed tomography.

1. Introduction

In recent years, there have been cases of the integrated use of different enhanced oil recovery methods (EOR), including combining thermal and chemical types. In order to achieve high oil recovery efficiency, acidic treatment of the carbonate reservoir can be used in thermal EOR, which is the dominant practice in the industry to increase the productivity of wells in low-permeability. For example in work [1], to enhance oil recovery during the process of thermal-steam stimulation (TSS) in high-viscosity reservoirs, acid system was injected into low productivity production wells.

Despite many years of experience in the application and a large amount of research conducted aimed at improving and increasing the efficiency of the method, a significant part of the treatments does not give good results. The heterogeneity of carbonate rock formations significantly influences to the filtration characteristics. The success of acid treatment is affected by the structure, length and branching of the filtration channels. In this regard, an important task is the formation of optimal channels of dissolution, leading to an increase in reservoir permeability. At the experimental level, the presence of an optimal injection mode has been proven. It depends on the dimensionless of Damköhler number and is associated with the formation of "wormholes" during the acid treatment [2]. The Damköhler number for channels of maximum efficiency has established in the interval 0.21-0.57 [3]. To determine the optimal characteristics, it is necessary to conduct experiments aimed at studying the pore space of the reservoir and its changes during the interaction of rock with acid.

X-ray computed tomography is the most convenient way to study the shape of the formed filtration channel. It allows getting maximum information about the real three-dimensional geometry of the "wormhole". However, all the advantages of this method are offset by the need to select the boundaries of the filtration channel and use approximate values of diameter and length for its calculation. The diameter of the wormhole can vary along its tracking, the wormhole itself can wriggle, making it difficult to determine its length. The situation is complicated by the formation of numerous branches from the main channel and the possible presence of several large connected channels. A new method of measurement the Damköhler number using the abilities of X-Ray computed tomography is proposed in this article.

2. Methodology

A correlation is established between the geometry of the filtering channels and the Damköhler number N_{Da}, which is defined as the ratio of the rate of flow of chemical interaction to the rate of acid supply, [3]. For the particular case of formation of the acid dissolution channel in the reservoir model, the dimensionless Damköhler number N_{Da} is determined by the formula [4]:

$$N_{Da} = \pi dlk / Q, \tag{1}$$

where k is the total reaction rate constant, cm/min; Q is the flow rate of fluid filtration, cm^3/min; l is the length of the dissolution channel, cm; d is the diameter of the dissolution channel, cm.

The ratio πdl is the lateral surface area of the cylinder that is closest to the size of the filtration channel. However, the wormhole often has bends and additional branches; it is very difficult to approximate its shape with a cylinder, which can significantly distort the calculation of the Damköhler number N_{Da}.

Therefore, instead of the lateral surface area of the “wormhole” πdl in formula (1), it is proposed to use an alternative parameter, which would simply be derived from the digital model. To characterize the area, it can be used not only the main filtration channel, but the entire network through which the acid has passed [5]. It can be easily segmented, as the largest associated object in the void-pore space of the sample. Thus, the new "Damköhler number" will look like:

$$N_{CT} = Sk / Q, \tag{2}$$

where k is the total reaction rate constant, cm/min; Q is the flow rate of fluid filtration, cm^3/min; S is the surface area of the dissolution filtration network, cm^2.

To compare the results of the measurement of N_{Da} and N_{CT} on cylindrical carbonate samples with a diameter of 30 mm and a height of 30 mm, experiments with filtration of acidic compositions were performed. The flow rate for all samples was equal – 0,5 ml/min. After that, X-ray computed tomography was done for the obtained samples with wormholes. Further, the boundaries of the filtration channels were identified on the 3D rock models. The values of N_{Da} and N_{CT} were measured according to the method described in more detail in [5].

3. Results and Discussions

The tomography results in the form of 3D visualization of wormholes are presented in Figure 1. The morphology of the obtained wormholes shows difference of their possible shapes. The first wormhole has the widest channel with thick side branches. In the second sample, two main filtration channels were formed. The wormhole of the third sample has a tortuous shape with numerous tree-like side channels.

Based on the formulas (1) and (2) for the data of wormholes, the values of the standard Damkeller number N_{Da} and the new N_{CT} were calculated (table 1). As can be seen from the results, an increase in N_{Da} values corresponds to an increase in N_{CT}, however, N_{CT} values increase to a greater extent. This is due to the fact that only the change in the diameter and length of the channel affects the N_{Da} parameter, whereas the N_{CT} takes into account not only the change in the area of the main channel, but also the area of all the resulting side branches.

Table 1. Statistic data on porous space of the samples.

Parameter	1	2*	3
N_{Da}	0,26	1,94	2,54
N_{CT}	0,84	18,15	19,49

* - in the sample 2 N_{Da} and N_{CT} were counted only for the biggest wormhole (the left one in Fig. 1)

4. Conclusions

Despite the similarity of formulas (1) and (2), they demonstrate different rate of dynamics when changing filtration parameters. Thus, the value of the classical Damköhler number will increase, but the new ratio will increase faster due to the increase in the network surface area according to a power law. Nevertheless, it is proposed to use a new dimensionless parameter, as more accurate for the current level of technology development. In the course of further experimental work, it is planned to compare the values of these parameters for easy orientation between them.

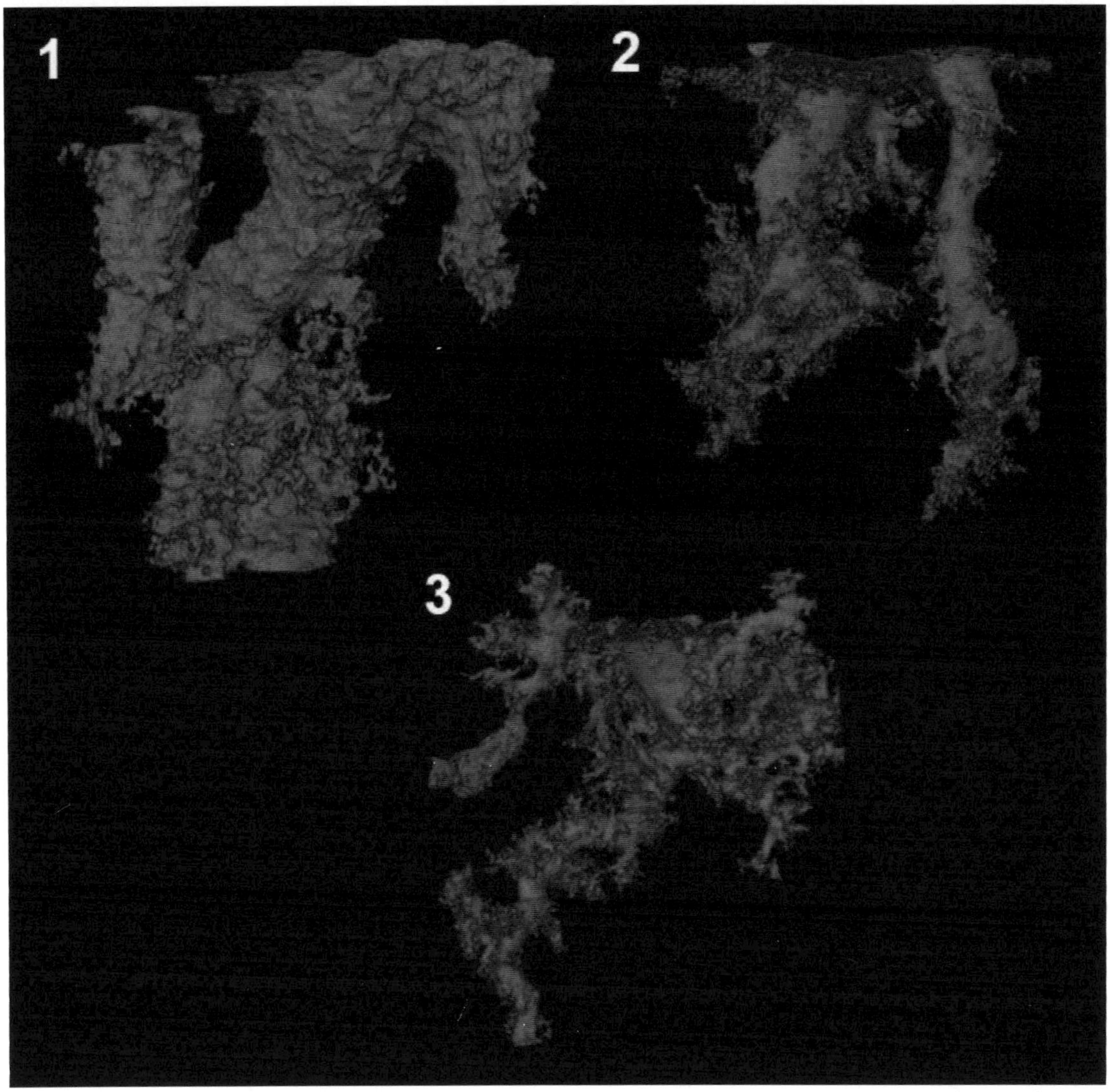

Figure 1. 3D visualization of wormholes, obtained for different acid compositions with flow rate 0.5 ml/min

Acknowledgements

The work is performed according to the Russian Government Program of Competitive Growth of Kazan Federal University.

References

[1] Altunina, L. K. *et al.* 2015 Pilot Tests of New EOR Technologies for Heavy Oil Reservoirs *SPE Russian Petroleum Technology Conference* 16

[2] Fredd, C. N. and Fogler, H. S. 1998 Alternative Stimulation Fluids and Their Impact on Carbonate Acidizing *SPE J.* **3**, 34–41

[3] Hoefner, M. L. and Fogler, H. S. 1988 Pore evolution and channel formation during flow and reaction in porous media *AICHE J.* **34**, 45–54

[4] Cohen, C. E. *et al.* 2008 From pore scale to wellbore scale: Impact of geometry on wormhole growth in carbonate acidization *Chem. Eng. Sci.* **63**, 3088–3099
[5] Kadyrov, R. I. *et al.* 2018 Structural Transformation of the Void-Pore Space of a Lime Reservoir During HCl Treatment *Chem. Technol. Fuels Oils* **54**, 307–318

IOP Conf. Series: Earth and Environmental Science **282** (2019) 012035 doi:10.1088/1755-1315/282/1/012035

Stages of thermal transformation of oil shales organic-mineral substance of the Middle Volga region during combustion

E Korolev[a], A Eskin[a], A Korolev[a], E Barieva[b]

[a] Institute of geology and petroleum technologies, Kazan Federal University, Kazan, Russia

[b]Kazan State Power-engineering University, Kazan, Russia

e-mail: eskin.aleksey@gmail.com

Abstract. The study of the staging of the thermal conversion of oil shale, common in the Middle Volga region, was carried out. It is established that the organic matter of rocks due to the polycomponent composition undergoes multi-stage pyrolysis in the temperature range 200-600°C. The most favorable temperature of pyrolysis, when up to 80% of the potential heat of shale is released, falls within the interval of 460-480°C. The main thermal transformations of the ash component are carried out at temperatures above 480°C. They are due to the release from the structure of clay minerals of the OH group, the oxidation of pyrite, the decomposition of calcite and the formation of spinels.
Summing up the obtained results, we can draw to the following conclusions:
1. Oil shale of the Middle Volga region is characterized by significant variations in the content of organic matter from 9 to 40%. The light isotopes C12 predominance in the composition of organic carbon indicates the predominantly biochemical origin of the organic matter.
2. The configuration of the DSC curve in the temperature range of 200–600°C indicates both the multicomponent composition of the organic compounds of oil shale and their multi-stage pyrolysis.
3. The most favorable pyrolysis temperature, when up to 80% of the potential heat of shale is released, is in the range of 460-480°C.
4. In the process of burning oil shale, the ash component of the rock also subjected to thermal transformation. The composition of rock-forming minerals determined the main temperature ranges of transformation in the areas of 480–600°C (extraction of hydroxyl water from clays, oxidation of pyrite with the formation of hematite), 680–800°C (destruction of calcite), 880–900°C (formation of spinels).

1. Introduction

The global trend of increasing energy consumption stimulates the development of research on the introduction of additional, local energy raw materials sources into the technology of the thermal electricity production. In the territory of the Middle Volga region one of such promising objects for energy needs is oil shale [1, 2, 3]. Shale strata confined to terrigenous-clayey complexes of Late Jurassic age can be traced in the Ulyanovsk and Saratov regions, as well as in the western part of the Republic of Tatarstan. The shale rocks are deposited at an insignificant depth and formed layers with a thickness of up to 7.0 m sustained along the strike [4].

IOP Conf. Series: Earth and Environmental Science **282** (2019) 012035 doi:10.1088/1755-1315/282/1/012035

Such geological conditions make the development of oil shale profitable by the cheapest open-pit (quarry) method. Despite the fact that oil shale is a low-grade energy resource, they have several advantages compared to higher-quality but imported coal. These are proximity to consumers, a well-established infrastructure in their areas of occurrence, the possibility of utilizing ash waste in the production of building materials in nearby factories. In this regard, the issue of considering the possibility of using the Volga oil shale in the energy sector becomes urgent.

2. Methodology

Taking into account the prospects for the industrial use of oil shale, in this work we studied the peculiarities of rocks organic component pyrolysis and their mineral substances transformation. Optical microscopic analysis of thin sections was carried out on an optical polarization microscope the Axio Imager research class (Carl Zeiss, Germany).The main research method was thermal analysis performed on an NETZSCH STA 449 Jupiter F3 instrument. The burning ranged from 30 to 1000 ° C, the heating step – 10 deg. / min. with constant air blowing. SEM, X-ray, and isotope analysis methods were used as auxiliary methods.

X-ray phase analysis. The X-ray phase analysis was performed on the diffractometer D2 Phaser (Bruker, Germany) used for measuring the powder products in the Bragg-Brentano geometry, with the use of monochromatic CuKα-radiation (λ = 1.54178 Å), in the step-scan mode. Measuring and recording modes: X-ray tube voltage-30 kV, current-30 mA. Scanning step – 0.02°; speed – 1 deg/min. The range of scanning angles in the Bragg-Brentano geometry 3-40°. Mineral types are determined by comparing experimentally obtained values of interplanar distance (d, Å) and relative intensity (Irel) of reflections to the standard XRD data from the International Database PDF-2 ICDD.

In order to establish the morphological appearance of minerals and their interrelations with each other, the samples were studied using a scanning electron microscope XL-30 ESEM, which equipped with an energy-dispersive analyzer (EDAR). From the most interesting areas were made small fragments, which were covered with carbon coating. After this, the objects were scanned and a selective survey of their surface was made.

The ratio of stable isotopes of light elements 13C / 12C and 18O / 16O was determined on a Delta V Plus isotope mass spectrometer (Thermo Fisher Scientific, Germany). The analysis of the substance was carried out in the gas phase.

3. Results and Discussion

In the outcrops, oil shales are characterized by a dark gray color, with black areas, cryptomerous structure, poorly-defined layered texture and well-pronounced foliation. According to optical microscopic and X-ray analysis data the shales are composed mainly of mixed-layer clay minerals illite-montmorillonite composition. The chlorite and muscovite flakes, as well as allothigenic grains of quartz, albite and microcline up to 0.1 mm in size are present as impurities [2]. Shales are enriched with limestone fragments of marine animals and pyrite aggregates. Coal shales are split by numerous sub-vertical cracks, some of which are filled with black organic matter. The latter indicates that in the process of lithogenesis, an active redistribution of the organic component took place. At the lithification time there were mobile carbonaceous compounds in shale, capable to migration and secondary concentration. To determine the genesis of carbon on some samples were analysis for the ratio of the C12 and C13 isotopes in them, as well as the content of the total amount of organic and inorganic carbon. Negative values indicate about less C13 contain in sample than the standard (lighter than the standard), positive values are heavier than the standard. PDB1 - the belemnite rostrum Belemnitella americana served as a standard (Table 1).

Table 1. The isotopic composition of organic and inorganic carbon oil shales.

IOP Conf. Series: Earth and Environmental Science **282** (2019) 012035 doi:10.1088/1755-1315/282/1/012035

Samples	Organic carbon		Inorganic carbon	
	total amount C, %	δC^{13}	total amount C, %	δC^{13}
1	36.0581	-2.74813	9.6213	0.09721
2	31.4912	-2.86547	10.9463	0.16138
3	22.2466	-10.34781	14.5334	0.10058
4	39.7480	-3.74303	38.0599	0.78185

The results showed that the content of organic carbon in oil shale reaches 40%. The light isotopes of C12 predominate, indicating it's predominantly biochemogenic origin. Considering the anoxide geochemical environment during the sedimentation and lithification period of carbonaceous-clay sediments, it can be assumed that the formation of organic matter was carried out with the active participation of anaerobic microbial communities. The latter, in the course of their vital activity, processing plant and animal biomass, could easily enrich the rocks with light C12 isotopes. The content of "heavy" inorganic carbon in the shells of marine animals is conditioned on the one hand by the forced extraction of C13 by floating organisms directly from sea water; on the other hand, by the ability of skeletal forms for a long time remain buried without significant changes. Due to the outer protective layer of organic-mineral matter, the shells retain the original ratio of light and heavy carbon isotopes, acquired during the life of mollusks.

On the basis of the data obtained, it can be assumed that the scattered dispersed organic matter of combustible shale is represented mainly by the products of biochemical transformation of plant-plankton residues by colonies of anaerobic microorganisms. Indirectly, this is confirmed by a significant amount of pyrite (2-3%), common in the rocks.

In order to detail the fractional composition of carbonaceous compounds and the ash component transformation stages, a study of oil shale was carried out using thermal analysis. The thermograms data showed that, with a consistent increase in the firing temperature, very diverse thermal transformations occur in the rocks, due to their multicomponent composition (Fig. 1).

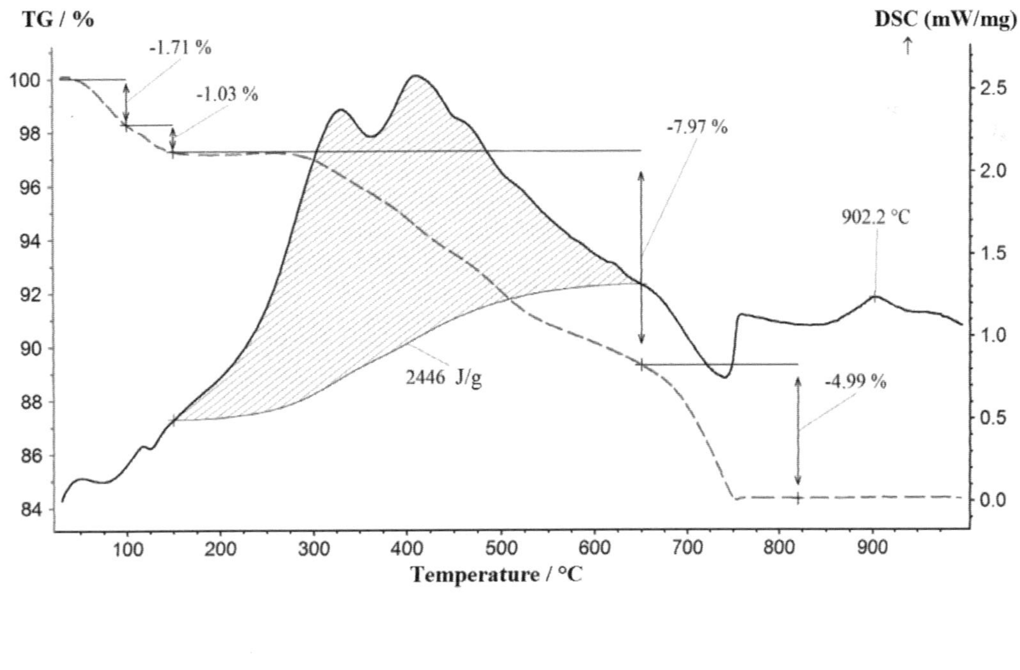

IOP Conf. Series: Earth and Environmental Science **282** (2019) 012035 doi:10.1088/1755-1315/282/1/012035

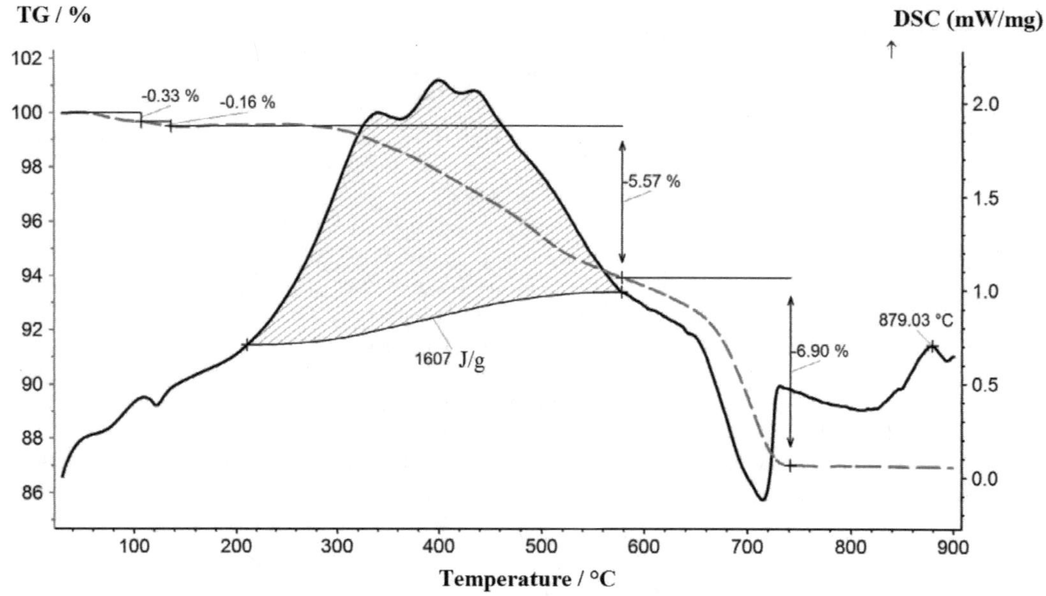

Figure 1. Synchronous thermal analysis data of the Middle Volga region oil shale deposit. The solid line is the differential scanning calorimetry (DSC) curve; the dotted line is the mass loss curve (TG).

The first changes occur already in the range of 60-150°C. Here on the thermogravimetric curve (TG) there is a two-step mass loss of the sample by 2-3%, accompanied by the appearance on the curve of differential scanning calorimetry (DSC) of two consecutive endo-effects. Considering the presence of montmorillonite, muscovite and chlorite in combustible shales, these effects can be attributed to the successive release of capillary-condensed and then molecular (interlayer) water from clay minerals of the shale.

The following changes are noted in the temperature range of 260-600°C. In this range occurs organic matter thermo-oxidative destruction. The complex configuration of the DSC curve indicates that the thermal decomposition of organic compounds proceeds in several stages. At the first stage, in the range of 260-320°C, boiling up of light hydrocarbon fractions occurs [5]. Considering the high content of methane components in the oil shales composition (up to 49.6% according to [4]), we can assume that the first exothermic effect is due to the release of CH4 from rocks. At the same time, at temperatures above 300°C, resinous substances condense with their transition to a semi-liquid state, that is, the bituminization process takes place. The main mass of shale resin is formed in the temperature range 320-450°C. Under conditions of rapid temperature increase in the heating chamber, the resulting resin evaporates and leaves in the form of steam. The mass restructuring process of kerogen carbon skeleton during its transition to a plastic state takes place at 380-400°C and fully ends at 450°C. The boiling up of shale resin is accompanied by the release of heat and, as a result, by the appearance of the following exo-effects on DSC in the interval of thermo-oxidative destruction of medium fractions hydrocarbons [5]. At temperatures above 400°C, along with resin, heavy hydrocarbons start to be released from shale, the amount of which in rocks according to [4] reaches 1.3%. The peak of their boiling is at 450-480°C, which is noted by the following exo-effect on DSC. Starting at 500 ° C, the pyrolysis of organic matter enters its final phase. Beginning processes of semi-coking the heavy resin and other relics of thermal bitumen. Due to the low content of hydrogen and oxygen in the products of carbonaceous clay shale pyrolysis, the subsequent increase in the firing temperature of rocks above 500 ° C does not lead to the significant gaseous substance formation. As a result, on the general decline of the DSC curve, only two small-intensity exo-effect in the region of 510–520 ° C and 560–580 ° C are distinguished, accompanied by a loss of mass. Their appearance is caused by the formation of coke and the decomposition of kerite.

Thus, the complex configuration of the DSC curve in the temperature range of 150-600 ° C indicates both the multicomponent composition of carbonaceous compounds in coal shale and organic matter multistage pyrolysis. The predominant primary components are the lipid constituents in kerogen, as indicated by the carbon isotope composition, as well as light and heavy oil-like fractions; the secondary ones are shale resins, which are the main constituents of the thermal bitumen; coke and kerite. The yield of volatile substances during the combustion of shales is 8-12%. Of these, light hydrocarbons - up to 49.6%, shale resin - up to 10%, heavy hydrocarbons - up to 1.3%. The significant yield of oil-like fractions in the process of burning coal shale is due to the presence high hydrogen content (up to 10.6% according to [4]) in the organic mass of rocks. It is his presence that contributes to the liquefaction of organic carbon compounds with the formation of various by compositions and boiling points of shale resins. Successively formed hydrocarbons, boiling in different, albeit overlapping, temperature ranges, create conditions for the continuous formation processes of the vapor-gas mixture, which determines the energy value of the carbonaceous-clay shale of the Middle Volga region. The vapors of hydrocarbons and non-condensable gases (CO_2, CO, H_2, N_2, H_2S) generated during the pyrolysis of organic matter, make conditions a relatively high heat of combustion of the rocks (2.5-6.8 MJ / kg). The most favorable pyrolysis temperature, when up to 80% of the potential heat of shale is released, is in the range of 460-480°C.

Further mass change on the TG curve to a temperature of 680°C is associated with lost hydroxyl water (OH-) from clay minerals structure. The implicitly pronounced nature of the endo-effect on the DSC curve appears to be due to the superimposed exo-effect caused by the oxidation of pyrite to form hematite. Mass loss in the temperature range of 680–800°C, accompanied by endo-effect, corresponds to the thermal destruction of calcite to CaO. The exo-effect on the DSC curve with a maximum at 880-902°C, which passes without a change in mass, corresponds to the recrystallization of amorphous products of destruction of clay minerals, accompanied by the formation of new mineral phases of the spinel type [5].

4. Conclusions

Given the above, we can draw the following conclusions:

1. Oil shale of the Middle Volga region is characterized by significant variations in the content of organic matter from 9 to 40%. The light isotopes C12 predominance in the composition of organic carbon indicates the predominantly biochemical origin of the organic matter.
2. The configuration of the DSC curve in the temperature range of 200–600°C indicates both the multicomponent composition of the organic compounds of oil shale and their multi-stage pyrolysis.
3. The most favorable pyrolysis temperature, when up to 80% of the potential heat of shale is released, is in the range of 460-480°C.
4. In the process of burning oil shale, the ash component of the rock also subjected to thermal transformation. The composition of rock-forming minerals determined the main temperature ranges of transformation in the areas of 480–600°C (extraction of hydroxyl water from clays, oxidation of pyrite with the formation of hematite), 680–800°C (destruction of calcite), 880–900°C (formation of spinels).

Acknowledgments
This study was performed in the context of the Russian Government Program of Competitive Growth of Kazan Federal University.

References:

[1] E.M. Atoyan, B.A. Semenov and S.S. Nazarov, "Pyrogasification of combustible shales of the Volga region in order to use the final products to produce electrical and thermal energy", SSTU Bulletin, vol. 54, pp. 101-105, 2011.
[2] E.A. Korolev, E.R. Barieva and A.A. Eskin, "Perspective usage estimation of Volga region combustible shale as a power generating fuel alternative", IOP Conference Series: Earth and Environmental Science, vol. 155, art. № 012011, 2018.

[3] A.I. Blokhin, M.I. Zaretsky, G.P. Stelmakh and T.S. Eyvazov, “New technologies for processing high-slate shale” (Thermal analysis of minerals and rocks Nauka), p 192, 2001.
[4] V. A. Cherdyntsev and E. I. Tikhvin, "The geology of the Tatar ASSR and neighborhood within 109 sheet” (Moscow, Publishing house «GONTI»), p 204, 1939.
[5] V.P. Ivanova, B.K. Kasatov, T.N. Krasavina and E.L. Rozinova, “Thermal analysis of minerals and rocks” (Moscow, Publishing house Nedra), p 399, 1974.

IOP Conf. Series: Earth and Environmental Science **282** (2019) 012036 doi:10.1088/1755-1315/282/1/012036

Heterogenous catalysts based on Fe-pillared clays for catalytic aquathermolysis processes of heavy oil

V Starshinova[1,2], Ya Abdelsalam[2], V Gorelysheva[2], A Oparkin[2], S Gnevashev[2], A Pyataev[1], R Khamidullin[2], E Karalin[2], and A Shinkarev[2] (jun.)

[1] Kazan Federal University, Russia

[2] Kazan National Research Technological University, Russia

E-mail: starshinovavl@gmail.com

Abstract. Aquathermolysis process is one of the key technologies of reducing the viscosity and exploitation of huge deposits of heavy crude oil. That process combines thermal and catalytical effects in the presence of water. There are a lot of types of catalysts used in aquathermolysis process, such as water and oil soluble, heterogeneous catalysts and minerals. One of the type of heterogeneous catalysts are pillared clays that used in catalytic, sorption and separation processes. Pillared clays (PILC) are the class of two-dimensional micro-mesoporous materials that have a high specific surface area, constant porosity, developed texture and active compaunds in the structure, i.e. combine both the carrier and the catalyst. In this study the physico-chemical properties of Fe-pillared clays and its application in the "reconnaissance" heavy oil catalytic aquathermolysis experiments of has been investigated.

1. Introduction

The catalytic aquathermolysis process becomes more significant research area for investigation of heavy crude oil upgrading since the end of last century.

Among well-known recovery techniques, thermal recovery by catalytic aquathermolysis process is the simple and effective method for heavy oil upgrading. The aquathermolysis has various advantages like 1) reduction of viscosity, 2) desulfurization, and 3) hydrogenation of heavy oils. Also the important benefit of aquathermolysis process is *in situ* viscosity reduction [1].

In general the catalysts used for aquathermolysis process are mineral, watersoluble, oil soluble, and dispersed catalysts. The viscosity reduction with these types of catalysts is in the order of mineral <water-soluble catalyst<oil-soluble catalyst<dispersed catalyst. Also during aquathermolysis, the saturates and aromatics increase while the amount of asphaltene and resin decreases.

There are a lot of investigation about reducing of heavy crude oil viscosity with modified hematite nanoparticles [2, 3], iron organic and non organic salts [4], iron catalyst and hydrogen donors [5], catalysis by keggin heteropoly acid salt [6], iron oxide and iron hydroxide ultradispersed colloidal particles [7], iron(III) tris(acetylacetonate) [8], amphiphilic metallic chelate catalyst-aromatic sulfonic iron [9], oil soluble catalytic solution (organic-metallic compounds of Mo, Fe, Co, Ni) [10], and mineral compounds with and without catalysts [11-13]. When the mineral is used with the metals, this mineral acts as support for

IOP Conf. Series: Earth and Environmental Science **282** (2019) 012036 doi:10.1088/1755-1315/282/1/012036

metals providing a more effective catalyst [14]. The use of different hydrogen donors on aquathermolysis also improves the quality of the heavy crude oil [15].

Investigation of mineral compounds modified by metal particles is interesting field in catalytical processes.

Pillared interlayered clays (PILC) are microporous materials synthesized by intercalation of oligonuclear complexes of different types metal cations into layered aluminosilicates interlayered spaces with following calcinations [16]. Large surface area, high pore volume and presence of catalytic active substrates make PILCs a promising material in heterogeneous catalysis. Moreover, varying the complex composition and metal type allow to adapt pillared systems for various catalytic processes.

Majority of domestic and overseas papers traditionally focus on synthesis of pillared aluminosilicates by method based on alkaline hydrolysis of metal salts with varying of initial reagents and process conditions [17]. In this case depending on process conditions in the pillaring solution various hydroxocations with different number of metal nuclears (charge and size) can be present. This circumstance doesn't allow obtaining catalyst with stable properties. It's worth to notice several works related to optimization of pillared clay obtaining methods including manipulations with parent aluminosilicates, polyhydroxocations, intercalation and pillaring conditions [18].

The particular research works describing hydrothermal synthesis of polyhydroxocations [19, 20] and research works [21, 22] where complex organic cations used as a pillaring agent should be noticed.

2. Objects and methodology

The precursor of layered compounds for PILC catalyst synthesis was bentonite clay with montmorillonite content about 95 wt. %.

Fe-PILC catalyst has been prepared by multi-stage saturation of 1% smectite suspension by iron acetate solution in ratio with 5 mmol of Fe to 1 g of clay. After preparation the samples was calcined in oxidation and reduction conditions at 450 °C during 4 h.

Chemical analysis of catalyst samples was determined with a universal XRF spectrometer SUR RENOM-FV.

X-ray diffraction (XRD) patterns of the samples were recorded by D2 Phaser diffractometer (Bruker Axs GmbH) with a CuKα radiation. Diffraction patterns processing was carried out by DIFFRAC.SUITE package software.

The textural properties of the samples were determined by nitrogen adsorption-desorption at 77 K by Quantochrome NOVA 2200e analyzer after degassing of samples at 378 K during 3 h. Specific surface area values were calculated by the Brunauer–Emmett-Teller (S_{BET}). The t-plotmethod was used to determine external surface area ($S_{external}$) and the micropore volume ($V_{micropore}$). The Barrett-Joyner-Halenda (BJH) method was applied to determine the mesopore volume ($V_{mesopore}$) and mesopore volume distribution as a pore size function.

Mössbauer spectra were recorded by MS1104EM spectrometer N40-12 with Co^{57} source in the Ro matrix transmission geometry and NaJ crystal scintillation counter detector. Spectrometer speed scale calibration was carried out by α-Fe spectra at the room temperature.

Scanning electron microscopy (SEM) microphotographs were obtained with powdered samples on a Carl Zeiss Merlin microscope operating at an accelerating voltage of 5 kV and a probe current of 300 pA for minimal exposure on the sample.

Transmission electron microscopy (TEM) microphotographs were collected on a Hitachi HT 7700 Exalens microscope operating at an accelerating voltage of 100 kV.

The object of the study was ultra-viscous oil from Ashalchinskoye oilfield (Volga-Ural basin, Republic of Tatarstan, Russia) with 0.965 g/cm^3 of density. Ashalchinskoye oilfield is one of the experimental objects for testing and optimization of high-viscosity oil extraction technologies located in the territory of the Republic of Tatarstan [23].

The aquathermolysis process experiments were conducted in a 0,25 L autoclave for 2 h at 250 °C at the vapor pressure 3 MPa. The mass ratio was 4:1 for oil:water and 200:1 for oil:catalyst into the catalytic system.

3. Results and Discussions

Saturation of smectite clay by iron pillaring solution led to iron content increasing in 10 times according to the chemical analysis. In the Tab. 1 the chemical composition of the smectite clay sample and the Fe-PIC sample is presented.

Table 1. Chemical composition of smectite clay and synthesised Fe-PILC

Sample	Chemical composition (wt.%)					
	SiO_2	Al_2O_3	Fe_2O_3	MgO	CaO	Na_2O
Smectite	61,3	18,5	4,1	2,2	0,9	1,9
Fe-PILC after calcination in the oxidation conditions	39,1	12,5	42,8	1,5	-	-

The textural properties of Fe-PILC samples are given in Table 2. Calcination of Fe-PILC samples in oxidation and reduction air conditions lead to the differences in surface areas and pore sizes. Chemical reduction of iron-containing clusters in the Fe-PILC lead to decrease the specific surface and micropore disappearance.

Table 2. Textural characteristic of Fe-PILC samples

Sample	S_{BET}, m^2/g	$V_{mesopore}$, cm^3/g	$V_{micropore}$, cm^3/g	$D_{average}$, nm
Fe-PILC after calcination in the oxidation conditions	320	0.21	0,04	2,77
Fe-PILC after calcination in the reduction conditions	134	0.26	–	7,96

X-ray diffraction results for Fe-pillared clay are shown the presence of the first basal spacing (d_{001}) of 2,1 nm that confirms successful intercalation iron compounds between aluminosilicate layers. After calcination the basal spacing has decreased to 1,9 nm due to remove organic ligands and water. Calcination in the reduction conditions leads to partial interlayer spaces collapse and to basal spacing displacement to 1 nm. The most possibly reason explain this behavior connected with decrease of particle sizes iron oxidic form to lower degrees of oxidation form in reducing process.

SEM and TEM images of clay and Fe-PILC samples before and after calcination were shown on Fig. 1.

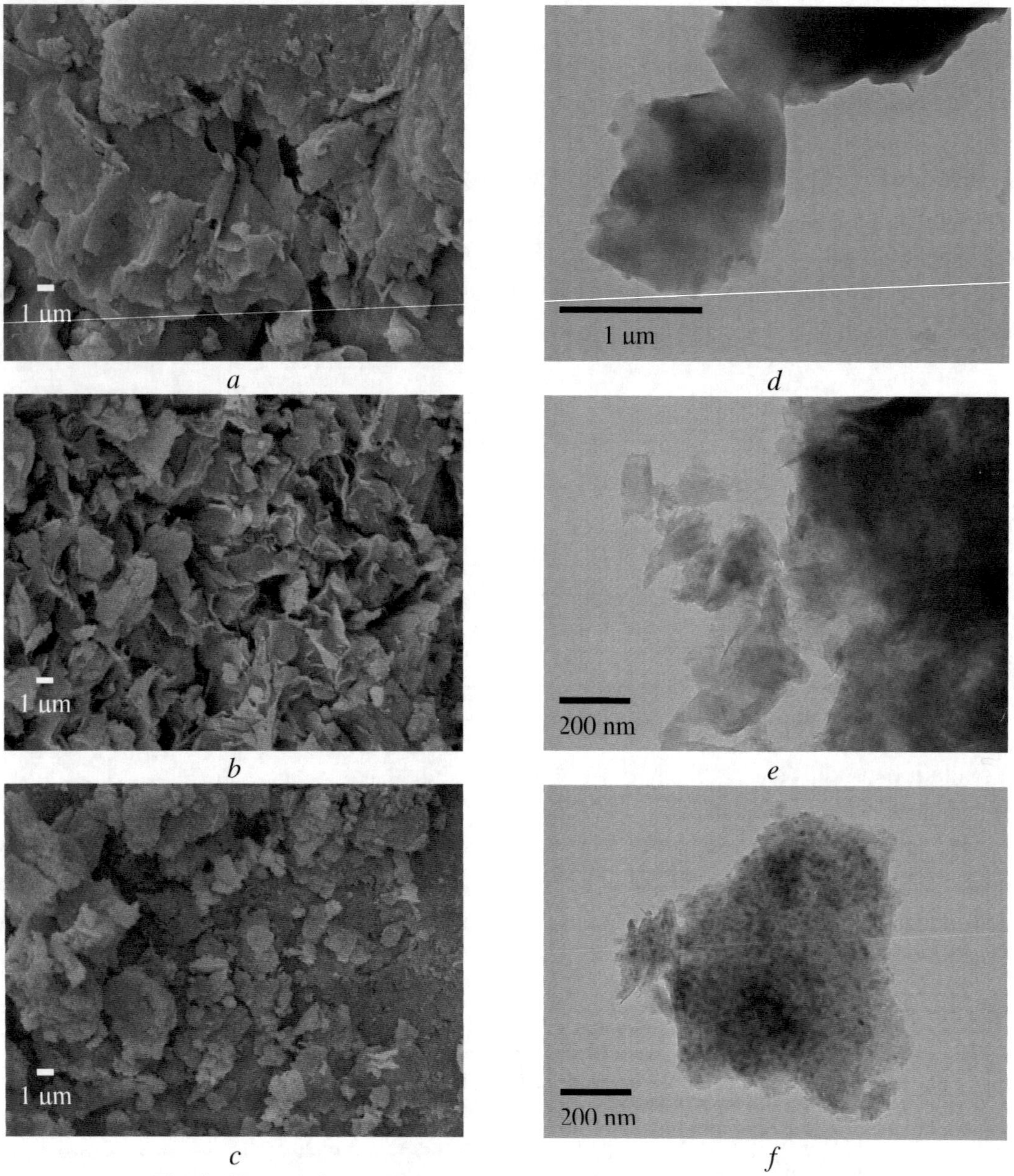

Fig 1. SEM (left) and TEM (right) images of Fe-PILC samples:
a,d – smectite; *b,e* – Fe-PILC after calcination in the oxidation conditions;
c, f – Fe-PILC after calcination in the reduction conditions

According to SEM images (*a,b,c*) the smectite clay sample has morphology with an extended silicate plates packed to tactoids. After pillaring processes the aggregate size decreases and it packed as house of cards. This effect is more expressed for calcined in the reduction conditions sample. On the TEM images (*d,e,f*) aluminosilicate plates are observed. For smectite sample (*d*) aluminosilicate plate has no any dark inclusions that appear in the Fe-PILC after calcination in the oxidation conditions and clearly marked for the sample calcined in the reduction conditions. That includes has a rounded shape and size about 10-20 nm. According TEM images with elemental mapping showed on fig. 2 the small size particles located inside and on the aluminosilicate sheets consist Fe.

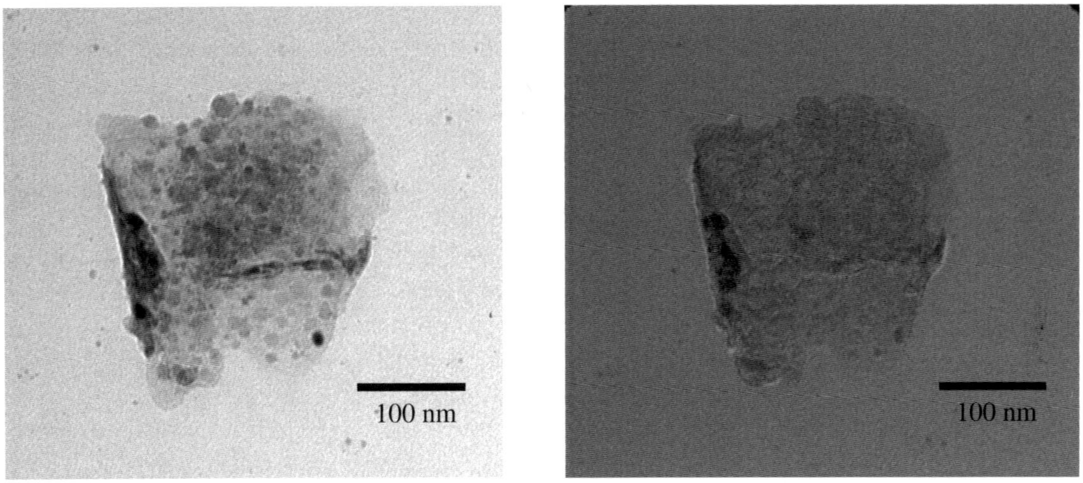

Fig 2. TEM images of Fe-PILC sample after calcination in reduction conditions

Mössbauer spectroscopy data of Fe-PILC samples show the presence of superparamagnetic iron and magnetite-like oxides forms particles separately distributed into clay structure and bounded with clay aluminosilicate matrix.

The aquathermolysis process results using different catalysts are given in the table 3.

Table 3. Results of catalytic aquathermolysis experiments

Reaction mixture	Density (g/cm^3 at 20 °C)	Boiling point (b.p.) (°C)	Dynamic viscosity (Pa·s at 20 °C)	Fraction output (wt.%)		Tars and asphaltenes content (wt.%)	Sulfur content (wt.%)
				200 °C	200-300 °C		
Initial oil	0,9650	170	2,68	1,00	12,00	34,4	4,70
Oil with water	0,9541	130	2,91	2,70	37,90	31,6	4,22
Oil with water and Fe-PILC after calcination in oxidation conditions	0,9578	100	2,61	2,63	15,37	30,6	4,16
Oil with water and Fe-PILC after calcination in reduction conditions	0,9437	80	1,23	5,00	20,55	20,6	3,56

4. Conclusions

Fe-pillared clays contain nano-clusters of iron oxides in different oxidation degrees which separately located in their structure. The results for aquathermolysis processes showed that viscosity, density, sulfur and tar-asphaltene content had been decreased and content of light oil fraction had been increased. Thus, Fe-pillared clay can be promising catalyst in the aquathermolysis process.

References:

[1] Maity S K, Ancheyta J, Marroquin G 2010 Catalytic aquathermolysis used for viscosity reduction of heavy crude oils: A review *Energy & Fuels.* **5** 2809-2816.
[2] Khalil M, Liu N, Lee R 2017 Catalytic aquathermolysis of heavy crude oil using surface-modified hematite nanoparticles *Industrial & Engineering Chemistry Research.* **15** 4572-4579.
[3] Khalil M, Lee R, Liu N 2015 Hematite nanoparticles in aquathermolysis: A desulfurization study of thiophene *Fuel. 145* 214-220.
[4] Yi Y, Li S, Ding F, Yu H 2009 Change of asphaltene and resin properties after catalytic aquathermolysis *Petroleum Science.* **2** 194-200.
[5] Zhong L, Liu Y, Fan H, Jiang S 2003 Liaohe extra-heavy crude oil underground aquathermolytic treatments using catalyst and hydrogen donors under steam injection conditions *SPE international improved oil recovery conference in Asia Pacific.*
[6] Chen Y, Wang Y, Lu J, Wu C 2009 The viscosity reduction of nano-keggin-$K_3PMo_{12}O_40$ in catalytic aquathermolysis of heavy oil *Fuel.* **8** 1426-1434.
[7] Nassar N, Husein M 2010 Ultradispersed particles in heavy oil: Part I, preparation and stabilization of iron oxide/hydroxide *Fuel Processing Technology.* **2** 164-168.
[8] Galukhin A, Erokhin A, Osin Y, Nurgaliev D 2015 Catalytic aquathermolysis of heavy oil with Iron Tris (acetylacetonate): changes of heavy oil composition and in situ formation of magnetic nanoparticles *Energy & Fuels.* **8** 4768-4773.
[9] Chen Y, Wang Y, Wu C, Xia F. 2008 Laboratory experiments and field tests of an amphiphilic metallic chelate for catalytic aquathermolysis of heavy oil *Energy & Fuels.* **3** 1502-1508.
[10] Fixari B, Peureux S, Elmouchnino J, Le Perchec P, Vrinar M 1994 New developments in deep hydroconversion of heavy oil residues with dispersed catalysts. 1 Effect of metal and experimental conditions *Energy & Fuel.* **8** 588-592.
[11] Merissa S 2013 Preliminary study of natural zeolite as catalyst for decreasing the viscosity of heavy oil In *AIP Conference Proceedings.* **1** 131-134.
[12] Fan H, Zhang Y, Lin Y 2004 The catalytic effects of minerals on aquathermolysis of heavy oils *Fuel.* **14-15** 2035-2039.
[13] Ovalles C, Vallejos C, Vasquez T, Rojas I, Ehrman U, Benitez J L, Martinez R 2003 Downhole upgrading of extra-heavy crude oil using hydrogen donors and methane under steam injection conditions *Petroleum science and technology.* **1-2** 255-274.
[14] Fan H F, Liu Y J, Zhong L G 2001 Studies on the synergetic effects of mineral and steam on the composition changes of heavy oils *Energy &Fuels.* **6** 1475-1479.
[15] Hyne J et al. 1982 Aquathermolysis of heavy oils *Revista Tecnica Intevep.* **2** 87-94
[16] Bergaya F, Lagaly G 2013 General introduction: clays, clay minerals, and clay science. In *Developments in Clay Science. 5* 1-19.

[17] Mandalia T, Crespin M, Messad D, Bergaya F, 1998 Large interlayer repeat distances observed for montmorillonites treated by mixed Al–Fe and Fe pillaring solutions *Chemical Communications*. **19** 2111-2112.
[18] Rightor E G, Tzou M S, Pinnavaia T J 1991 Iron oxide pillared clay with large gallery height: Synthesis and properties as a Fischer-Tropsch catalyst *Journal of Catalysis*. **1** 29-40.
[19] Ding M, Zuo S, Qi C 2015 Preparation and characterization of novel composite AlCr-pillared clays and preliminary investigation for benzene adsorption *Applied Clay Science*. **115** 9-16.
[20] Li J Zuo S, Yang P et al. 2017 Study of CeO2 Modified AlNi Mixed Pillared Clays Supported Palladium Catalysts for Benzene Adsorption/Desorption-Catalytic Combustion *Materials*. **10** 1-14.
[21] Doff D, Gangas N, Allan J et al. 1988 Preparation and characterization of iron oxide pillared montmorillonite *Clay Minerals*. **23** 367-377.
[22] Yamanaka S, Hattori M 1988 Iron oxide pillared clay *Catalysis Today*. **2** 261-270.
[23] Muslimov R H 2005 Unconventional oil deposits are a significant potential for the further development of the old oil-producing regions *Georesources*. **1** 2-8

Forecast estimation of sandstone collectors transformation of Lower-Karmalinsky bituminous deposit during its development by the SAGD method

E Korolev[a], A Bakhtin[a], A Eskin[a], E Barieva[b] , A Korolev[a]

[a] Institute of geology and petroleum technologies, Kazan Federal University, Kazan, Russia

[b]Kazan State Power-engineering University, Kazan, Russia

e-mail: eskin.aleksey@gmail.com

Abstract. In order to increase hydrocarbon reserves in recent years, begun to actively involve natural bitumens in the development. On the territory of Tatarstan, one of the experimental sites for testing various technologies for the extraction of bitumen is the Lower-Karmalinsky bitumen deposit. Currently, on the field is testing the SAFD method. According to this technology for the extraction of hydrocarbons, steam is pumped in large volumes at a temperature of about + 180°C. Superheated steam reduces the viscosity of hydrocarbons, making them mobile. However, focusing on the reduction of bitumen viscosity, the problem of mineral phase's transformation due to anthropogenic impact in reservoir is not taken into account. In this work, using thermodynamic modeling, it was shown that the injection of superheated steam into the formation of bitumen-saturated sandstone will lead to the dissolution of calcite cement. Moreover, depending on the salinity of the injected water in the form of steam, the dissolution of carbonates will occur at different temperatures. To improve the technological characteristics and to increase the recoverability of bitumen from reservoir, it is recommended to pump hot water with as little mineralization as possible into sandstone layers. Temperature intervals were established, which should be followed in the process of technological development of bitumen collectors at the stage of heating the reservoir and at the stage of extracting water-bitumen solutions.

1. Introduction

One of the priorities in the strategy for the development of oil regions is the involvement in the development of unconventional hydrocarbon resources. As part of this strategy, on the territory of the Tatarstan Republic, began to actively explore the possibilities of extracting shallow lying natural bitumen in the sediments of the Ufa low Permian tier [1,2]. Since 2006, in the pilot industrial development finding Ashalchinskoye and Mordovo-Karmalyinskoye bituminous deposits. In recent years, one of the experimental sites is the Lower-Karmalinsky bitumen deposit.

Upheaval is included in the group of similar syn-sedimentary morphostructure having a common spatial orientation. All the structural ridges are located within the western slope of the South Tatar arch, which is a monocline, stepping in steps plunging down to the Melekess depression. According to one of the hypotheses, sedimentation-accumulative uplifts are bodies of sand bars elongated along the coastline of the Ufa Paleo Sea.

Currently, for development of the Lower-Karmalinsky bitumen deposit, steam assisted gravity drainage technology (SAGD) is being tested [3,4]. In the productive sandstone layer two horizontal wells are drilled one above the other. In the productive sandstone layer two horizontal wells are drilled one above the other. To the upper injection well is supplied water steam with a temperature of about +180°C, which heats the layer of bituminous sandstone. Condensing water enriches by hydrocarbons and enters the producing underlying well. Thus, between the two wells, a constant circulation of the oil-water fluid is created. Over time, the heat chamber increases in size, reaching 10.0-15 m. According to the simulation, the heat-affected zone can extend over distances of 25–30 m from the injection well.

The superheated water vapor injection into the bitumen deposit will drastically change the hydrogeological conditions of the reservoir, which can lead to changes of filtration-capacity properties of sandstone at developing. Under the influence of temperature, a wide variety of chemical reactions can be activated, contributing to the dissolution and subsequent deposition of mineral phases within the sand layer. Given this, in the proposed article, thermodynamic modeling of possible hydrochemical processes was carried out with an assessment of their effect on the bituminous reservoir being developed.

2. Object

Lower-Karmalinsky bituminous deposit tectonically is confined to the western slope of the South Tatar arch. Bituminous deposit belongs to the layer-uplifted deposit type. The bitumen-saturated rocks are sandstones of the Sheshminsky horizon of the Ufa low Permian tier (P_1ss) with an average thickness of about 26 m [2]. The reservoir is confined to the Lower-Karmalinsky brachianticlinal structure, which stretch in the northeast direction. Its length is about 13 km, width - 2-3 km, height - up to 45 m.

In thermodynamic modeling, the data of the mineral composition of sandstones and the chemical composition of interstitial waters in the bitumen reservoir are used. As shown by optical studies, the bitumen-saturated reservoir is represented by greywacke sandstones [5,6], which are composed of the Ural magmatic complexes destruction products. In them, the debris components are represented by rounded fragments of mafics effusive rocks (55-60%), subrounded metamorphic and siliceous rocks (20-25%), subrounded quartz grains (15%), angular fragments of plagioclase and microcline (5%), and rare scales of muscovite and chlorite. Clastic fragments of minerals and rocks are cemented by clay-calcite cement, which constituting 10-30% of the rock. In the productive part of the bitumen deposits there is a cement of two types: pore-filling and argillaceous lump cement. In the most bituminous part of the sandstone pack, organic-mineral cement accounts for 10-35% of the rock. The organic components are represented by medium and heavy oil, as well as fragments of syngenetic organics of flora [7]; the mineral components are calcite and analcime. The sandstone reservoir porosity is not constant, varies from 10 to 30% [6].

The bituminous reservoir of the Lower-Karmalinsky field is coinciding with low yield aquifer and locally with water bearing stratum terrigenous complex of the Sheshminsky horizon of the Ufimian stage. Water-bearing rocks are porous sandstones and siltstones with a thickness of 4 to 29 m. The underground waters of the Ufimian sediments are pressure water, piezometric levels are set above the top of aquifer layer. The magnitude of the head flow above the top of the reservoir - from 50 to 60 m. Piezometric levels in wells are often set at absolute elevations from 75 to 80 m. Well water output (discharge) varies from 1.4 to 150 m^3 per day. The recharge of the Sheshminsky horizon is carried out by infiltration from above and ascending filtration from below. Recharge of groundwater into the bitumen deposit occurs through cracks or tectonically weakened zones. The main direction of underground water flow is westerly direction towards to the Sheshma River valley. In the cross-section of the Lower-Karmalinsky bitumen deposit there are several aquifers, which are practically unrelated. The groundwater of deposit is sodium hydrocarbonate waters with TDS (Total dissolved solids) of 1.2-5.7 g/L, containing a significant amount of hydrogen sulfide [7]. Waters of the underlying sandstones and edge waters are sulphate-calcium-magnesium and hydrocarbonate-calcium type with TDS up to 1.0 g/L.

3. Results and Discussion

According chemical analysis data of bitumen deposit groundwater's, which planned for development, hydrocarbonate ions are predominate in them, sulphate ion and chlorite ion are in small amount. From cations, in accordance with the hydrochemical type, sodium ions prevails (Table 1). Potassium, calcium and magnesium ions are present in insignificant amounts. TDS of the formation water varies from 1.5 to 1.9 g/L, pH corresponds to weak acid medium with the temperature +8°C.

Table 1. The chemical composition of groundwater Lower-Karmalinsky bitumen deposit.

Sample	The content of major elements, mg/L							mg/L	pH
	Ca^{2+}			K^+	HCO_3^-	SO_4^{2-}	Cl^-		
1	12,82	16,84	953,20	9,68	470,21	95,25	99,00	1657	6,3
2	7,27	16,27	1097,95	8,17	620,34	33,08	92,92	1876	6,1
3	8,51	24,45	1082,19	9,70	608,70	14,05	125,40	1873	6,2
4	6,31	20,09	922,78	9,18	557,79	36,16	109,69	1662	6,3
5	9,23	22,47	855,29	13,88	411,15	146,89	104,09	1563	6,5

Given the temperature difference between stratal and injected waters is necessary to consider how increasing temperature in the bituminous reservoir will effect on the changes of groundwater chemical composition and their interaction with matrix minerals. Analysis of bituminous sandstone mineral composition showed that the most unstable phase will be calcite cement. Therefore, in this work we carried out a thermodynamic simulation of calcite interaction process with waters having different temperatures. Since the initial process water injected into the layer of bitumen-saturated sandstone has a temperature of +180°C, all calculations were carried out in the temperature range of 25-200°C. To calculate the temperature dependence of the Gibbs energy change, we used the carbonate equilibrium equation:

$$CaCO_3 + H_2O + CO_2 = Ca^{2+} + 2HCO_3^- \quad (1)$$

The calculation was made according to the formula:

$$\Delta rG(T) = \Delta rG°(T) + R*T*lgk \quad (2)$$

where **R** is the universal gas constant (8.3143 J * deg-1 * mol-1);

T – the absolute temperature;

k – the equilibrium constant of the reaction (1):

$$k = \frac{a_{Ca^{2+}} \times a^2_{HCO_3^-}}{a_{CO_2}}$$

in which **a** – the activity of the reaction components (1).

The calculation of ΔrG°(T) was made according to formulas 2.19-2.23 from [8] using the corresponding thermodynamic constants. The calculation was done for the two types of water: 1) stratal water, containing 8.83 mg/L of Ca^{2+} ions and 533.64 mg/L of HCO^{3-} ions; 2) river water injected to well and containing 100.00 mg/L Ca^{2+} ions and 375.00 mg/L HCO^{3-} ions. The results of the calculations are presented in the table 2.

For clarity of the process, using obtained data the plots were drawn. Figure 1 shows the dependence of the Gibbs energy change of equation (1) depending on the temperature of stratal water and heated river water injected to well. Analysis of calculations and graphs shows the following. In the heating stratal groundwater of bitumen reservoir the carbonate equilibrium (ΔrG(T) = 0) is established at temperatures of about +120°C. Calcite dissolution of cement occurs below this temperature; above, on the contrary, calcite precipitates. In the injected hot river water into the reservoir, carbonate equilibrium is established at a temperature of +90°C. Calcite dissolves below this temperature and precipitates above.

Table 2. Change of thermodynamic parameters [ΔfG°(T)] components, logarithm of equilibrium constant (lgk°) and Gibbs energies [ΔrG°(T) and ΔrG(T)] of reaction (1) at different temperatures for stratal and river waters.

T,°C	Thermodynamic parameters of the reaction component, (ΔfG°(T))					ΔrG°(T), kJ/mol	lgk°	ΔrG(T), kJ/mol	
	$CaCO_3$	H_2O	CO_2	Ca^{2+}	HCO_3^-			stratal	river
25	-1128,9	-237,1	-394,4	-552,8	-586,86	33,83	-5,9	-12,27	-8,0
50	-1122,3	-233,1	-394,4	-553,6	-578,09	39,97	-6,5	-10,00	-5,4
100	-1109,1	-224,9	-394,6	-555,3	-559,82	53,63	-7,5	-4,07	+1,3
150	-1096,1	-216,8	-394,7	-557,3	-540,61	69,07	-8,5	+3,64	+9,7
200	-1083,1	-208,6	-394,8	-559,1	-52,30	86,81	-9,6	+13,65	+20,4

A question may arise whether calcite will dissolve simply by the action of water without the participation of carbon dioxide by the reaction:

$$CaCO_3 + H_2O = Ca^{2+} + HCO_3^- + OH^- \qquad (3)$$

since the initial components of this reaction are common for the system described by the carbonate equilibrium equation of reaction (1). We calculated the amounts of the Gibbs energy change ΔrG°(T) and amount of the logarithm of the equilibrium constant lgk° of reaction (3) for temperatures of 25–200°C. The results are shown in the table 3 and figure 1.

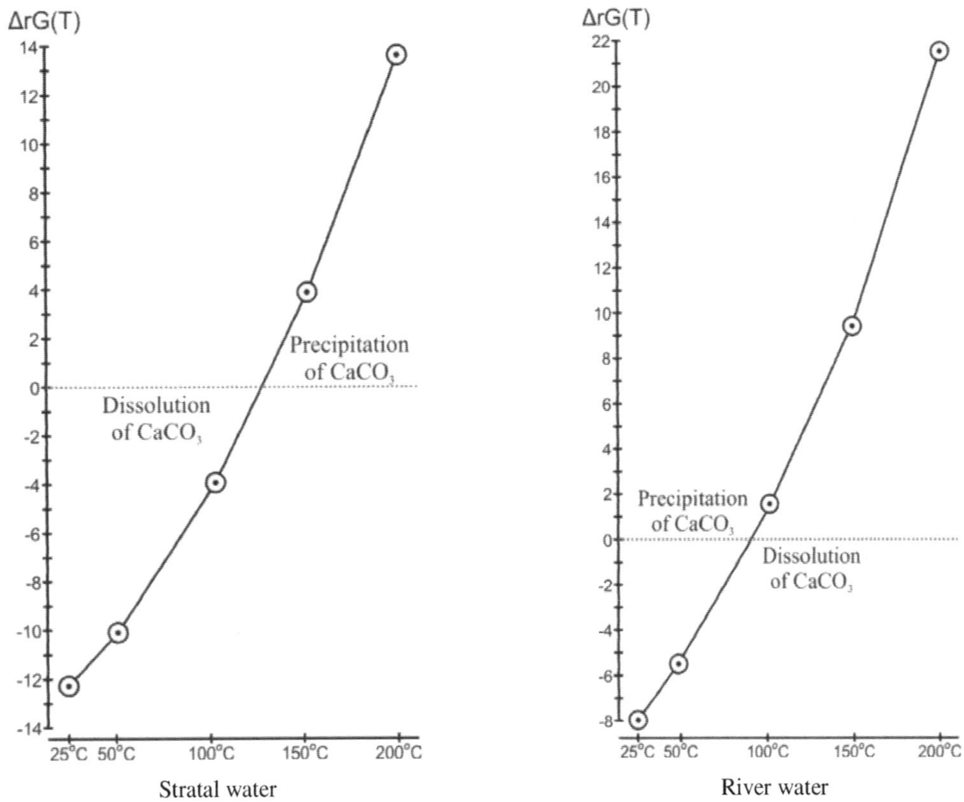

Figure 1. Synchronous thermal analysis data of the Middle Volga region oil shale deposit. The solid line is the differential scanning calorimetry (DSC) curve; the dotted line is the mass loss curve (TG).

Table 3. Thermodynamic parameters of ΔrG°(T) and lgk° at various temperatures.

T,°C	ΔrG°(T), kJ/mol	lgk°
25	+69,06	-12,10
50	+72,67	-11,75
100	+81,34	-11,36
150	+91,82	-11,33
200	+104,65	-11,55

These results show that the equilibrium constant of the reaction (3) is a million times less likely in comparison with reaction (1) and therefore it can be ignored.

4. Conclusions

Given the above, we can draw the following conclusions:

1. Depending on the mineral composition of water in the reservoir, when stationary conditions of develop are reached, it is necessary to maintain a temperature in the range from +90 to +120°C. In this case, a decrease in the bitumen viscosity will not be accompanied by processes of the "dissolution-precipitation" of calcites.

2. If it is necessary to improve the filtration-capacity properties of the reservoir, it is possible at the initial stages to maintain the temperature in the thermal chamber at the level of +70-90°C. This will lead to partial dissolution of calcite cement in sandstones.

3. To improve the technological characteristics of the reservoir layer and increase the recoverability of bitumen, it is recommended to inject hot water into the strata with low TDS and less mineralization.

Acknowledgments

This study was performed in the context of the Russian Government Program of Competitive Growth of Kazan Federal University.

References:

[1] E.A. Korolev, S.A. Usmanov, D.S. Nikolaev and R.R. Gabdelvaliyeva, “Effect of lithological heterogeneity of bitumen sandstones on SAGD reservoir development”, IOP Conference Series Earth and Environmental Science, vol. 155, art. № 012019, 2018.

[2] E.A. Korolev, A.I. Bakhtin, A.A. Eskin and R.R. Hanipova, " Diagenetic changes of sandstone reservoir of Ashalchinskoye bitumen deposit”, Neftyanoe Khozyaystvo – Oil Industry, vol. 10, pp. 26-28, 2016.

[3] R.S. Khisamov, V.G. Bazarevskaya, T.I. Tarasova, A.A. Kostina, R.R. Abusalimova and S.A. Panina, “Classification of geologic cross-section types of Sheshminskian P_1ss_2 sand sequence”, Neftyanoe Khozyaystvo – Oil Industry, vol. 6, pp. 27-29, 2017.

[4] E.A. Korolev, M.G. Khramchenkov, E.M. Khramchenkov, A.A. Eskin, R.R. Gabdelvalieva and A.N. Garaeva, “Modeling of the suffusion cavities development in bitumen saturated sandstones of Ashalchinskoye deposit under SAGD technology application”, Neftyanoe Khozyaystvo – Oil Industry, vol. 1, pp. 55-57, 2018.

[5] A.I. Mullakaev, R.R. Khasanov and B.M. Galiullin, “Mineralogy of sandstones and localization of oil matter in productive horizons of high-viscosity oil in permian deposits of the Volga-Ural region (Russia)”, 17th International Multidisciplinary Scientific GeoConference, vol. 17(11), pp. 353-358, 2017.

[6] R.R. Khasanov and A.I. Mullakaev, “Paleogeographic factors of the formation of Permian reservoir rocks of bitumen deposits in the east of the Russian plate (Russia)”, 16th International Multidisciplinary Scientific GeoConference, vol. 1(1), pp. 469-474, 2016.

[7] A.I. Mullakaev, R.R. Khasanov, O.R. Badrutdinov, N.M. Khasanova and N.M. Nizamutdnov, “Features of bitumen-containing sandstones of the Volga-Ural oil and gas province according to

Electron paramagnetic resonance and Gamma spectrometry data", 17th International Multidisciplinary Scientific GeoConference, vol. 17(11), pp. 133-139, 2017.
[8] M. Borisov, Yu. Shvarov, Thermodynamics of geochemical processes (Moscow: MSU Publishing House), p 256, 1992.

IOP Conf. Series: Earth and Environmental Science **282** (2019) 012038 doi:10.1088/1755-1315/282/1/012038

Influence of formation conditions on reservoir thickness and reservoir properties

B V Uspensky[1], S V Khaliullina [1], S E Valeeva[1] , R F Vafin
[1]Kazan Federal University, Kazan, Russia

E-mail: borvadus@rambler.ru

Abstract. Authors of the article identified the influence of sedimentation conditions on reservoir thickness and the patterns of change in reservoir properties of superviscous oil field from the following three zones of the sandstone sequence of the Sheshminsky horizon: Northern, Southern and Central identified by PJSC Tatneft and prepared for initial development. The following relationships have been investigated: open porosity relation with bitumen content (by weight and by volume), open porosity relation with carbonate content, open porosity relation with bulk density, carbonate content relation with bulk density, bitumen content (by weight and by volume) relation with carbonate content, and also bitumen content relation with their depth. The revealed patterns of changes in reservoir properties and bitumen content within the sandstone sequence can be the result of not only changing conditions of sedimentation, but also the result of post-sedimentation processes resulting in calcitization or redistribution of carbonate cement in the rock under the influence of aggressive products (oxidation, biodegradation) of oil deposits destruction.

1. Introduction

The study of reservoirs formation processes and patterns of the reservoirs distribution as well as study of associated with them traps of oil and bitumen-bearing rocks traditionally involves the forecast and the search for oil and gas fields. In this work, the main emphasis is made on identifying the influence of sandstone-aleuritic sediments formation conditions on the structure and reservoir properties of bitumen (oil) containing sediments. The object of the study are sandstone-aleuritic sediments of the Sheshminian horizon, Ufimian Stage. The zone consisting of hydrocarbon-saturated rocks is at least 200 km in length. The conditions of the hydrocarbons formation are polyfacial, however this question is interpreted ambiguously by different researchers (Forsh N.N., Miropolsky L.M., Ellern S.S., Shalin P.A., Uspensky B.V., Petrov G.A. and others [1, 2]. This is due to the fact that in the areas of oil and bitumen-containing sediments distribution there are channel, delta (distal and proximal parts) and bar facies, as a rule, without specific boundaries of mutual transitions of one facies zone to another. Most of the authors identify only one type of facies. However, each type corresponds to specific and particular sizes of traps, thickness of sediment containing hydrocarbon deposits, particle size distribution and reservoir properties, varying both along strike and along the vertical section. In this regard, knowledge of the detailed geological structure of reservoirs of specific fields, their relation to certain facies zones is extremely important when developing predictive criteria for the search for hydrocarbons, as well as when choosing methods to increase oil recovery using secondary and tertiary methods.

2. Methodology

Analysis of the applied Russian and international technologies of ultra-viscous oil (UVO) and natural bitumen (NB) production showed that there are no clear criteria for their applicability to specific oilfields. Each technology should be individually adapted taking into account the geological and geochemical conditions of the ultra-viscous oil and natural bitumen reservoirs formation. Examples of recommended criteria for the applicability of technologies for the production of ultra-viscous oil and natural bitumen, depending on their physicochemical properties and the reservoir geological characteristics from [3] are given in Table 1.

Table 1. Criteria for the selection of objects recommended to develop by in-situ combustion[1] and steam injection[2]

Parameters	Recommended Values[1]	Recommended Values[2]
Depth, m	< 2100	up to 1200
Reservoir thickness, m	> 3	> 6
Porosity,% for terrigenous rocks for carbonate rocks	 > 18 > 12	 > 18 -
Oil saturation to the start of the process, %	> 40	> 40
Permeability, μm^2	> 0.1	> 0.1
Viscosity of reservoir oil (bitumen), MPa·s	> 10	> 1000
Density of reservoir oil (bitumen), kg/m^3	> 870	> 900

From the above data, it follows that most of the recommended parameters (reservoir thickness, porosity, permeability, bitumen content, viscosity and bitumen density) for selecting development criteria by these methods depend on the reservoirs and hydrocarbon deposits formation conditions. In order to ensure proper reservoir development, construction of horizontal wells, use of enhanced oil recovery methods, besides the listed criteria, it is important to know the morphology and size of traps, direction of change in reservoir properties, which is also a consequence of sedimentation processes.

In order to develop and improve the concepts of the search for, production and refining of ultra-viscous oil and natural bitumen, this article summarizes and interprets the results of actual geological, geophysical and production data, laboratory petrophysical and geophysical research data performed at the Department of Oil and Gas Geology of Kazan (Volga Region) Federal University. Based on the fundamental studies of the reservoirs' geological structure and their formation conditions, according to the authors of [2, 4], there are four facial zones in this area: channel, delta, avandelta and bar.

3. Results and Discussions

There were studied a sandstone (Ashalchinskaya) sequence of the Sheshminsky horizon, Ufimian stage and associated ultra-viscous oil and natural bitumen reservoirs. The core material analyzed in this section describes individual deposits of super-viscous oil and natural bitumen confined to three main groups: the northern, central and southern Cheremshan-Bastryk zone [3] located on the western slope of the South Tatar arch (Fig.1.). Specialists of PJSC Tatneft performed the classification of the ultra-viscous oil reservoirs by zones in order to prepare these reservoirs for pilot commercial development.

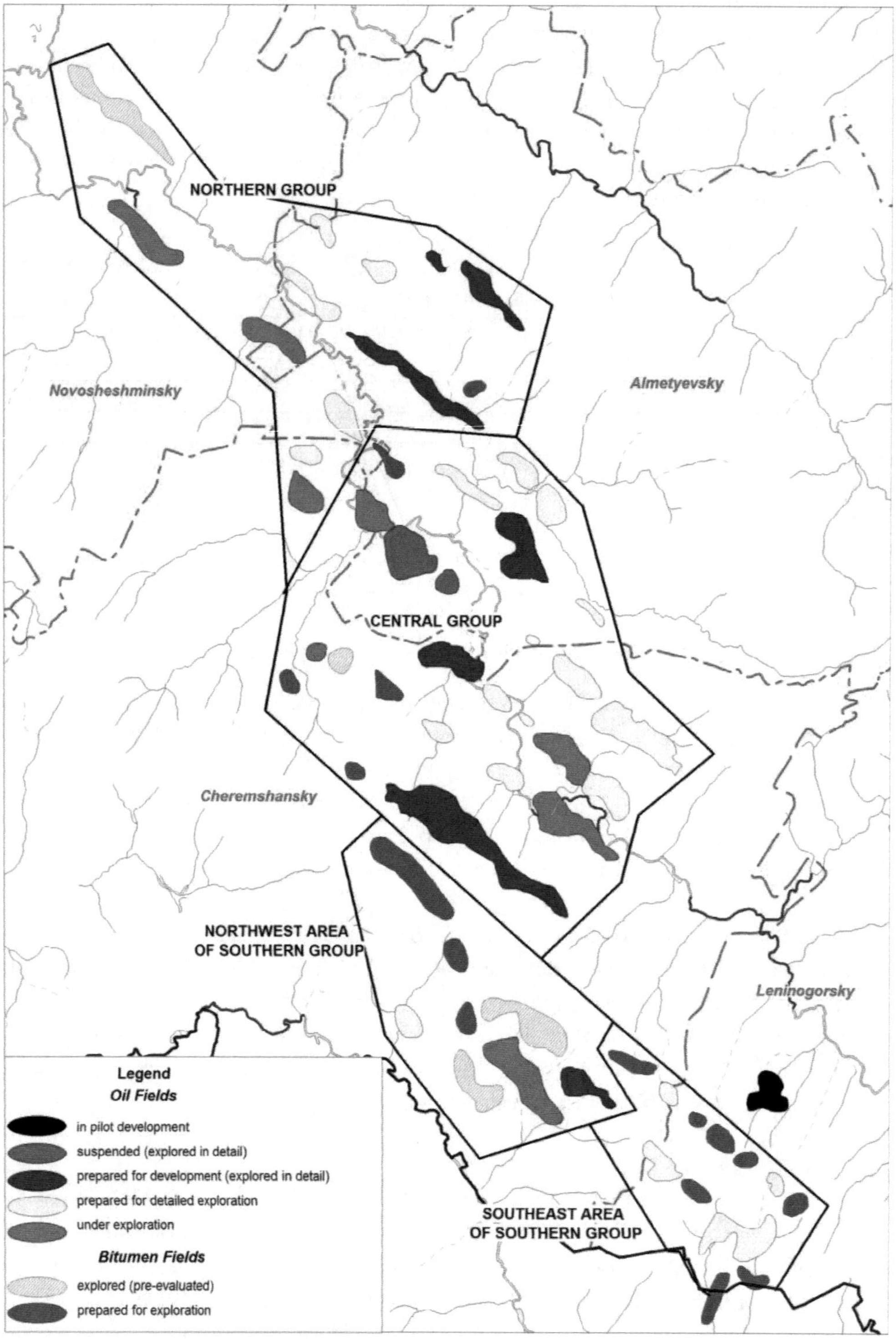

Figure 1. Cheremshan-Bastryk zone of super-viscous oil and natural bitumen

The Northern Group includes ultra-viscous oil reservoirs located between the Ashalchinsky and Arkhangelsky deposits. The Central group includes ultra-viscous oil reservoirs located between the Upper Karmalsky and South Ashalchinsky oil deposits. The Southern group includes oil reservoirs located between the Oikino-Altuninskoye and Sofievskoye oil deposits.

Reservoir properties and bitumen content of the rocks vary widely within the selected zones. Thus, in the fields of the Northern group, the open porosity of rocks varies from 3.3 to 33.9% (an average of 13.8%), carbonate content varies from 0 to 33% (an average of 16.5%), mineralogical density ranges from $2.5 \cdot 10^{-3}$ to $3.6 \cdot 10^{-3}$ kg/m^3 (on average $2.7 \cdot 10^{-3}$ kg /m^3), bulk density varies from $1.7 \cdot 10^{-3}$ to $3.3 \cdot 10^{-3}$ kg/m^3 (average $2.3 \cdot 10^{-3}$ kg/m^3), bitumen content by weight varies from 0.1 to 16.3% (average 5.5%), bitumen content by volume varies from 4.4 to 88.6 % (average 41%).

In the areas of the Central Group, the open porosity of rocks varies from 2.4 to 44.5% (average 23%), carbonate content varies from 0.6 to 46.2% (average 15.9%), mineralogical density varies from $2.3 \cdot 10^{3}$ to $3.3 \cdot 10^{-3}$ kg/m^3 (on average $2.7 \cdot 10^{-3}$ kg/m^3), bulk density varies from $1.4 \cdot 10^{-3}$ to $3 \cdot 10^{-3}$ kg/m^3 (on average $2.2 \cdot 10^{-3}$ kg/m^3), bitumen content by weight varies from 0.1 to 14.9% (on average 5.0%), bitumen content by volume varies from 2.5 to 96.8% (average 39.2%).

In terms of reservoir properties and bitumen content, the rocks from the productive reservoirs of the southern group of deposits are close to the rocks of the northern group of deposits. The open porosity of rocks varies from 0.9 to 38.7% (17.4% on average), carbonate content varies from 0.0 to 46.5% (average 16.7%), mineralogical density ranges from $2.2 \cdot 10^{-3}$ to $3.5 \cdot 10^{-3}$ kg/m^3 (on average $2.7 \cdot 10^{-3}$ kg/m^3), bulk density ranges from $1.6 \cdot 10^{-3}$ to $3.4 \cdot 10^{-3}$ kg/m^3 (on average $2.3 \cdot 10^{-3}$ kg/m^3), bitumen content by weight varies from 0 to 15.5% (average 4.5%),bitumen content by volume varies from 0 to 97.9% (average 40%).

Relationships between various reservoir parameters of the sandstone sequence within oil deposits from all three selected groups were investigated for a more in-depth study of the reservoir properties of Sheshminsky sediments. The following relationships have been studied: open porosity relation with bitumen content (by weight and by volume), open porosity relation with carbonate content, open porosity relation with bulk density, carbonate content relation with bulk density, bitumen content (by weight and by volume) relation with carbonate content [5, 6]. This paper presents the most informative correlation dependencies. The northern group is represented by the Olympiadovsky oil field, the Central group is represented by the Melnikovskoye oil field, and the Southern group is represented by the Verkhne-Karmalskoye oil field. The reservoir properties of the Olympiadovsky oil field are presented in table 2.

The correlation dependences between the parameters characterizing the reservoir properties of the Olimpiadovsky field, as well as of the Melnikovsky, Verkhne-Karmalsky and other fields referred to the Central and Southern groups are presented in Figure 2. Analysis of reservoir properties in the well section of individual oil fields showed a good direct correlation between the following indicators: porosity and bitumen content (by weight and by volume) with a correlation coefficient from 0.5 to

0.78. A clear inverse relationship appears between porosity and bulk density with a fairly high correlation coefficient from minus 0.71 to minus 0.95. There is a good inverse correlation of open porosity with carbonate content (correlation coefficient varies from minus 0.6 to minus 0.7) There is also a reverse correlation between carbonate and bitumen content (by weight and by volume) with a correlation coefficient from minus 0.55 to minus 0.8. In general, the obtained correlation dependences indicate that the concentrations of hydrocarbons in the Permian sediments are confined to highly porous reservoirs, which are characterized by rather high bitumen content and low carbonate content.

Table 2. Reservoir properties of Olympiadovsky super-viscous oil field (Northern group): min, max, x are the minimum, maximum and average values of the parameter, respectively

Objects of research		Porosity %		Bitumen content, % by weight		Bitumen content, % by volume		Carbonate content, %	
		min/max	X	min/max	X	min/max	X	min/max	X
Rocks from individual wells	246	4.9/25.8	10.5	0.5/8.6	27.0	27.0/72.0	40.7	2.8/22.7	14.5
	248	7.9/22.1	12.4	1.2/3.2	25.2	25.2/66.2	42.5	9.2/33.0	19.7
	251	4.1/23.8	15.4	0.5/8.5	6.6	6.6/74.2	39.6	3.0/20.0	12.6
	247	4.3/23.2	11.2	0.5/7.5	12.5	12.5/83.5	47.8	3.0/30.6	14.3
	256	3.3/20.5	11.4	1.3/10.5	29.4	29.4/97.6	67.8	3.9/24.5	15.7
	748	4.9/26.0	12.9	0.3/11.6	9.0	9.0/69.3	44.3	6.4/32.4	23.0
	790	7.0/24.0	15.9	0.3/9.3	7.7	7.7/68.1	42.5	23.6/30.0	26.7
	258	5.1/32.6	13.8	0.2/16.3	4.1	4.1/93.0	43.1	0.0/26.6	9.9
Formtion in general		**3.3/32.6**	**12.9**	**0.2/16.3**	**4.6**	**4.1/97.6**	**46.5**	**0.0/33.0**	**18.1**

Earlier [4], a connection was found between the thickness of a sandstone pack and its lithological features. It has been identified that the degree of the reservoir's clastic rocks homogeneity (the increase in degree of sorting and increase in average grain size) also increases with an increase in the reservoir thickness, which, in turn, is related to the dynamics of sedimentation. The areas of increased sandstone sequence thickness correspond to more coarse-grained rocks accumulated in the center of the stream (channel), rocks from these areas have better sorting and minimal clay and carbonate content.

The vertical section of the sandstone sequence of the Ufimian Stage is characterized by a strong pattern of changes in reservoir properties with depth.

The open porosity as well as the bitumen content of the rocks, both by weight and by volume decreases with an increase in the depth of the sandstone sequence. The lower parts of the section within deposits are usually composed of tightly and moderately cemented weakly porous (5-12%) sandstones. Weakly and moderately cemented porous (porosity from 25 to 42%) rocks are developed in the upper part of the sequence. This leads to a higher degree of their oil saturation. The middle part of the reservoir is represented by compacted calcareous sandstones with a porosity of 12-24%. In this part of the section, changes in open porosity with depth are not so sharp, although the tendency to decrease in its values is maintained, while bitumen content of rocks decreases with depth similarly to the upper and lower parts of the formation.

4. Conclusions

Thus, the conditions of sandstone sequence rocks formation were decisive in the distribution of their reservoir properties and the degree of their saturation with hydrocarbons. Relationships between various reservoir parameters of the sandstone sequence within oil deposits from three groups of fields (Northern, Southern and Central) identified by PJSC Tatneft and prepared for initial development were investigated for a more in-depth study of the reservoir properties of Sheshminsky sediments. The following relationships have been identified: open porosity relation with bitumen content (by weight and by volume), open porosity relation with carbonate content, open porosity relation with bulk density, carbonate content relation with bulk density, bitumen content (by weight and by volume) relation with carbonate content, and also bitumen content relation with their depth. The revealed patterns of changes in reservoir properties and bitumen content within the sandstone sequence can be the result of not only changing conditions of sedimentation, but also the result of the influence (perhaps even more) of post-sedimentation processes resulting in calcitization or redistribution of carbonate cement in the rock under the influence of aggressive products (oxidation, biodegradation) of the initial oil deposits destruction.

Acknowledges

This work was supported by the Russian Foundation for Basic Research and the Russian National Science Foundation.

References

[1] Muslimov R H *et al* 1995 *Locations and development of natural bitumen resources (Geology of oil and gas* **2**) pp 7-10
[2] Uspensky B V *et al* 2008 *Geology of natural bitumen deposits of the Republic of Tatarstan* (Kazan: PF Gart LLC Publishing house) p 349
[3] Muslimov R H *et al* 2012 *Complex development of heavy oils and natural bitumens of the Perm system of the Republic of Tatarstan* (Kazan: Fen Publishing house of the Academy of Sciences of the Republic of Tatarstan) p 396
[4] Uspensky B V *et al* 1998 *Geological and geochemical basis for the development of bituminous deposits of the Middle Volga* (Kazan: Publishing house of Kazan University) p 148
[5] Uspensky B V *et al* 2015 *Influence of sedimentation conditions on the structure and reservoir properties of oil bitumen-containing rocks* (*Georesources* **3**) pp 17-23
[6] Uspensky B V *et al* 2016 *Dependence of the reservoir properties of the Ashalchinskaya sequence on the conditions of its formation* (*Oil industry* **7**) pp 69-71

IOP Conf. Series: Earth and Environmental Science **282** (2019) 012039 doi:10.1088/1755-1315/282/1/012039

Problems of shale oil hydrocarbons, ultra-viscous oil reservoirs and natural bitumen deposits development in Tatarstan

B V Uspensky[1], V P Morozov[1], S E Valeeva[1]

[1]Institute of Geology and Petroleum Technologies, Kazan Federal University, Kazan, Russia

E-mail: borvadus@rambler.ru

Abstract. The area of the Volga-Ural oil and gas basin belongs to the "old" oil and gas bearing regions. The most part of easily recoverable reserves of large and unique oil fields already have been developed within the area. Therefore, super-viscous oil reservoirs, natural bitumen deposits and shale hydrocarbons seized the attention of local petroleum geologists. Tatarstan is a classic example in that sense. The Republic of Tatarstan is a leader in the development of heavy oil resources, a leader in the degree of knowledge of natural bitumen accumulations, as well as a leader in technological readiness for their development. In the depths of Tatarstan, there are more than 450 fields of ultra-viscous oil and natural bitumen with reserves ranging from 2 to 7 billion tons of this type of hydrocarbons. However, their industrial development due to low profitability is slow. To solve the problem, it is necessary to search for and commercialize economically viable technologies for the production, transportation and processing of hydrocarbons, which take into account its features and do not require increased energy and other material costs. Therefore, the search for the right unconventional hydrocarbons production concept is an urgent problem of our time.

1. Introduction

Using the fundamental knowledge and accumulated Russian and international experience in the study and development of super-viscous oil reservoirs, natural bitumens and shale hydrocarbons, one can approach the effective industrial development of heavy hydrocarbons.

Most of the foreign programs on the use of natural bitumens and super-viscous oils, ranging in value from a few million to several billion dollars, have been compiled taking into account all the features of the mineral resource, the size of deposits, extraction conditions, etc. Despite the expected significant growth in extraction of natural bitumen (from 78 million tons in 2010 to 240 million tons in 2035 mainly due to Canada) and biofuels from 75 million tons in 2010 to 145 million tons in 2035 (mainly in the USA and Brazil), the most favorable conditions for increasing the production of unconventional hydrocarbons are associated with shale oil reservoir development. United States are leading in this area. There are more than 20 large shale oil basins in the US, the largest of which are located in the west of the country in the states of Utah, Wyoming and Colorado. Theoretically, among the countries which have shale gas and shale oil, Russia can take a leading position because it has a developed infrastructure, great oil and gas potential, and can quickly master the technology of multi-stage hydraulic fracturing in horizontal wells achieving its high profitability in areas of high-carbon formations (Bazhenites and Domanikites), the potential sources of shale oil and gas. In addition to shale hydrocarbons, huge reserves of heavy oils and natural bitumen are concentrated in Russia. The Republic of Tatarstan is a leader in the development of this type of hydrocarbons. Downhole and mine development methods are the main methods of natural bitumen production. Using the downhole methods, natural bitumens are produced after a preliminary increase in their mobility in the reservoir by heating. Heating can be carried out by steam injection, steam gas injection, in-situ combustion, etc. Mining Methods include Mine Development Method and Open Pit Development Method. In this methods, the rock is delivered to the surface, and then the natural bitumens and high-viscosity oils contained in the rock are extracted with solvents, hot water, steam with surfactant additives, alkalis, etc. The selection of a system for influencing natural bitumens, super-viscous oils and shale hydrocarbons should be based on global criteria for the applicability of enhanced oil recovery methods.

At present, Canada took the leading position in terms of oil production from bituminous sandstones. Over 150 million tons of synthetic oil per year is produced from the Athabasca sandstones of Alberta province.

Bituminous sandstones of Alberta are being developed [3] using surface mining methods by such companies as Syncrude, Suncor, Shell, etc., as well as using downhole methods ("in situ production") by such companies as Petro-Canada, Nexen-Opti, Imperial Oil, etc. The level of oil recovery for Mining Development methods averages 90% and reaches 70% for the downhole methods.

Some new promising methods and technologies are being tested at the pilot stage. In particular, the following technologies should be noted: Solvent Aided Process (SAP), the combination of in-situ combustion (vertical well) and oil production from a horizontal well (THA1 - Toe to Heel Air Injection), Electro-Thermal Dynamic Stripping Process (ET-DSP), Sonics-Pressure Pulsing and technology with injection of nitrogen, nitrogen and carbon dioxide or carbon dioxide only.

2. Methodology

There are a number of different and considering successful ultra-viscous oil and natural bitumen production technologies. Various development approaches have been tested in the process of pilot works at the Mordovo-Karmalsky and Ashalchinsky fields of viscous and highly viscous oils (Tatarstan) in the sandstones of the Ufimsky Stage [2, 3]. The development technologies using horizontal wells and the technologies combining impact on the formation by use of thermowave emitter and a steam-gas generator have proved themselves most effective. Significant practical developments have been made in the study and exploration of the bituminous sandstones of the Ufimian Stage. Positive results of hydrocarbons extraction were achieved during a number of pilot works. In September 2018, ultra-viscous oil production amounted to 165,210 tons, and reached 1,432,719 tons from the beginning of the year (for 9 months) resulting in more than 5 million tons produced over the entire period of ultra-viscous oil reservoirs development. Most part of ultra-viscous oil was extracted by horizontal wells, the flow rates of which are 8-10 times higher than flow rates of vertical ones.

The development of shale hydrocarbons, super-viscous oil and natural bitumen in Tatarstan is complicated not only by the lack of currently tested cost-effective technologies for their production, refining, transportation, but also by the unresolved individual issues of subsoil use and scientific design of unconventional hydrocarbon development schemes which would be based [4] on the study of the reservoirs geological structure, knowledge of the basic laws in changing developing site selection criteria in accordance with specific lithofacial zones [1, 5, 6, 7]. Thus, the new scientific approach presented in the article for solving the problem of unconventional sources of hydrocarbons development is based on a system analysis: from the detailed geological structure of strata containing unconventional hydrocarbons, studying the formation conditions and patterns of hydrocarbon accumulations to the preparation of oilfields for development.

3. Results and Discussions

The study of the reservoirs' geological structure saturated by unconventional hydrocarbons allows [4] to systematize groups of oilfields with a common lithological and petrophysical composition of reservoirs, depth, shape and size of hydrocarbon traps, screening capacity of caprocks, physical and chemical properties of saturating fluids, etc. The identified features will contribute to the development of geological and geochemical criteria for ranking and determining rational production methods (based on the world experience of heavy oil reservoirs and natural bitumen deposits development as well as on criteria for the application of foreign and Russian technologies) and the sequence of launching oilfields into operation [1].

The Domanikites containing shale hydrocarbons have the following features: 1) anomalous material composition 2) limited stratigraphic prevalence, 3) confinement to the eastern part of the Russian Plate.

Reason of the anomalous material composition is the presence in the rocks of the following main rock-forming components: quartz (chalcedony), calcite and organic matter. The content of the latter can vary widely from the first percent to 20-30%, or even more in rare cases.

The limited prevalence of these rocks is due to their limited stratigraphic position. Such rocks are developed only in the sediments of the Semiluksky (Domanik) horizon of the Fransian stage of the Upper Devonian. Their thickness is usually the first tens of meters. Such sediments are no longer found in the section of the Volga-Ural region sedimentary cover, the thickness of which is about two thousand meters. However, they are very widespread in the shape of a meridional strip along the western slope of the Urals and cover large areas. The Semiluksky (Domanik) horizon deposits look like a belt that stretches along the Urals from north to south from the Pechora Sea to the Caspian syneclise through Bashkortostan and Tatarstan.

Based on the available data, it has been established that such formations form a cyclical pattern of different ranks, which is associated with hydrothermal activity, eustatic fluctuations of sea level and climatic changes. According to [8], an important role in the domanikites formation belongs to hydrothermal fluid systems or systems related to them.

Our own studies [10] showed that the studied sections were composed of three following lithotypes: carbonate rocks, carbonate-siliceous rocks with a high content of organic matter and carbonate breccias comprised of carbonate fragments cemented by a carbonate-siliceous material enriched in organic matter.

It should be noted that there is another feature of the core material studied: the presence of feldspars and mica in the carbonate-siliceous lithotype [11]. The presence of these minerals in the rocks cannot be explained by the terrigenous introduction of the material, since Domanikites are developed among carbonate (not terrigenous) rocks (limestones). Therefore, carbonate breccias, given the irregular position of such rocks in the studied sections and the impossibility of their correlation, suggest their lenticular shape.

Another important fact, in our opinion, that allows a more rigorous discussion of the domanikites sedimentogenesis conditions is the established relationship of organic matter with the siliceous component of sediments. According to the data based on the study of the samples mineral composition and their organic matter content, there was identified the dependence indicating that the organic matter content increases with the increase in proportion of quartz in the rocks.

Endogenous fluids that supplied silica acted as a source of substance for the formation of quartz (chalcedony) of carbonate-siliceous rocks. Their periodic activity ensured the cyclicity in sedimentation.

As part of endogenous fluids, there were also some biophilic elements, which ensured high bio-productivity and accumulation of organic matter having a sapropelian genesis.

The following additional facts that do not contradict the previously established concept, and which require an explanation appeared in the process of studying core samples:

- the relationship of silica accumulation with accumulation of organic matter;
- the presence in the sediments of not only limestone and carbonate-siliceous rocks, but also carbonate breccia;
- connection of biophilic chemical elements with carbonate-siliceous rocks enriched with organic matter;
- Domanikites development among carbonate rocks (limestones).

The relationship of silica accumulation and accumulation of organic matter can be completely explained by the fact that the formation of sediments took place under the conditions of the so-called fluid lithogenesis, when endogenous fluids served as one of the material suppliers for the formation of sediments. This led to the formation of the siliceous component of the domanikites and also the feldspathic and mica components of the domanikites, which are only found in carbonate-siliceous layers. At the same time, the fluids introduced trace elements with themselves, which, being biophilic, contributed to the development of biota and the formation of sapropelium substance.

The material of hydrothermal, and, perhaps, volcanogenic origin could provoke a rapid increase in the development of biota. Thus, released CO_2, silica, methane and trace elements could lead to the flowering of certain organisms and the oppression of others. Thus, the abundance of silica may be associated with the appearance of masses of silicon-forming organisms, while methane and other components could cause a rapid increase in the development of biota.

Such an interpretation of the sedimentation process does not indicate the formation of domanikites in the conditions of a long-existing reducing environment, since the preservation of organic matter may be due to its high content in the sediment. Complete oxidation of organic matter becomes impossible under conditions of a very high rate of its accumulation. However, such a statement can be contested and requires additional confirmation.

Carbonate breccias formed due to turbiditic flows (underwater-landslide flows), which according to our opinion have arisen due to the changes in sea level within the transgressive marine basin. The latter should probably be considered as the most promising objects for the possible development of domanikites. However, the assumption made needs thorough justification.

4. Conclusions

It was identified by many researchers that Domanic deposits were formed during an extensive marine transgression from the Paleoural, the maximum of which was observed in the Semiluksky time. Therefore, taking into account the ubiquitous and irregular presence in the section of the Domanik carbonate breccia deposits, it is possible to propose the most likely way of their formation. According to our understanding, under the conditions of the marine basin transgression, a certain "barrier" is formed on the seabed separating the carbonate accumulation area from the accumulation area of carbonate and carbonate-siliceous material (domanikites itself) with periodic inflow of endogenous fluids to the bottom. At the same time, endogenous fluids, in addition to the main component of silica, also supplied biophilic chemical elements that led to the mass formation of biota. The "barrier" formed during the transgression serves as a factor in the formation of turbidites or similar gravitational flows, and, as a result, the introduction of detrital carbonate material into the forming domanikites. This method of carbonate breccias formation is in many ways analogous to one of the hypotheses of the anomalous Bazhenov sections formation by underwater-landslide penetration into it of terrigenous aleurite-sand size material. Similar rocks in some publications are attributed to oil-bearing ones.

Acknowledges

This work was supported by the Russian Foundation for Basic Research and the Russian National Science Foundation.

References

[1] Uspensky B V *et al* 2015 *Geological criteria and geophysical methods of natural bitumen deposits preparation for development* (Kazan: Ikhlas Publishing house) pp 294-296

[2] Kayukova G P *et al* 2015 *Properties of Permian heavy oils and bitumens of Tatarstan in natural and technogenic processes* (Moscow: GEOS Publishing) p 343

[3] Muslimov R H *et al* 2012 *Complex development of heavy oils and natural bitumens of the Perm system of the Republic of Tatarstan* (Kazan: Fen Publishing house of the Academy of Sciences of the Republic of Tatarstan) p 396

[4] Uspensky B V 2015 *A methodology for studying oil and bitumen-bearing territories and objects* (*Oil industry* **4**) pp 30-33

[5] Uspensky B V *et al* 2016 *Dependence of the reservoir properties of the Ashalchinskaya sequence on the conditions of its formation* (*Oil industry* **7**) pp 69-71

[6] Uspensky B V *et al* 2016 *Formation conditions and geophysical methods for forecasting complicating factors for the development of deposits of natural bitumen* (*Oil Industry* **11**) pp 75-77

[7] Uspensky B V *et al* 2015 *Influence of sedimentation conditions on the structure and reservoir properties of oil bitumen-containing rocks* (*Georesources* **3**) pp 17-23
[8] Kolchugin A N *et al* 2018 *Depositional environments and diagenesis of Devonian black shales (case study Volga-Ural region, Russia)* (SGEM – Bulgaria) pp 367-379
[9] Kayukova G P *et al* 2018 *Modeling the extraction of shale oil from high-carbon Domanik rocks using hydrothermal technologies* (Kazan: Slovo Publishing) pp 225-228
[10] Liang S *et al* 2018 *The mechanism of modeling the influence of volcanic activity on the conditions of high-carbon sequences formation* (Kazan: Slovo Publishing) pp 248-249
[11] Morozov V P *et al* 2018 *Geological model and reconstruction of the Volga-Ural region Domanikites formation conditions* (Kazan: Slovo Publishing) pp 271-272

Application of thermal methods to increase the efficiency of coalbed methane production

R R Khasanov[1], M A Varfolomeev[1,2], D A Emel'yanov[2], A I Rakhimzyanov[1], A I Mullakaev[1]

[1] Institute of Geology and Petroleum Technologies, Kazan Federal University, Kazan, Russia

[2] Alexander Butlerov Institute of Chemistry, Kazan Federal University, Kazan, Russia

E-mail: rinat.khassanov@kpfu.ru

Abstract
In recent years, the expansion of the base of energy raw materials occurs through the involvement of new unconventional sources into operation. The development of new technologies made it possible to extract methane and its homologues from sedimentary formations that were not previously considered as sources of natural gas. One of these formations are coal-bearing strata. In some cases, the development of coal seams using traditional methods (quarry and mine) is impossible due to mining, geological, economic, technical and other reasons. The volume of industrial production of coalbed methane in foreign countries is beginning to have a significant impact on global gas markets. Industrial extraction of methane is conducted in the USA, China, Canada and other industrialized countries. It should be noted that the potential of unconventional resources of methane is huge - the resources of coal in the world reach 15 trillion tons. Coals are a promising source of coalbed methane. Gas output can be increased by thermal exposure to the coal seam. We have carried out studies of residual gas components from samples of fossil coal by the method of combined thermogravimetry (TG) and IR spectroscopy (IR). The method of thermogravimetry allows us to estimate the amount of gas sorbed in the coals, and IR spectroscopy to identify the composition of the emitted gases in each temperature range. Data on the composition of gas fractions in various temperature ranges was obtained.

1. Introduction

The expansion of the base of energy raw materials occurs due to the involvement of new sources of hydrocarbons in the operation, one of them is coal seams gases. Fossil coals are high-gas-capacity rocks. They have a great resource potential of coalbed methane. It should be noted that the potential of unconventional methane resources is huge. The world's coal resources in sediment exceed the amount of any other energy carriers. Coals can be a practically inexhaustible source of coalbed methane. Currently, industrial production of coalbed methane is growing in the world, which is beginning to have a significant impact on gas markets. The largest amounts of coalbed methane are mined in the United States, China and Canada [1,2]. In Russia, the extraction of methane from coal seams is carried out at the pilot site by the company Gazprom Mining Kuznetsk [3]. Natural gas is one of the most environmentally friendly types of hydrocarbons. The development of coal seams is carried out using inclined drilling technologies. Produced gas is an analogue of natural gas. It is used for domestic needs and for the working of local power plants. Significant untapped coal resources are associated with the Volga-Ural oil and gas province, where coal-bearing strata are distributed in Tatarstan, Udmurtia and Bashkortostan. Oil reserves here are greatly depleted as a result of years of development. In this regard, in the future, coal-bearing strata of the region can be considered as a promising source of hydrocarbon raw materials. To date, only in Tatarstan their total resources exceed 3.5 billion tons [4, 5, 6]. In total, 95 deposits were identified and assessed in Tatarstan, which in their scale correspond to small and medium-sized deposits. The coal seams are confined to the fill-in section of the Visean terrigenous formation. The thickness of coal deposits varies from 1 to 30 m. In terms of scale, they all belong to small and medium-sized deposits with resources ranging from the first to several hundred million tons. The main structural forms to which the most powerful coal seam is confined are erosion-karst downcuts. The contours of the coal deposits are controlled by the boundaries of the downcuts. The coal seam thickness reaches 10-29 m. The Visean coals have the following quality parameters: ash content - 15-26%, volatile matter yield - 41- 48%, sulfur content - 3.1- 4.2%, heat of combustion 29.9- 31.4 MJ / kg, content of toxic elements low. The mining and geological conditions of the Visean coal deposits are extremely complex, due to the deep occurrence of coal seams (about 900-1400 m), unfavorable groundwater regime, gas content and poor stability of coal-bearing rocks. Currently, coal deposits are not being developed, but they can be used as a source of coalbed methane. Gas output from coal seams is significantly lower than in natural gas fields. However, it can be increased as a result of thermal effects on the coal seam, which can be carried out through wells.

2. Methodology

Hard coal and brown coal from Visean coal beds in the territory of the Volga-Ural oil and gas province was used as the samples under study. By nature, the Visean coals are humits. By the degree of metamorphism, the Visean coals are mostly hard, sometimes brown. The degree of metamorphic transformation have great importance for the gas content of coal and for the composition of hydrocarbon gases sorbed in the coal. In the present work, a study was made of the composition of the gas components of coal and the characteristics of their release during heating. For this, the combined thermogravimetry (TG) and IR spectroscopy (IR) method was used. The thermogravimetric experiment was described in detail in previous works [7, 8]. The method of thermogravimetry allows to estimate the amount of gas sorbed in the coals. Using IR-spectroscopy is determined by the composition of the evolved gases in each temperature range. The analysis was performed on a TG 209 F1 Libra (Netzsch) thermogravimeter combined with an Alpha FTIR-spectrometer (gas cell) (Bruker Inc.). The measurements were carried out in a dynamic stream of nitrogen at a purge rate of 75 ml / min in the temperature range from room temperature of 30 ° C to 800 ° C in open corundum (45 μl) crucibles at a heating rate of 20 K / min. The temperature of the transit line to the spectrometer was 250 ° C, the temperature of the optical part of the spectrometer — 200 ° C. To analyze the composition of exhaust gases during thermal exposure (EGA or Evolved Gas Analysis), the IR absorption spectrum was taken in the range of wave numbers from 4000 cm^{-1} to 400 cm^{-1}. Gas purity was improved to 99.99999% and water vapor was removed using a Click-On filter (SGT-A-flow Moisture trap). The device was calibrated by temperature according to 9 melting points of high-purity metals (Bi, In, Sn, Bi, Pb, Zn, Al, Ag, Au).

Samples of coal were weighed on a Sartorius precision scale with an internal calibration function.

3. Results and Discussions

The existing technology for producing methane from coal seams and coal-bearing strata is based on drilling wells (vertical, inclined or horizontal) in the coal deposit (field) and destruction of the coal seam using the hydraulic fracturing method. As a result, additional coal degassing occurs. Excess gas in the free or dissolved state in water is brought to the surface through wells and after a certain cleaning can be used for its intended purpose. The experience of the methane-coal field in Kuzbass showed [3] that the volume of gas production depends mainly on the filtration parameters of coal seams (porosity) and structural and geodynamic features of the rock mass, and not on the thickness of the coal seams and their total methane content. Visean coals occur at a depth of about 900-1400 meters, where high pressure and temperature prevail. Under these conditions, gas, coal and water in the coal seam are in a complex interaction. Unlike traditional deposits of natural gas, where natural gas is concentrated in the hollow-pore space of reservoir rocks [9], coal seams with sorbed methane are complex natural systems. They consist of organic matter of coal, sorbed in the coal substance of gas, gas-saturated waters and natural cracks in the coal. In coal, methane is present as free gas, sorbed gas and solid solution [10]. Up to 75% of methane in coal is present in adsorbed form [11], where its amounts are adsorbed on the surface of coal particles. At the considered depths, the main amount of methane is in the sorbed state. With increasing pressure, the amount of sorbed methane in coal increases, and decreases with increasing temperature. With the existing technologies of hydraulic fracturing extracting that part of the gas, which is in the pores and adsorbed on the surface of coal particles. This gas is easily removed from the coal. The remaining quantities of methane are associated with the substance of coal and can be extracted as a result of the process of its destruction. The most effective way to destroy the structure of a coal substance is thermal exposure, which leads to the removal of gaseous components. As shown in [12], during thermal exposure, volatile components are released, their proportion reaching 30–40 wt. %. The coals studied are hard, sometimes brown [13]. Hard coal has the greatest gas content. Our studies of Visean coals by the method of thermogravimetry (TG) and IR spectroscopy (IR) allow us to estimate the amount and composition of the gas sorbed in the coals. The test samples were subsequently heated to 800 ° C with the determination of the weight loss resulting from the release of gaseous components.

The release of gaseous components occurs in several intervals, each of which is characterized by its composition of gases. At the first stage, in the temperature range from room temperature to 150 ° C, the adsorbed water evaporates. Then, in the temperature range from 250-270 ° C to 500-550 ° C, a pyrolysis zone of polycyclic aromatic structures is observed, which is accompanied by the release of methane and some cyclic methyl-substituted cycloalkanes. At the final stage, in the temperature range of 550-800 ° C, a pyrolysis zone of higher thermostable cyclic compounds is observed. This zone is characterized by the release of partially heavy hydrocarbon gases and further by simplifying the composition of the gas components until the organic matter is completely burned out.

4. Conclusions

As a result of the research, it was established that during the pyrolysis of polycyclic aromatic structures with the release of the main mass of methane occurs in the temperature range from 250 to 550 ° C. Further heating of the sample to 800 ° C leads to the release of heavy hydrocarbon gases and up to the complete burning out of organic matter. Thus, the use of modern technologies for the development of "shale gas" in combination with thermal methods will increase the volume of gas extracted from coal seams. The obtained results can be useful for choosing the temperature regime of impact on the coal seam during its development.

This study was funded by the subsidy allocated to Kazan Federal University as part of the state program for increasing its competitiveness among the world's leading centers of science and education.

References

[1] Moore T.A. Coalbed methane: A review, International Journal of Coal Geology, vol. 101, pp 36-81, 2012

[2] Bartis, James T.; LaTourrette, Tom; Dixon, Lloyd; Peterson, D.J.; Cecchine, Gary. Oil shale development in the United States: prospects and policy issues. The RAND Corporation. 2005. 68 p.

[3] Dmitrievskaya T.V., Ryabukhina S.G., Zaitsev V.A. Problems of methane production from coal beds and recent geodynamics with reference to Taldinskoye field (South Kuzbas). The journal Oil and Gas Geology (4), pp. 85-91, 2012.

[4] Khasanov, R.R., Larochkina, I.A. Prospects for the joint development of oil and coal deposits in previsean depressions of Volga-Ural oil and gas province. Neftyanoe Khozyaystvo - Oil Industry (1), pp. 36 -39, 2013. (in Russian)

[5] Khasanov R.R., Gafurov Sh.Z., Mullakaev A.I. The paragenesis of coal and oil deposits in paleozoic sediments in the east of the Eastern European platform / 16th International Multidisciplinary Scientific GeoConference SGEM 2016, SGEM2016 Conference Proceedings, Book 1. Vol.1, 627-631pp. DOI: 10.5593/SGEM2016/B11/S01.079

[6] Khassanov, R.R., Islamov, A.F. and Bogomolov, A.Kh. Rare-earth mineralization in early Carboniferous coals of the Volga-Ural region, Moscow University Geology Bulletin, Russia, Vol. 69, No. 4, pp. 251–257, 2014.

[7] Grachev A.N., Varfolomeev M.A., Emelianov D.A., Zabelkin S.A., Gilfanov M.F., Nuriyakhmetov R.A. Joint thermal treatment of heavy oil and liquid products of fast wood pyrolysis for producing fuels and chemicals, Chemistry and Technology of Fuels and Oils, 2017, № 5, 638-645.

[8] Varfolomeev M.A., Grachev A.N., Makarov A.A., Zabelkin S.A., Emelianov D.A., Musin T.R., Gerasimov A.V., Nurgaliev D.K. Thermal analysis and calorimetric study of the combustion of hydrolytic wood lignin and products of its pyrolysis, Chemistry and Technology of Fuels and Oils, 2015, № 1, 140-145.

[9] Storozsky NM, Khryukin VT, Mitronov DV, Shvachko EV Nonconventional resources of coal bed methane, Russian Chemical Journal, 2008, T. LII, No. 6, pp 63-72.

[10] Skipochka S.I., Palamarchuk T.A. On the form of methane in coal. - Geotechnical Mechanics, 2008, No. 78, pp 43-47. (in Russian)

[11] Kreinin E.V. Nonconventional hydrocarbon sources. New technologies for their development. Moscow: Publisher: Prospekt. 208 p., 2016. (in Russian)

[12] Khasanov R.R., Varfolomeev M. A., Emel'yanov D.A., and Rakhimzyanov A.I., Investigation of Thermal Effect on Samples of Coals to Determine the Prospect of Their

Utilization as Sources of Gaseous Fuels, Chemistry and Technology of Fuels and Oils, 2(606), pp.3-7, 2018.
[13] Khasanov, R.R., Gafurov, Sh.Z., Rakhimzyanov, A.I. The degree of the epigenetic transformation of an organic matter in the Early Carboniferous sediments of the central part of the Volga-Ural oil and gas province. 2016. Neftyanoe Khozyaystvo - Oil Industry. 2016. (10), pp. 29-31. (In Russan).

Laboratory studies of the behavior of organic matter of the Bazhenov deposits during thermal effects

L Sitdikova, E Bondarev, N Khasanova, E Sidorova

Kazan Federal University, 18 Kremlevskaya Str., Kazan, Russia

E-mail: sitdikova8432@mail.ru

Abstract. Currently, the role of unconventional facilities for hydrocarbons is increasing. Such objects include the Domanik rocks (Republic of Tatarstan) and Bazhenov deposits (Western Siberia). The article presents the results of studying the type of organic matter of the Bazhenov suite and studying the process of pyrolysis of organic matter. It is established that the formation of new free hydrocarbon radicals occurs at temperatures up to 350°C (pyrolysis of organic matter). This indicates that the rocks of the Bazhenov suite did not fully realize their hydrocarbon potential. The deposits of the "normal" section of the Bazhenov suite not only contain a large amount of ready and hard to recover hydrocarbons, but also have an additional generation potential.

Introduction and Methodology

Currently, there is a decrease in the production of traditional oil in Russia. There is a need to identify, study and introduce into the development of hydrocarbons unconventional reservoirs. Bitumen deposits, carbonate reservoirs, etc. are among such objects. Among the promising for the search and extraction of hydrocarbons are deposits of "shale" formations. On the territory of the Republic of Tatarstan, a number of such objects are Domanik rocks [1, 2, 3], and in Western Siberia the Bazhenov suite is the most famous and promising shale formation [4]. The formation of the Bazhenov suite took place during the Volga and early Berriane centuries in the relatively deep-water epicontinental West Siberian Sea.

The object of research of this work is the rocks of the Bazhenov suite and the Achimov formation. To perform the work, a complex of studies was used: traditional macro-structural-texture analysis, optical methods (Leica microscope), X-ray phase analysis (diffractometer

IOP Conf. Series: Earth and Environmental Science **282** (2019) 012041 doi:10.1088/1755-1315/282/1/012041

DRON-3M), electron paramagnetic analysis (spectrometer CMS8400 3cm range), scanning electron microscopy (Philips XL-30).

Results and Discussions

Within the study area, two types of sections of the Bazhenov suite are distinguished: “normal” and “anomalous”. Rocks of the “normal” section are composed of homogeneous black, dense, and platy argillites, and the “anomalous” section differs sharply in structural and textural characteristics. Typical minerals of rocks of the Bazhenov suite are fine-grained quartz, plagioclase (albite), biotite, muscovite, chlorite. The rocks also include clay minerals: illite, kaolinite, and sites of ore mineral extraction.

In the rocks of the “anomalous” section, on the background of homogeneous gray, fine-grained sandstones, interlayers of black, strongly deformed, dense argillites stand out. The Achimov formation is more homogeneous and is composed of gray, fine-grained, quartz-feldspathic sandstones. At present, the main mining operations are carried out from sections of rocks corresponding to the “anomalous” type of Bazhenov deposits (Figure 1).

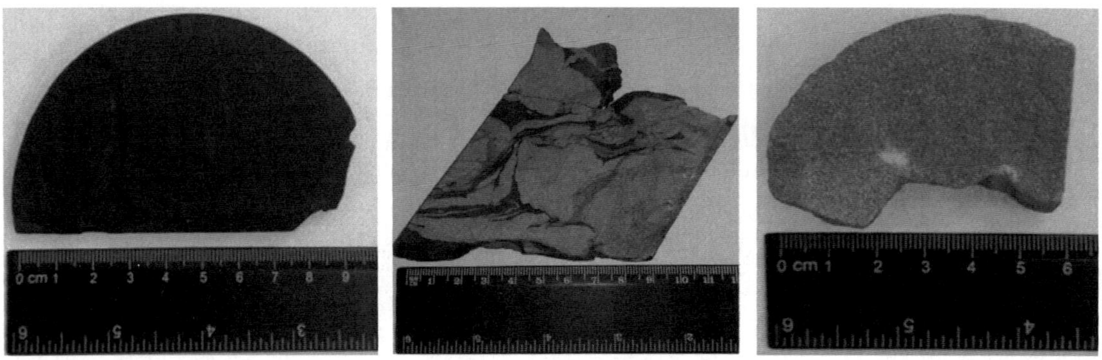

a b c

Figure 1. Structural and textural features of the studied rocks: a) the rocks of the “normal” section of the Bazhenov suite; b) typical rocks of "anomalous" section; c) rocks of the Achimov suite.

In the process of research, special attention was paid to the study of the type of organic matter, the characteristics of its distribution and behavior during thermal methods of exposure to the rock (Figure 2a, b).

The rocks the “normal” section of the Bazhenov suite are characterized by heterogeneity of distribution and difference in the shape and size of organic matter secretions: from diffuse, disseminated forms to contents in the form of large clusters, lenticular secretions with residual fragments of detritus. Fragments of detritus are replaced by ore minerals with preservation of the structure (Figure 2a).

IOP Conf. Series: Earth and Environmental Science **282** (2019) 012041 doi:10.1088/1755-1315/282/1/012041

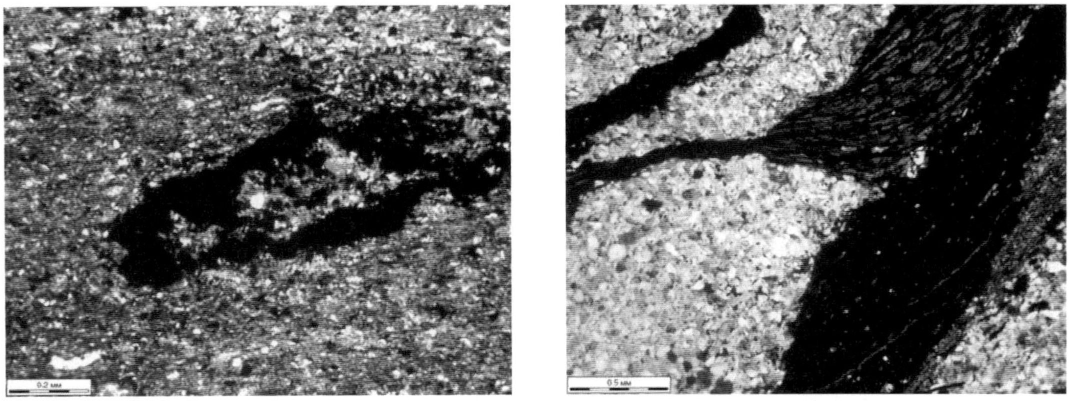

a b

Figure 2. Features of the distribution of organic matter in the Bazhenov suite: a) spotted accumulations of organic matter, the rocks of the “normal” section of the Bazhenov suite; b) typical rocks of "anomalous" section of the Bazhenov suite.

According to the results of scanning electron microscopy with microprobe analysis, it was found that the rocks of the Bazhenov suite within the studied wells have microporosity. Pores in different parts of the rock differ in size and shape. In samples of wells from deeper depth an interval, the presence of “felt” forms of illite is noted, which fulfills the void space, but the pores remain permeable. The presence of grains of the secondary quartz is noted in larger pores. In the rocks of the “normal” section of the Bazhenov suite in association with clay minerals in the void space, there are excretions of fine pyrite, the larger precipitates – framboids scattered throughout the rock.

a b

Figure 3. a) Features of the void space, “felt” forms of illite, the “normal” section of the Bazhenov suite; b) sample from "anomalous" section of the Bazhenov suite (sandstone - light part).

To establish the features of the distribution of organic matter and its behavior during thermal effects in samples of the “normal” and “anomalous” types of sections, laboratory studies of the samples by electron paramagnetic resonance (EPR) were carried out on a CMS8400 spectrometer 3 cm range. Experiments with samples of the Bazhenov suite were carried out at different temperatures: at room temperature and after heating for 30 minutes in an atmosphere of hydrogen at temperatures up to 350°C in a horizontal electric furnace (type SUOL).

Studies show that the spectra corresponding to the samples of a "normal" section of the Bazhenov suite emit intense lines of the organic radical, E'-quartz lines, Fe^{3+} lines and kaolinite (with g~4.3) (Figure 4a). After thermal exposure at a temperature of 350°C, the lines of the organic radical are more symmetrical and have an age intensity of the reflex, which indicates the process of pyrolysis of organic substances (Figure 4b).

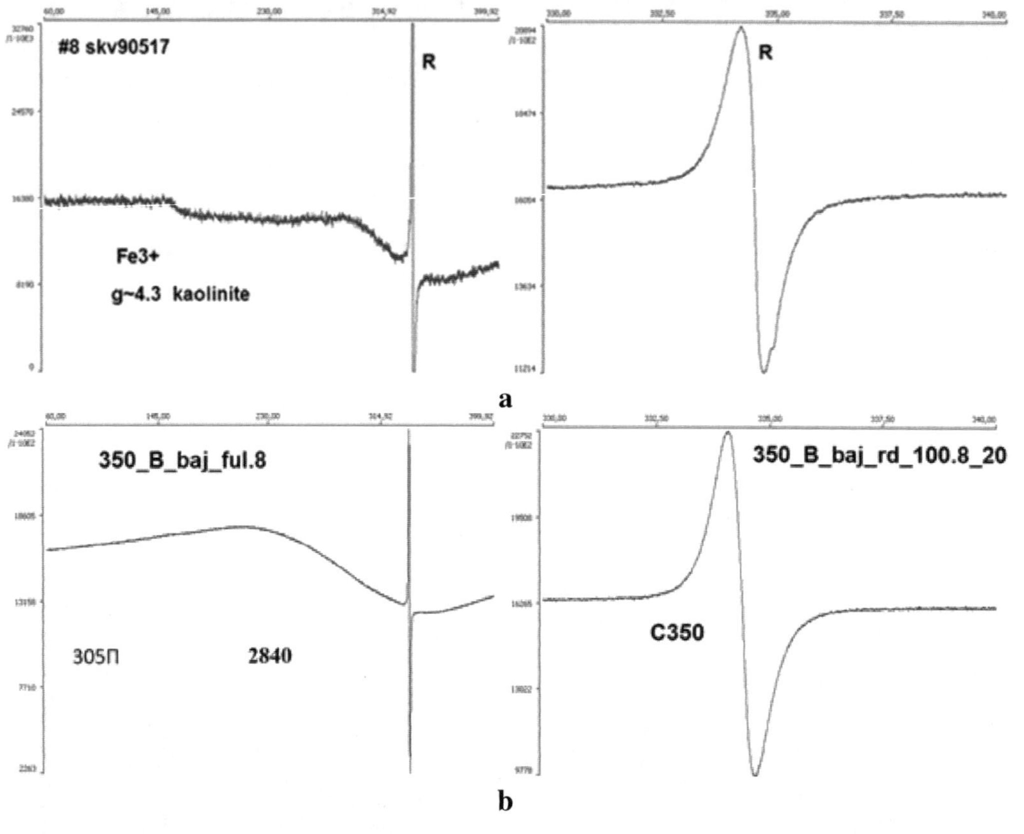

Figure 4. Characteristic EPR Spectra – sample analysis for rocks of the "normal" section of the Bazhenov suite. a) initial sample, at room temperature; b) after thermal exposure at a temperature of 350°C.

Figure 5 shows the EPR spectra of the sample of the "anomalous" section of the Bazhenov suite; the behavior of organic matter in the light part (sandstone) and clay-argillite part (black part) was analyzed. The largest amount of organic matter in the original sample is associated with the clay-argillite part of the rocks, where the line of the organic radical is clearly distinguished (Figure 5a). At temperatures of 350°C, pyrolysis of organic matter occurs, and the intensity of line of free organic radical increases (Figure 5b). Examination of the light part of the sample (sandstone) shows that the mineral composition of the rocks differs sharply from the black part: the lines corresponding to Mn^{2+} in calcite $CaCO_3$, the Fe^{3+} line and a distinct E'-quartz line are distinguished in the EPR analysis (Figure 5c). The organic radical line in the light part of the rock is less intense than the black part. After thermal impact on the sample, an increase in the intensity of the organic radical line was not observed in the EPR spectra.

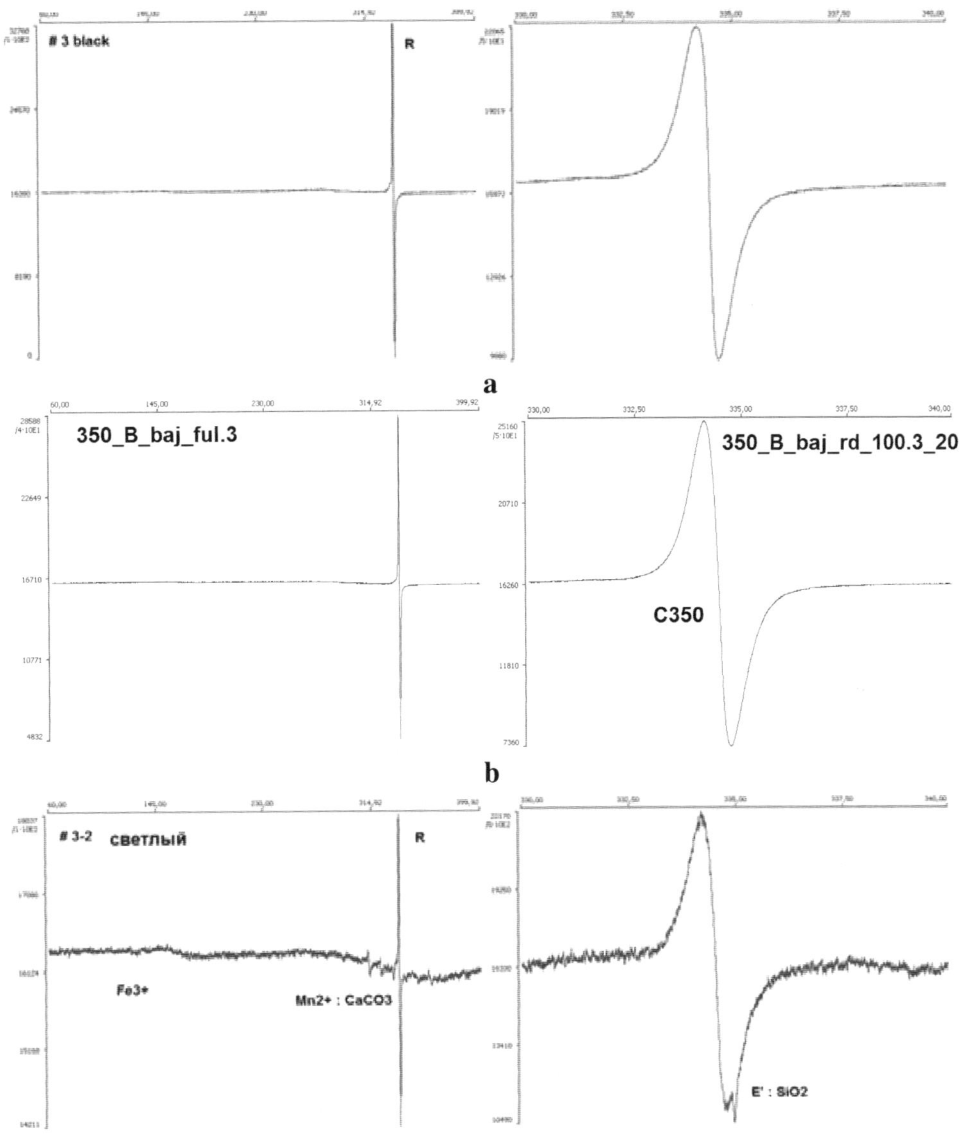

Figure 5. Characteristic EPR Spectra – sample analysis for rocks of black and light parts of the "anomalous" section of the Bazhenov suite. a) initial sample, at room temperature (black part); b) the black part of the sample after thermal exposure at a temperature of 350°C; c) the light part at room temperature.

Conclusions

1. The obtained results indicate that the value of the free organic radical in the rocks of the Bazhenov formation is significantly higher than that of the rocks of the Achimov formation. High values of the free organic radical indicate an increased content of organic matter in the original rock (Figure 4a).

2. In some of the studied samples, the presence of vanadium was recorded on the EPR spectra and it was noted that in samples belonging to the "normal" section of the Bazhenov suite, the content of vanadium increases with depth.

3. In order to determine the possible hydrocarbon potential of the rocks of the Bazhenov suite, studies have been conducted of the process of pyrolysis of organic matter, it was found that at a temperature of 350°C, the formation of new free hydrocarbon radicals was recorded. This indicates that the rocks of the suite in natural conditions did not fully realize their hydrocarbon potential.

4. According to the results of studies of the rocks of the Bazhenov suite, it can be concluded that the deposits of the "normal" section not only contain a large number of ready and hard to recover hydrocarbons, but also have an additional generation potential.

5. The extraction of hydrocarbons is carried out only from the "anomalous" section of the Bazhenov suite due to the clinoform complexes of the Achimov formation. Profitable extraction of hydrocarbons from the "normal" section requires more detailed study and application of new technologies for the development and production of hydrocarbons.

Acknowledgements

The work is performed according to the Russian Government Program of Competitive Growth of Kazan Federal University.

References

[1] Kayukova G P, Kiyamova A M, Mikhailova A N, Kosachev I P, Petrov S M, Romanov G V, Sitdikova L M, Plotnikova I N, Vakhin A V 2016 Generation of Hydrocarbons by Hydrothermal Transformation of Organic Matter of Domanik Rocks *Chemistry and Technology of Fuels and Oils* **52** 149

[2] Khasanova N M, Sitdikova L M, Morozov V P, Nizamutdinov N M, Khasanov R A, Nurgaliev D K 2016 Geochemical features of the Domanik deposits of the Republic of Tatarstan according to EPR studies *Geology, minerals and problems of geo-ecology of Bashkortostan, the Urals and adjacent territories* pp 189-191

[3] Vakhin A V, Onishchenko, Y V, Chemodanov A E, Sitdikova L M, Nurgaliev D K 2016 Thermal transformation of bitumoid of Domanic formations of Tatarstan *Neftyanoe Khozyaystvo - Oil Industry* **10** 32

[4] Sitdikova L M, Khasanova N M, Izotov V G, Vakhin A V, Sidorova E U 2017 Temperature dynamics of organic matter of the bazhenov suite from electron paramagnetic resonance *International Multidisciplinary Scientific GeoConference Surveying Geology and Mining Ecology Management SGEM* **17** 565

IOP Conf. Series: Earth and Environmental Science **282** (2019) 012042 doi:10.1088/1755-1315/282/1/012042

Analysis of the results of tracer tests for the monitoring of the development of super-viscous oil deposit

M. S. Shipaeva[1], V. A. Sudakov[1], R. R. Akhmadullin[2]

[1]Institute of Geology and Petroleum Technologies, Kazan Federal University, Kazan, Russian Federation
[2]PJSC «TATNEFT», Department of extra-viscous oil production, Almetyevsk, Russian Federation

E-mail: mariasipaeva@gmail.com

Abstract. The article describes the results of implementation tracer studies at one of the super-viscous oil deposits in the Republic of Tatarstan. Accurate information on the flow of liquid in oil deposits is absent, this is an obstacle to the successful development of such deposits. The information obtained as a result of tracer studies is an addition to the general history of field development and is used to reduce uncertainties in the construction of geological and hydrodynamic models of deposits. The main tasks were the determination of the hydrodynamic connection between the wells along channels with low filtration resistance, the determination of the filtration rates of the condensate in the reservoir and the heterogeneity of the reservoir in the inter-well space of the investigated section of the oilfield. In order to determine the quality of the hydrodynamic connection within the bituminous deposit, the geological structure, as well as the direction of the filtration streams, a solution of the chemical eosin indicator was pumped into one of the steam injection wells, followed by registration in the samples of the surrounding production wells. To detect fluorescent indicators in the wells, unique equipment was used that significantly reduces the previous sample processing and detection time.

1. Introduction

One of the informative and inexpensive ways for investigation of the filtration properties of reservoirs is the well to well tracer test. The results of this method reflect the characteristics of the reservoir in the entire interwell space. The technology of the indicator method is simple and based on the injection of a tracer agent into the injection well, followed by the selection of the products of the production wells surrounding it. These samples are measured in the laboratory for the presence of an indicator. The concentration and arrival time of the indicator are interpreted, and the results are used to create a picture of fluid movement through the formation, as well as to clarify the geological structure of the site [1].

By the time of the first registration of the indicator from exploration wells, zones of high dynamic permeability are determined. The concentration and quantity of the indicator extracted to the surface determine the conductivity of the filtration paths. The indicator method allows to establish control over the distribution of flows in directions, to identify areas of absence of hydrodynamic connectivity between wells.

Eosin Y was chosen as an indicator, because it is supposed to use markers in the development of extra-viscous oil deposits using steam and gravity drainage technology, that is, at pressures up to 10 MPa and temperatures up to 200 ° C. During laboratory studies, it was found that exposure to high pressure and high temperature led to a significant increase in fluorescence intensity. This effect is explained by an increase in the solubility of eosin Y at high temperatures. This feature of the selected indicator will be very useful in the real conditions of the production of SVO in the Republic of Tatarstan, because will positively affect the sensitivity of the method.

The object of research was the SVO deposit, associate with the Ufa bituminous complex, to the sand pack of the Sheshminsky horizon of the Permian system (Fig.1).

This territory is well studied, the productive horizon contains lithological heterogeneities represented by dense consolidated sandstones and lenses of water saturated sands above the oil-water contact. The deposit is characterized by irregular OWC over the entire area, in contrast to traditional deposits [2].

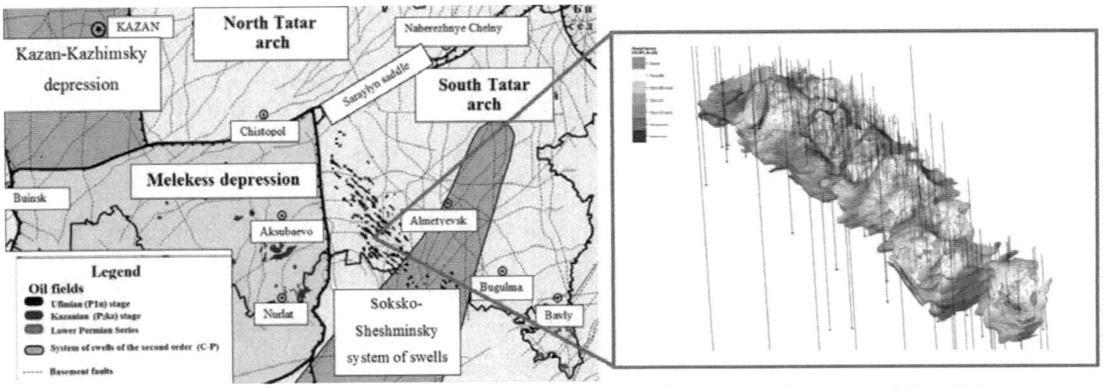

Figure 1. Fields of super-viscous oils and natural bitumens of the Republic of Tatarstan, object of research

2. Methodology

The methodology includes both field and laboratory work.

2.1 Field Work

To select wells for the study the specialists were guided primarily by the geology of the deposit and the tasks that were set. The deposit is a reservoir with several domes. It was necessary to find out whether there is a connection between the Eastern and Western domes, whether the water moves through the depression that separates them (Fig. 2). It was also necessary to find out the presence and direction of movement of the and the influence of the location of the pump on the flow of condensate inside the reservoir.

Before the indicator was injected, water samples were taken from wells for baseline survey.

The fluorescent indicator aqueous solution was injected through injection well № 61. This well located in the center of the East dome and is the highest in the entire field and has relatively high production rate. In order to realize a homogeneous distribution of the tracer in the wellbore and in the reservoir, it was decided to inject the indicator into the heel of the well №61 (Fig. 2).

Before the indicator was injected, well №61 was stopped to reduce the temperature in the column. Well downtime lasted 2 days. Well № 60 was stopped before the indicator was injected to ensure equidistribution of the indicator solution in the perforation interval of the production well №60. Well did not work for 1 day.

The aqueous solution contained 15 kg of the eosin indicator dissolved in 5 m^3 of warm fresh water. The final concentration of eosin in water was 3 g/l. Surveys were conducted for 10 days. The established monitoring fund is 24 wells.

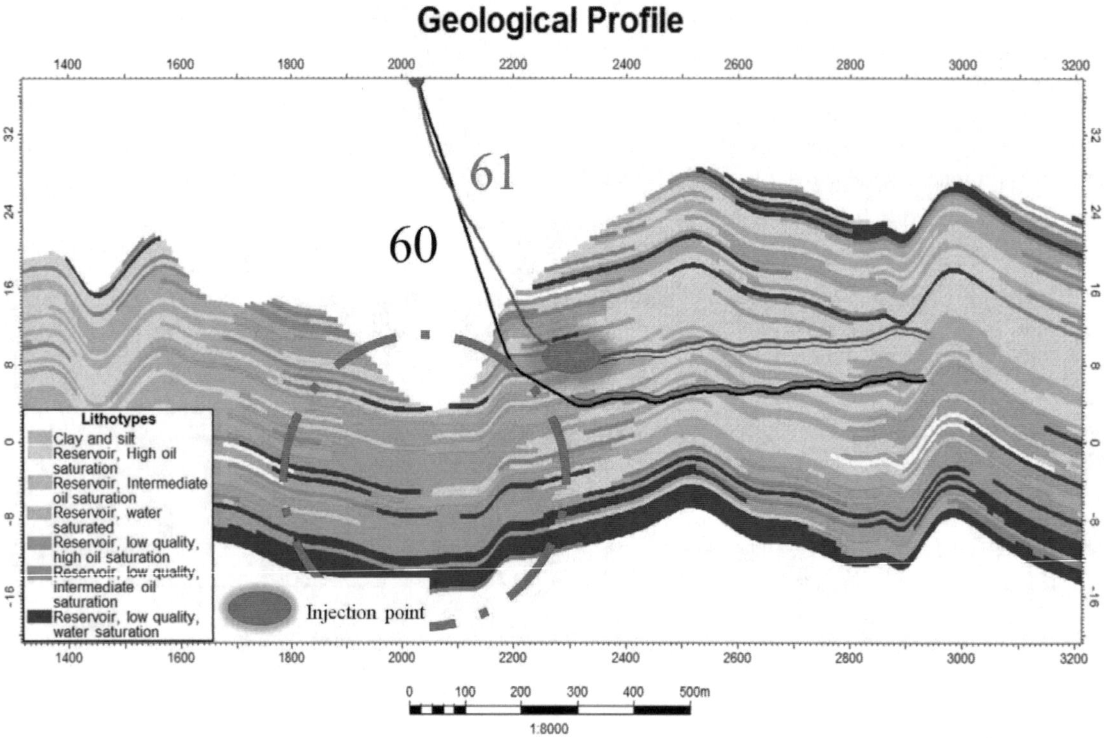

Figure 2. Geological profile of the Eastern and Western domes

2.2 Lab work

Experimental hydrodynamic studies of a technical fluid using hydrodynamic markers (in particular, fluorescent indicators) indicate the presence of a special highly sensitive optical system for their detection by fluorescence spectra in real conditions of oil production. For the study, an automated mobile field laser spectrofluorimeter was used (Fig. 3). The MPLS-2 was developed as part of a joint project of Kazan Federal University and Kazan E. K. Zavoisky Physical-Technical Institute of the RAS.

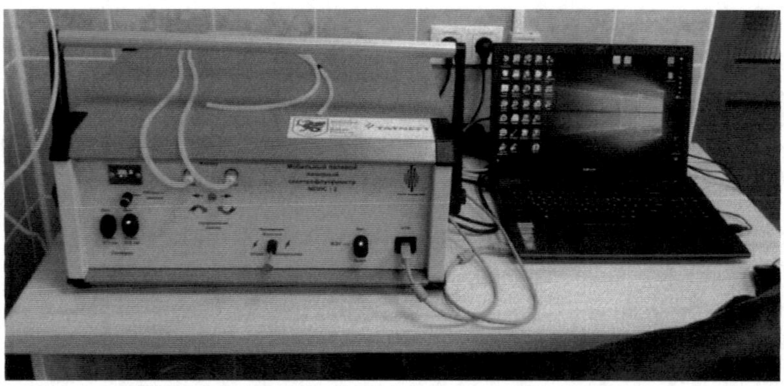

Figure 3. Mobile field laser spectrofluorimeter

First of all, it is necessary to investigate test solutions of eosin with a known concentration prepared in natural water wells, with the aim of calibrating the device MPLS-2. Samples were cleaned with a paper filter, as they contained oil contamination.

Results and discussion

As a result of the study, a map has been created showing the location of the horizontal wells, percentage extraction of indicator and arrival time (Fig. 4). All calculations were carried out at the location of the pump inside the producing horizontal well due to the predominant flow into this area. The tracer is confidently fixed in all investigation wells. The tracer enters in the wells mainly on the 1st and 2nd day.

The wells located near the injection well №61, failed to investigate due to the absence of water in the produced fluid. This may be due to the stop of the steam injection well №61 to reduce the temperature in the column by 2 days. Hence we can conclude about the existence of impact of the development of the steam chamber №61 for the extraction of nearby wells №58 and №47

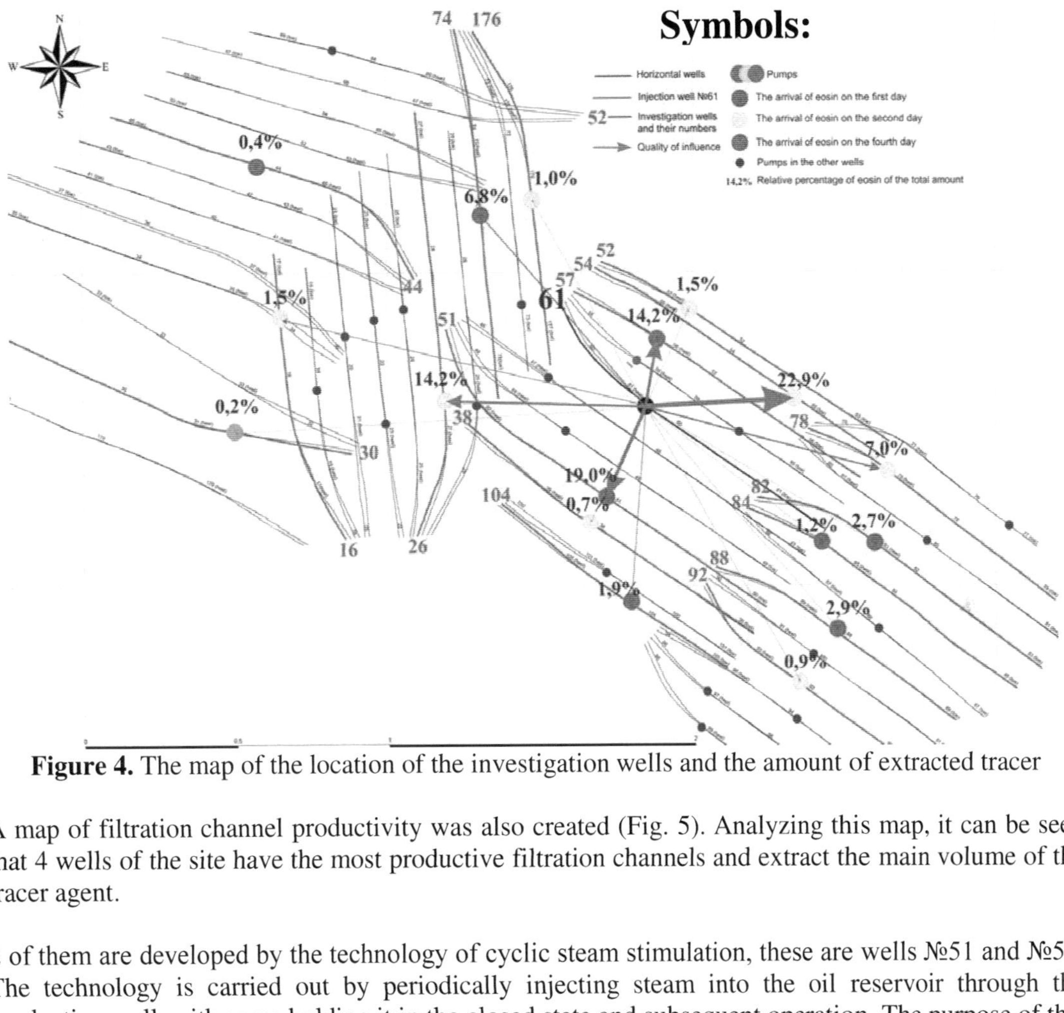

Figure 4. The map of the location of the investigation wells and the amount of extracted tracer

A map of filtration channel productivity was also created (Fig. 5). Analyzing this map, it can be seen that 4 wells of the site have the most productive filtration channels and extract the main volume of the tracer agent.

2 of them are developed by the technology of cyclic steam stimulation, these are wells №51 and №57. The technology is carried out by periodically injecting steam into the oil reservoir through the production well, with some holding it in the closed state and subsequent operation. The purpose of this technology is to increase the flow of oil to the wells by reducing the viscosity of the oil, increasing the bottomhole pressure, facilitating filtration conditions [3]. If we developed part of field in this way, a positive depression is created around the well, where condensate is easy to flow than the SAGD method where the steam chamber creates overpressure and repels condensate. Also during the tracer studies well №54 worked only for production, steam injection into the upper pair was suspended for it. This explains the high productivity of the channels in these 3 wells.

Also areas of wells with the same tracer behavior are identified (well №№78, 82, 88, 84). In these wells, there are 2 portions of the arrival of the tracer agent (Fig.6). It is assumed that the first portion is associated with the movement of the tracer through the wellbore of the injection well №61. Leaving the bottomhole tracer was established in the immediate vicinity of the well №61. And the second portion of the indicator came completely on the pore part of the collector later, on the 5th day of the research.

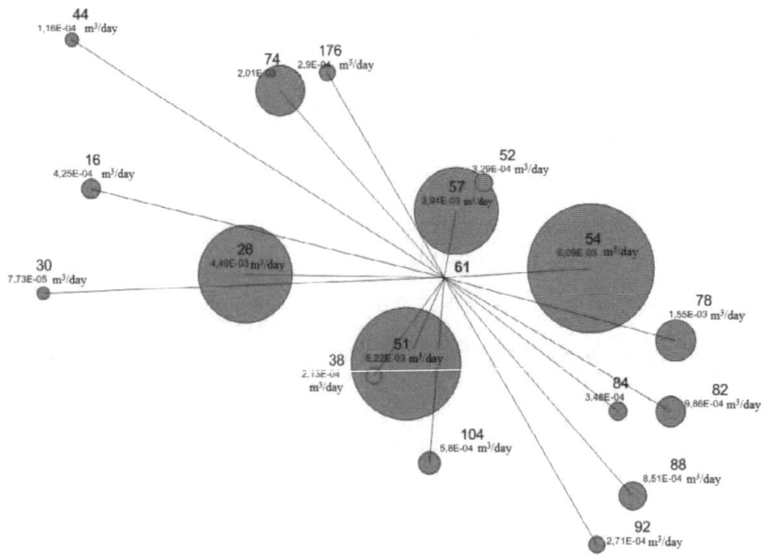

Figure 5. Filtration channel productivity map.

The tracer is fixed in all the studied wells, which indicates the presence of water flows inside the reservoir. But tracer came in different ways. It is also worth noting that the first portion comes in most wells on the 1st and 2nd days. During the study, the operation mode of some wells changed and this is reflected in the behavior of the tracer in the interwell space. In total, 15 g of eosin Y (out of 15 kg) was extracted. Such a small extraction is associated with the injection of the tracer into the reservoir with water that differs significantly in temperature from the reservoir conditions; therefore, main part of tracer is filtered in the underlying aquifer.

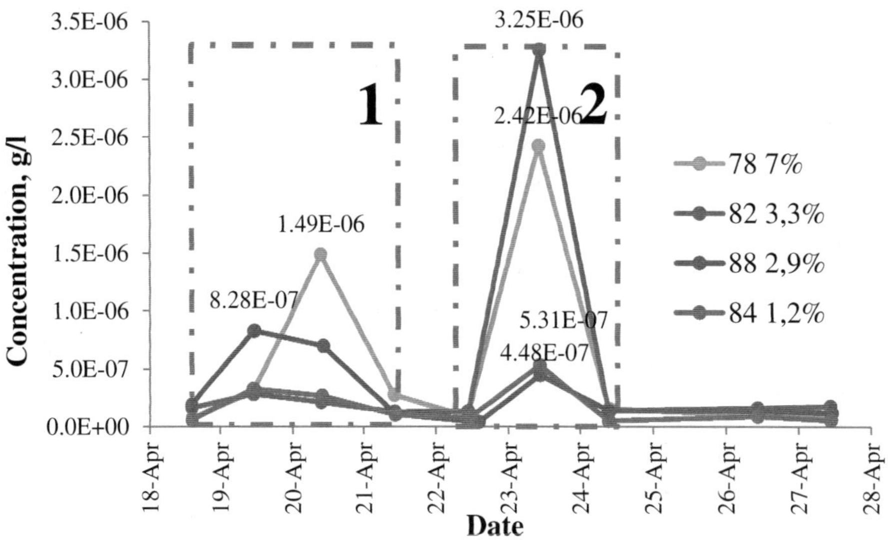

Figure 6. The graph "Tracer concentration - time"

IOP Conf. Series: Earth and Environmental Science **282** (2019) 012042 doi:10.1088/1755-1315/282/1/012042

3. Conclusions

The existence of water flows within the bitumen deposit, resulting from the contact of the steam chambers with each other is shown in this work. It was also found that the reservoir is a single hydrodynamic reservoir, the Eastern and Western domes of the investigated area are interconnected (~23% of the tracer left the Eastern dome to the Western). Water with a tracer dissolved in it moves through different aquifers and in different ways.

Acknowledgements

The work was supported by the Ministry of Education and Science of the Russian Federation (project No. 02.G25.31.0170) and by the subsidy allocated to Kazan Federal University as part of the state program for increasing its competitiveness among the world's leading centers of science and education.

References

[1] Kubarev P.N, Kamyshnikov A.G. Quantitative Estimation of Main Ways of Fluid Conductivity After Use of Technologies of Oil Refining Unit and Formation Hydraulic Fracturing by Tracer Method of Research. 2016. TatNIPIneft Institute, PC «Tatneft», Bugulma.

[2] Novikova S.P., Nurgaliev D.K., Sudakov V.A., Delev A.N., Nurgalieva N.G., Lyabipov M.R., Akhmadullin R.R. The Main Features of the Geological Modeling Process of a Shallow Deposit of Super-Viscous Oil in Aspect of Development Strategy Planning with the Use of Steam-Assisted Gravity Drainage Method. Georesursy = Georesources. 2017. V. 19. No. 4. Part 1. pp. 331-340.

[3] Sheng, J.J. Cyclic steam stimulation. In Enhanced Oil Recovery Field Case Studies; Elsevier: Amsterdam,The Netherlands, 2013; pp. 389–412.

Thermal Conversion of Heavy Crude Oil in the Presence of Formic Acid as a Hydrogen-donor Solvent

A. Almuntaser[1]*, M. Varfolomeev[1,2], M. Suwaid[1], R. Djimasbe[1], D. Garayeva[1], A. Ulbaev[1]

[1] Department of Petroleum Engineering, Kazan (Volga region) Federal University, Russian Federation

[2] Department of Physical Chemistry, Kazan (Volga region) Federal University, Russian Federation

E-mail: aalmuntaser@mail.ru

Abstract. In this study, thermal conversion experiments of heavy oil using steam and formic acid as a hydrogen donor were carried out in a batch reactor at T = 380 °C and P = 165 bar. Material balance and products distribution were calculated after the experiments. Properties of crude oil before and after thermal conversion were analyzed including: viscosity, API gravity, SARA and elemental analysis measurements. It was shown, that application of formic acid as a hydrogen-donor solvent leads to the reduction of coke and gaseous products formation and to the increase of liquid products yield. In addition, the viscosity of upgraded oil was decreased by 23.2% due to addition of formic acid in comparison with thermal conversion without the hydrogen-donor solvent.

1. INTRODUCTION

In last decades, research related to the development of new technologies for the production of heavy hydrocarbons, such as heavy and extra-heavy oils, natural bitumens, oil sands has been extensively conducted [1,2]. This is due to the reduction of the light oil production and the annual increase in the amount of heavy oil involved in recovery processes. Heavy oils have a high content of resins, asphaltenes and other heteroatomic compounds. Many studies were dedicated to the use of unconventional methods that could ensure the production of lighter, so-called synthetic oil from heavy feedstock, which can be further processed according to existing schemes without substantial changes in refinery processes. One of the promising ways to produce synthetic oil is thermal conversion of heavy hydrocarbon feedstock in the presence of water in the form of superheated steam or at critical or supercritical conditions [3]. It was reported that using of superheated steam in the process of cyclic steam stimulation is fostered for the thermal enhanced oil recovery including heavy oils and oil sands, like Athabasca oil sand and Orinoco heavy oil [4-6]. Another research done by Li et al. suggested that thermal conversion of heavy oil takes place in the oil reservoir when injected the superheated steam [7]. In addition, Song et al. offered a new thermal conversion project for conventional heavy oil by superheated steam. Authors observed [7] that the reaction will reach evenness after a confirmed time and will not be sensitive to reaction time. Katritzky et al. indicated that water in the form of superheated steam participates as a catalyst, reactant, and a good hydrogen-donor solvent in reactions of hydrocarbon compounds with superheated steam [8]. Under these conditions, water has the properties of a non-polar proton-donor solvent. In addition, physical properties of water under these conditions like density, dielectric constant, hydrocarbon solubility and ion dissociation constants, which can speed up the thermal conversion, increased significantly [9]. Consequently, it makes possible to increase the efficiency of thermal processes for the destruction of heavy oil feedstock and, as a consequence, to obtain "synthetic" oil with a low content of high molecular weight heteroatomic compounds and a high content of low-boiling fractions.

The main disadvantage for thermal conversion of heavy hydrocarbon resources in the presence of superheated steam is the significant formation of coke and gaseous products [10], which leads to the loss of the desired components of the liquid products (synthetic oil). However, this issue can be solved by using some additives such as hydrogen-donor solvents or any agent, which prevent the coke and gas formation.

In addition, thermal conversion of heavy oil using hydrogen-donor solvents along with superheated steam lead to an increase in the ratio of hydrogen and carbon, as a result of transfer an additional hydrogen from hydrogen-donor solvents, which participates in hydrogenation reactions (preventing the recombination of the formed radicals). It should be noted that, there are many studies and works showed that using additives such as hydrogen-donor solvents or even different catalysts with steam improve the efficiency of upgrading and conversion of heavy oil and reduce the coke and gas formations. Fan et al indicated that, the coke and gas formations can be decreased by adding hydrogen donor and catalysts [11, 12]. This technology successfully has been applied in the field with good results. For extra-heavy oil (Shengli oilfield, China) the viscosity reduction rate was about 70% after thermal-catalytic upgrading by adding of a hydrogen donor [12].

In Tatarstan region, Russia, there are many heavy and extra-heavy oil fields, which are developed by using steam based method for enhanced oil recovery (SAGD) [13,14]. Using of hydrogen-donor solvents as additive can help to improve the efficiency of their extraction and *in-situ* upgrading. In this work formic acid is using as hydrogen-donor solvent together with superheated steam, to reduce the formation of coke and gas products, and to prevent the recombination reactions during the thermal conversion process of heavy oil.

2. MATERIALS AND METHODS

Heavy oil (Tatarstan region, Russia) was used in this work. The main characteristics are shown in Table 1.

Table 1. Properties of studied heavy oil

Parameter	Value	Parameter	Value
°API gravity	14.1	Elemental analysis (wt. %):	
Viscosity at 25 °C, mPa·s	2073	Carbon	83.68
Saturates, % mas.	28.79	Hydrogen	11.44
Aromatics, % mas	44.32	Sulfur	4.52
Resin, % mas	20.98	Nitrogen	0.36
Asphaltenes, % mas	5.91		

Upgrading experiments were carried out in a stainless-steel 350-ml batch Parr reactor equipped with a heating device and a temperature controller. In a typical experiment, the reactor was loaded with crude oil and water at a weight ratio of 2:1, respectively. The reactor was heated at 5°C/min to 380°C, generating a final pressure of approximately 164 bars for 6 hours (2 bar initial N_2 pressure). The thermal conversion experiment of using steam and hydrogen-donor solvent (2% mas. formic acid) together was conducted at the same conditions as experiment without hydrogen-donor solvent.

Physical-chemical properties including viscosity, API gravity and SARA analysis of oil before and after thermal conversion in the presence and absence of formic acid were studied. Viscosity and API gravity of oil were measured by SVM-3000 Stabinger viscometer. The elemental composition of oil before and after conversion was determined by Analyzer Perkin Elmer 2400 Series II. SARA analysis was done according to ASTM D 4124.

3. RESULTS AND DISCUSSIONS

3.1. *Material Balance and products distributions for the process of thermal conversion*

The mass balance (Table 2) of the three products i.e., liquid, gas, and coke for two experiments was calculated as percentage of the mass of feed oil into the reactor using Eqs. (1) and (2):

$$\text{Yield (wt.\%)} = Wi/WF \times 100 \quad (1)$$

$$\text{Yield of gas (wt\%)} = (WF - WA)/WF \times 100 \quad (2)$$

where Wi is the weight of component i and WF is the weight of the feed oil (heavy oil downloaded into the reactor) and WA is sum of weights of coke and liquid phase in autoclave after the experiment.

Table 2. Material balance for thermal conversion of crude oil at 380 °C.

Reaction system	Temperature, °C	Liquid yield (wt, %)	Coke yield (wt, %)	Gas yield (wt, %)
oil + superheated steam + N_2	380	73.04	16.63	7.33
oil + superheated steam + formic acid + N_2	380	80.10	13.89	6.01

From table 2 it can be noted high yield of coke and gas after thermal upgrading with superheated steam, due to polymerization and condensation of high molecular weight components. Quite the contrary, adding the formic acid to the reaction system provides decreasing in the yield of the coke and gases and increasing of liquid phase content.

3.2 Viscosity and API gravity of heavy oil before and after thermal conversion at 380 ^0C

The results of studies of crude oil properties (viscosity, API gravity and SARA analysis) before and after upgrading under superheated steam conditions are given in Table 3.

Table 3. Viscosity, API gravity and SARA composition of oil before and after thermal conversion

Reaction system	viscosity at 25 °C, mPa·s	^{0}API, gravity	mass content,%			
			Saturates	Aromatics	Resins	Asphaltenes
initial oil	2073	14.1	28.79	44.32	20.98	5.91
oil + superheated steam	15.73	21.6	55.17	29.83	12.45	2.47
oil + superheated steam + formic acid	12.08	22.9	57.84	31.04	9.77	1.35

From the table 3, we can observe the increase of saturates fraction content and the decrease the amount of resins fraction due to thermal conversion of heavy oil with and without of formic acid as a hydrogen-donor solvent. However, the conversion degree in presence of formic acid is higher. Application of hydrogen donor helps to improve destruction of high-molecular components (resins and asphaltenes) of heavy oil [10,11]. In comparison with the initial oil, the content of saturates significantly increases and the amount of resins and asphaltenes significantly decreases in the same conditions with superheated steam and formic acid. These changes are reflected in the increasing of ^{0}API gravity from 21.6 to 22.9 by adding the hydrogen-donor solvent and reduction viscosity from 2073 to 15.73 without formic acid and 12.08 with it.

3.3 Elemental analysis of oil sample before and after thermal conversion

Table 4 shows the elemental analysis of oil before and after thermal conversion, we can see a noticeable decrease in the heteroatoms, especially sulfur content from 4.52 for initial oil to 2.96 with superheated steam and 2.77 with superheated steam and formic acid. At the same time we can see the decreasing in the percentage of carbon and hydrogen due to their conversion to coke and gas after process of the polymerization and condensation. It should be noted that, adding of formic acid as a hydrogen-donor solvent leads to an increasing the percentage of hydrogen comparing to the upgrading without formic acid.

Table 4. Elemental analysis of crude oil before and after thermal conversion

System	Content, wt %			
	C	H	N	S
Initial oil	83.68	11.44	0.36	4.52
Oil + superheated steam	83.65	8.65	0.19	2.96
Oil + superheated steam + formic acid	83.01	10.87	0.34	2.77

4. CONCLUSION

Thermal conversion of heavy crude oil using a superheated steam together with formic acid as hydrogen-donor solvent gave promising results. The content of high-molecular heteroatom substances such as resins and asphaltenes was decreased in contrast with thermal treatment using just superheated steam. The viscosity of heavy oil was improved by adding formic acid, compared with thermal conversion without the hydrogen-donor solvent (viscosity reduction rate is equal to 23.2%). Additionally, using the formic acid as a hydrogen-donor solvent leads to the reduction of obtained coke and gaseous products, increase the amount of

IOP Conf. Series: Earth and Environmental Science **282** (2019) 012043 doi:10.1088/1755-1315/282/1/012043

liquid products (upgraded oil) yield. In addition, the reduction of sulfur content was observed due to adding the formic acid as a proton donor for the thermal conversion process.

Acknowledgments

The work is performed according to the Russian Government Program of Competitive Growth of Kazan Federal University.

References

[1] J. B. Joshi *et al.*, "Petroleum residue upgradation via visbreaking: A review," *Ind. Eng. Chem. Res., vol. 47, no. 23, pp. 8960–8988, 2008.*

[2] Ameen A M Al-Muntaser, M. A. Suwaid, I. Z. Rakhmatullin, and M. A. Varfolomeev, *Int. Multidiscip. Sci. GeoConference SGEM Surv. Geol. Min. Ecol. Manag.*, vol. 17, no. 1.5, pp. 75–82, 2017.

[3] P. Arcelus-Arrillaga, J. L. Pinilla, K. Hellgardt, and M. Millan, *Energy & Fuels*, vol. 31, *no. 5, pp. 4571–4587, 2017.*

[4] Maity, S. K.; Ancheyta J.; *Marroquín G. Energy Fuels 2010, 24, 2809-2816.*

[5] Speight, J. G., 4th ed.*; CRC Press/Taylor & Francis Group: Boca Raton, FL, 2007.*

[6] Huang, Shijun, Meng Cao, and Linsong Cheng. *Energy & fuels 32, no. 4 (2018): 4850-4858.*

[7] Li, Jian, Yanling Chen, Huachao Liu, Pujian Wang, and Feng Liu. *Energy & Fuels 27, no. 5 (2013): 2555-2562.*

[8] Katritzky, A. R.; Alin, S. M.; Siskin, M. *Chem. Res. 1996, 29(8)399-406.*

[9] Ameen A M Al-Muntaser, C. Yuan, M. A. Suwaid, and M. A. Varfolomeev, *Int. Multidiscip. Sci. GeoConference SGEM Surv. Geol. Min. Ecol. Manag.*, vol. 18, pp. 335–341, 2018.

[10] X. Z. Yuan, J. Y. Tong, G. M. Zeng, H. Li, and W. Xie, *Energy & Fuels*, vol. 23, no. 6, pp. 3262–3267, 2009.

[11] Salih I Sh S, Mukhamatdinov I I, Akhmetov B R 2018 *18th International Multidisciplinary Scientific GeoConference - SGEM 18 735.*

[12] Fan, Ze-xia, Fu-lin Zhao, Jie-xiang Wang, and Yong-gang Gong. *Journal of Fuel Chemistry and Technology 34, no. 3 (2006): 315.*

[13] Mohammad, A.A.A. *PhD diss., Texas A&M University, 2008.*

[14] Suwaid, Muneer A., Chengdong Yuan, Ameen A. Almuntaser, and Mikhail A. Varfolomeev. *International Multidisciplinary Scientific GeoConference: SGEM: Surveying Geology & mining Ecology Management 18 (2018): 283-289.*

IOP Conf. Series: Earth and Environmental Science **282** (2019) 012044 doi:10.1088/1755-1315/282/1/012044

Comparison of upgrading of heavy oil and vacuum distillation residues by supercritical water

Richard Djimasbe*, Ameen A. Al-muntaser, Muneer A. Suwaid, Mikhail A. Varfolomeev

Department of Petroleum Engineering, Kazan Federal University, Kazan 420008, Russia

E-mail: oiltchad91@gmail.com

Abstract

The purpose of this work is to test the upgrading technology by supercritical water of two different raw materials heavy oil and vacuum distillation residue. The experiments were carried out under the same conditions (temperature 400 °C, pressure 236 bar). The products obtained after upgrading were analyzed. Density, viscosity, SARA-composition, elemental composition and composition of saturates fraction were determined. The results revealed that after upgrading at 400 °C during 60 minutes raw materials heavy oil and vacuum residues yielded light oil 85.6 % and 64.7 %, coke at level of 11.3 % and 13.7 % and gaseous products 3.1 % and 21.6 %.

1. INTRODUCTION

The problem of shortage of light oil is increasing day by day because of the excessive consumption of energy. This question interest many scientists and researchers for some decades and a lot of research has been done to find an alternative source for fossil energy. One of the options is the development of unconventional crude oil reserves such as heavy oil, shale oil, oil sand [1]. Heavy oil and natural bitumen are mainly composed of high molecular weight hydrocarbons, resins and asphaltenes. The exploitation of these resources is very difficult due to high viscosity and bad transport properties of heavy oil, where asphaltenes and resins are the main factors.

The process of upgrading of heavy oil under supercritical water (SCW) conditions is an approach that can be used nowadays in the field of heavy oil exploitation or shale oil extraction and residues refining. Apart from heavy oil and oil sands production methods like steam-assisted gravity drainage (SAGD) and cold heavy oil production with sand (CHOPS), the supercritical fluid technology can be alternative to improve recovery, transportation and refining [2]. Cheng et al. found [3] that asphaltenes can be partially dissolved and dispersed under the effect of supercritical water.

One of the common problems encountered during the production or refining of heavy oil is the presence of asphaltenes in the oil. Asphaltenes are defined as one of the highest molecular weight components of crude oil. They present solid substances of black or brown color with average molecular weight 1000-5000 and density 1.1 g/cm^3 [4]. They can be softened only in an inert atmosphere and at high temperatures 200-300 °C.

Regarding the supercritical conditions (T = 374 °C and P = 221 bar), the water decreases its dielectric constant by about 78 to 10, where it has a solvent-like property such as acetone, benzene, toluene, etc. [5-6]. The water in the supercritical state is capable to supply the hydrogen to saturate the radicals formed from the oil component molecules at the time of thermal cracking and the hydroxyl group can cut off the bridge of the naphthenic alcohol in the oil and this prevents the formation of coke [7-8]. The asphaltenes are the main target in upgrading of heavy oil under the supercritical water conditions.

The process of upgrading of heavy oil using supercritical water was studied by Timko et al. [9]. Authors observed high level of heavy oil upgrading and desulphurization. Kozhevnikov et al.[5] studied the conversion of asphaltenes in a supercritical water environment at 380 °C and 226 bar during 3 hours. They indicated the formation of maltenes, gaseous products and coke from the hydrothermal destruction of asphaltenes. Gao et al. [10] used the D_2O as a isotope tag to study the hydroconversion of the Tahe residue in the SCW treatment. The results showed the role of water in the transformation of hydrocarbons. They determined [10] that the proportion of water molecules involved in hydrogen exchange was about 40% when the water content in the reaction mixture was 10% by weight. Olukcu et al. [11] suggested in their study that the absence of *n*-alkenes and the presence of alcohol in the product could be due to the transfer of OH from water to organic free radicals generated from an oil shale when extracting by SCW. Fedyaeva et al. [12] also confirmed in their study that the involvement of water contributed to the upgrading of bitumen in an up flow reactor with the proof of the appearance of oxygen and the increase in the concentration of hydrogen in the products and the conversion residue.

The interest of this research work is to study in details the treatment of heavy oil and vacuum distillation residue from Tatarstan Republic by supercritical water for the upgrading purpose. Currently, Republic of Tatarstan has large reserves of heavy oil and bitumen's. So, new technologies are required for their extraction and refining. To study the mechanism of upgrading process different physical-chemical methods has been used in this study. They helped to identify changes in composition and properties of heavy oil and vacuum residue.

2. EXPERIMENTAL SECTION

2.1 Materials

Experiments were carried out on two raw materials. One of them is heavy oil from Tatarstan oilfield. Second sample is the vacuum residue from Refinery Company of Tatarstan Republic. The studied crude oil is considered as heavy oil because of its density and viscosity (971 kg/m^3 at 20 °C, API gravity 14.1 and 2073 mPa∗s at 25 °C) and elemental composition (C 83.68%, H 11.44%, S 4.52%, and others 0.36%). The vacuum residue sample has density 1.2 g/cm^3 with a viscosity of 26056 mPa∗s. It contains 5.97 % of paraffin's.

Table 1. Density and viscosity of studied heavy oil and vacuum residue.

Sample	Density, kg/m^3	Viscosity, mPa*s at 25 °C
Tatarstan heavy oil	971	2073
Vacuum residue	1200	26056*

*The viscosity of initial vacuum residue determined at 40 °C because of its solid state

Table 2. SARA composition of heavy oil and vacuum residue.

	Saturate	Aromatic	Resins	Asphaltenes
Heavy oil	28.79	44.32	20.98	5.91
Vacuum residue	28.67	40.14	26.78	4.39

2.2 Methods

The experiment was carried out at 400 °C and at initial pressure 236 bar. The heavy oil or vacuum residue sample was charged into the autoclave-reactor model 4848 (Parr Instruments, USA) together with water in a ratio of 1:1 (75 g of distilled water and 75 g of heavy oil or residue). The reactor was hermetically closed. To remove the low volatile impurities and provide the initial pressure in the reactor, nitrogen was injected at 10 bars before starting the experiment. During the experiment, the heavy oil and vacuum residue were stirred with a rotation speed 140 rpm (rotation per minute). The reactor has a load capacity of 500 ml. The reactor is equipped with an automatic control system. The average heating rate of the reactor at the first stage was (dT/dt = 8 °C/min) and when the temperature reaches 380 °C the heating rate was decreased to 4 °C/min. After reaching the desired temperature of 400 ^{0}C, the experiment was continued at this temperature for 60 minutes and then stopped. M. Hosseinpour et al. reported that it is important to use fast cooling system of the product to stop the reaction [13]. The same procedure was used in our work. After cooling the sample, the liquid products were separated on upgraded oil and water using the centrifuge. If the emulsion was stable and did not separate by centrifugation, vacuum evaporator was used to

IOP Conf. Series: Earth and Environmental Science **282** (2019) 012044 doi:10.1088/1755-1315/282/1/012044

remove water. Viscosity was measured by Brookfield DV-II Pro viscometer according to GOST 3900-85. Density was determined by the standard method of the relative density test (pycnometer method) according ASTM D 2320-98 method [8].

The SARA analysis of the two samples (heavy oil and vacuum residue) after upgrading was done according the ASTM D-2007. The asphaltenes for each upgraded liquid products were precipitated by *n*-heptane with a ratio (oil:*n*-heptane) of 1:40. After 24 hours, the precipitated asphaltenes were separated from the maltene fraction by paper filtration, and the asphaltenes remaining in the paper were prepared and washed in the soxhlete extractor. After extraction, the solvent was evaporated and the residue in the flask was weighed for the balance calculation. At the same time, the maltene fraction was dried and the residue was prepared for adsorption in the glass chromatographic column. The column was loaded with aluminum oxide and then the residues of the maltenes were added to the aluminum oxide in the column for the adsorption of the maltenes. Then the saturates, aromatics and resins fractions were separated by respective solvents (180 ml of n-heptane for saturates, 180 ml of toluene for aromatics and a mixture of toluene with isopropyl alcohol in a ratio of 1: 1 200 ml for the resins). The respective 3 fractions are then evaporated and stabilized for material balance calculation.

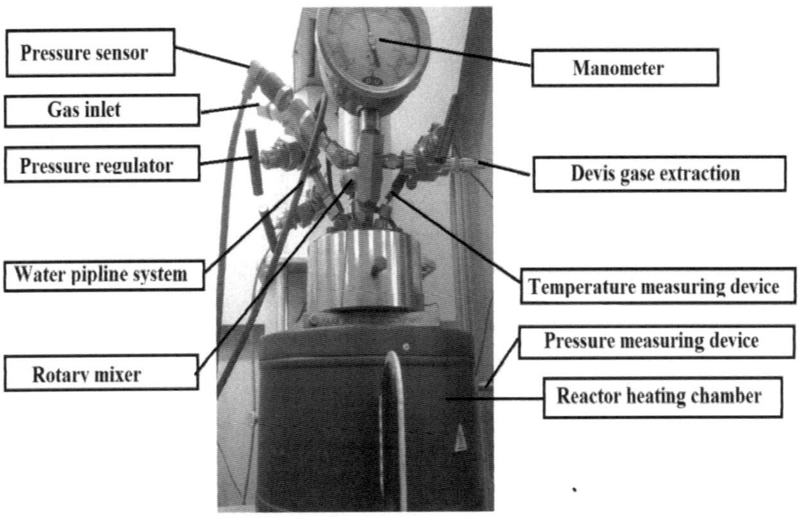

Fig 1. The autoclave reactor Parr 4848

3. RESULTS AND DISCUSSION

3.1. Products yields

Two experiments for thermal treatment of heavy oil and vacuum residue by supercritical water were carried out at 400 ^{0}C. After the experiment the liquid products and coke were collected and their mass was determined. Based on these data the material balance of upgrading process was calculated. The mass of gaseous products was determined by the subtraction of the coke and upgraded oil masses from the mass of initial heavy oil. The mass fraction of each product was calculated by the following formula:

$$(x_1, \%) = \frac{M2*100}{M1}, \quad (1)$$

(x_1) is the mass fraction of upgraded oil; M_1 is the initial mass of crude oil, (g); M_2 is the mass of upgraded oil after experiment, (g).

$$(x_2, \%) = \frac{M3*100}{M1}, \quad (2)$$

(x_2) is the mass fraction of the coke; M_1 is the initial mass of crude oil, (g); M_3 is the mass of coke after experiment, (g).

$$(x_3, \%) = \frac{(M_1 - M_2 - M_3)}{M_1}, \quad (3)$$

(x_3) is the gaseous mass fraction.

Table. 3 Material balance of products after supercritical water treatment

Sample	Products yields , %		
	Gases	Oil	Coke
Heavy oil	3.1	85.6	11.3
Vacuum residue	21.6	64.7	13.7

3.2. Physical properties of the upgraded oil

Viscosity and density of upgraded oil obtained after SCW treatment were measured. Results are presented in Table 4. Viscosity of upgraded heavy oil and upgraded oil from vacuum residue significantly decreased, 209,4 and 1350 times respectively. Density of liquid products also decreased from 971 to 918 kg /m^3 for heavy oil and from 1200 to 995 kg/m^3 for vacuum residue. These changes induced by supercritical water effect on the heavy oil, which results in conversion of heavy fractions to light components [14].

IOP Conf. Series: Earth and Environmental Science **282** (2019) 012044 doi:10.1088/1755-1315/282/1/012044

Table. 4 Density and viscosity of upgraded products

Sample	Density, kg/m^3	Viscosity, mPa*s at 25 °C
Tatarstan heavy oil	918	9.90
Vacuum residue	995	19.3

3.3. Analysis of SARA- fractions

In order to understand the chemical transformation of heavy oil and vacuum residue induced by SCW effect, the SARA composition of liquid products was determined. Comparison on Fig. 2 shows, that saturates fraction increased significantly after upgrading from 28.79 to 78.51 % for heavy oil sample and from 28.67 to 57.24 % for vacuum residue sample. At the same time the fraction of aromatics, resins and asphaltenes decreased respectively from (44.32 % to 16. 45 %, 20.98 % to 2.65 %, 5.91 to 2. 39 % for upgraded heavy oil and (40.14 % to 24.20 %, 26.78 % to 15.43 %, 4.39 % to 3.13 % for upgraded vacuum residue. The increase in saturated hydrocarbons and the decrease of aromatics, resins, and asphaltenes in the products is due to the hydrothermal reactions. Previously it was reported [14] that thermal treatment of vacuum residue at 400 °C during 60 minutes led to decrease of density by 0.2 g/cm^3 and remarkably improved the quality of product yields. After 1 hour of treatment, the fraction of asphaltenes decreased by 86 % and the aromatics increased from 20 to 63.3% [14].

Table. 5 SARA composition of the upgraded liquid products

Nomination	SARA analysis, (%)			
	Saturate	Aromatic	Resins	Asphaltenes
Tatarstan heavy oil	78.51	16.45	2.65	2.39
Vacuum residue	57.24	24.20	15.43	3.13

*The viscosity of initial vacuum residue determined at 40 °C because of its solid state

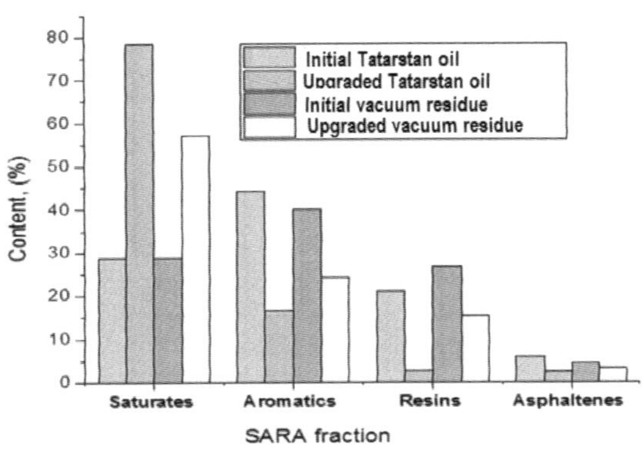

Fig 2. The SARA fraction of upgrade oil

3.4. Analysis of Saturates fraction by Agilent Technologies 7890B GC Gas Chromatograph System

Fig.3. presents the results of gas chromatographic analysis of saturates fraction before and after SCW treatment of heavy oil and vacuum residue. In the initial sample of heavy oil the saturated hydrocarbons from **C_{19}** to C_{27} were not detected, and other hydrocarbons as **C_{28}** to **C_{30}** were detected but in small amount. In the initial vacuum residue, the saturated hydrocarbons were detected from **C_{21}** to **C_{31}.** However, after upgrading, the both samples were enriched by new aliphatic hydrocarbons (dealkylation reaction of heavy components). In the product of the heavy oil after upgrading at 400 °C, a wide range of saturated hydrocarbons were detected from **C_{11}** to **C_{29}**, these increase of saturated hydrocarbons was due to thermal cracking reactions of resins and asphaltenes of heavy oil. In the liquid product of the vacuum residue after supercritical water treatment at 400°C, the saturated hydrocarbons concentration was considerably increased at **C_{11}-C_{31}** (fig. 3). The increase in concentration of saturated hydrocarbons is due to the intensification of conversion of vacuum residue under hydrothermal cracking in the supercritical water medium.

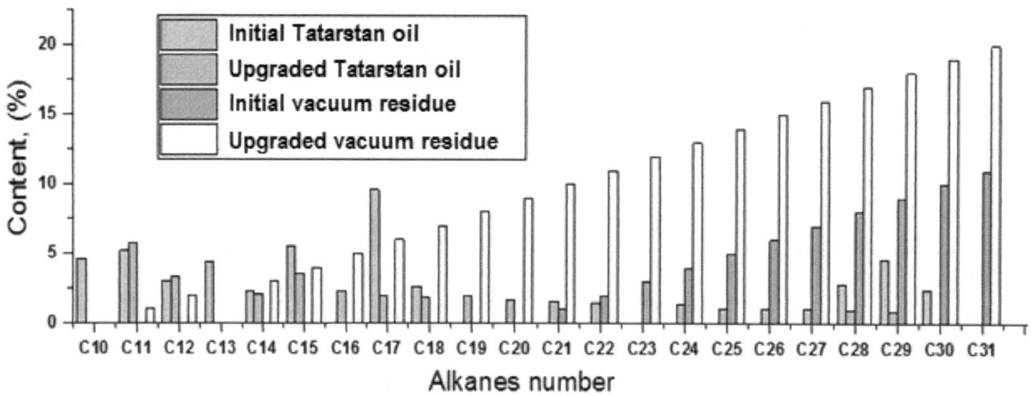

Fig 3. Curve of saturates hydrocarbons of initial and upgraded products

IOP Conf. Series: Earth and Environmental Science **282** (2019) 012044 doi:10.1088/1755-1315/282/1/012044

CONCLUSION

The process of upgrading of heavy oil and the vacuum residues at 400 °C and 60 min was carried out in this work. The composition and properties of upgraded products were compared with the initial samples. It was found that viscosity significantly decreased from 2073 to 16.6 mPa*s for heavy oil and from 26056 to 9.9 mPa*s for vacuum residues. At the same time the density increased from 971 kg/m^3 to 918 kg/m^3 for heavy oil and 1200 kg/m^3 to 995 kg/m^3). The conversion rate of heavy oil reached respective liquid products 85.6% and 64.7%, gases 3.1% and 21.6% and coke 11.3% and 13.7%. Also, the SARA fraction of crude oils (heavy oil and vacuum residue) was compared. The fraction of saturated hydrocarbons increased intensively in both products from 28.79 % to 78.51% and from 28.67 % to 57.24 %, while aromatics decreased from 44.32 to 16.45% and from 40.14 % to 24.20 % due to hydrothermal destruction and hydrogenation, the resins decreased from 20.98 % to 2.65 % and 26.78 % to 15.43 %. The asphaltenes, due to the supercritical effect of water decreased from 5.91% to 2.39% and 4.39% to 3.13%. The treatment of heavy oil and vacuum residue by supercritical water showed a very good performance on the conversion of heavy components into light liquid products and this technology can be perspective for industrial application.

Acknowledgments

This work was performed according to the Russian Government Program of Competitive Growth of Kazan Federal University.

REFERENCES

[1] R. M. Mills, *The myth of the oil crisis: overcoming the challenges of depletion, geopolitics, and global warming*. Greenwood Publishing Group, 2008.
[2] S. Akin, "Mathematical modeling of steam-assisted gravity drainage," *Comput. Geosci.*, vol. 32, no. 2, pp. 240–246, 2006.
[3] Z.-M. Cheng, Y. Ding, L.-Q. Zhao, P.-Q. Yuan, and W.-K. Yuan, "Effects of supercritical water in vacuum residue upgrading," *Energy & fuels*, vol. 23, no. 6, pp. 3178–3183, 2009.
[4] S. E. Moschopedis, J. F. Fryer, and J. G. Speight, "Investigation of asphaltene molecular weights," *Fuel*, vol. 55, no. 3, pp. 227–232, 1976.
[5] O. N. T. of petroleum asphaltenes in Kozhevnikov, I.V.; Nuzhdin, A.L.; Martyanov and 217–222. supercritical water. J. Supercrit. Fluids 2010, 55, "Transformation of petroleum asphaltenes in supercritical water," *J. Supercrit. Fluids*, vol. 55, no. 1, pp. 217–222, 2010.
[6] C. Ma, R. Zhang, and J. Bi, "Upgrading of coal tar in supercritical water," *J. Fuel Chem. Technol.*, vol. 31, no. 2, pp. 103–110, 2003.
[7] C.-C. Zhu, C. Ren, X.-C. Tan, G. Chen, P.-Q. Yuan, Z.-M. Cheng, and W.-K. Yuan, "Initiated pyrolysis of heavy oil in the presence of near-critical water," *Fuel Process. Technol.*, vol. 111, pp. 111–117, 2013.

[8] P. R. Filgueiras, C. M. S. Sad, A. R. Loureiro, M. F. P. Santos, E. V. R. Castro, J. C. M. Dias, and R. J. Poppi, "Determination of API gravity, kinematic viscosity and water content in petroleum by ATR-FTIR spectroscopy and multivariate calibration," *Fuel*, vol. 116, pp. 123–130, 2014.

[9] M. T. Timko, A. F. Ghoniem, and W. H. Green, "Upgrading and desulfurization of heavy oils by supercritical water," *J. Supercrit. Fluids*, vol. 96, pp. 114–123, 2015.

[10] L. Gao, Y. Liu, L. Wen, W. Huang, X. Mu, B. Zong, H. Fan, and B. Han, "The effect of supercritical water on the hydroconversion of Tahe Residue," *AIChE J.*, vol. 56, no. 12, pp. 3236–3242, 2010.

[11] N. Olukcu, J. Yanik, M. Saglam, M. Yuksel, and M. Karaduman, "Solvent effect on the extraction of Beypazari oil shale," *Energy & Fuels*, vol. 13, no. 4, pp. 895–902, 1999.

[12] O. N. Fedyaeva, A. V Shatrova, and A. A. Vostrikov, "Effect of temperature on bitumen conversion in a supercritical water flow," *J. Supercrit. Fluids*, vol. 95, pp. 437–443, 2014.

[13] M. Hosseinpour, S. Fatemi, S. J. Ahmadi, M. Morimoto, M. Akizuki, Y. Oshima, and E. Fumoto, "The synergistic effect between supercritical water and redox properties of iron oxide nanoparticles during in-situ catalytic upgrading of heavy oil with formic acid. Isotopic study," *Appl. Catal. B Environ.*, vol. 230, pp. 91–101, 2018.

[14] Y. Liu, F. Bai, C.-C. Zhu, P.-Q. Yuan, Z.-M. Cheng, and W.-K. Yuan, "Upgrading of residual oil in sub-and supercritical water: An experimental study," *Fuel Process. Technol.*, vol. 106, pp. 281–288, 2013.

AUTHOR INDEX

AUTHOR INDEX